JOURNEES DE GEOMETRIE ALGEBRIQUE D'ANGERS

ALGEBRAIC GEOMETRY ANGERS 1979

Journées de géométrie algébrique d'Angers (juillet 1979)

Algebraic Geometry Angers 1979

Variétés de petite dimension

edited by

Arnaud Beauville
Université d'Angers

SIJTHOFF & NOORDHOFF 1980
Alphen aan den Rijn, The Netherlands
Rockville, Maryland, USA

ISBN 90 286 0500 2

Library of Congress Catalog Card Number: 80–83268

Printed in The Netherlands

TABLE DES MATIERES

Introduction VII

Liste des exposés non publiés dans ce volume IX

Liste des participants XI

Fibrés vectoriels 1

A. van de Ven
Twenty years of classifying algebraic vector bundles 3
Robin Hartshorne
Four years of algebraic vector bundles: 1975–1979 21
Wolf Barth
Kummer surfaces associated with the Horrocks-Mumford bundle 29

Périodes des variétés algébriques 49

James A. Carlson and Phillip A. Griffiths
Infinitesimal variations of Hodge structure and the global Torelli problem 51
James A. Carlson, Eduardo H. Cattani and Aroldo G. Kaplan
Mixed Hodge structures and compactifications of Siegel's space 77
James A. Carlson
Extensions of mixed Hodge structures 107
A. Conte and J.P. Murre
The Hodge conjecture for Fano complete intersections of dimension four 129
Ron Donagi and Roy Smith
The degree of the Prym map onto the moduli space of five dimensional abelian varieties 143
Frans Oort and Joseph Steenbrink
The local Torelli problem for algebraic curves 157

Surfaces 205

ARNAUD BEAUVILLE
Sur le nombre maximum de points doubles d'une surface dans $\mathbf{P}^3$ ($\mu(5) = 31$) 207
S. BLOCH
On an argument of Mumford in the theory of algebraic cycles 217
J.-L. COLLIOT-THELENE et J. SANSUC
La descente sur les variétés rationnelles 223
YOICHI MIYAOKA
On the Mumford–Ramanujam vanishing theorem on a surface 239
C.A.M. PETERS
On automorphisms of compact Kähler surfaces 249

Variétés de dimension 3 269

YUJIRO KAWAMATA
A characterization of an abelian variety 271
MILES REID
Canonical 3-folds 273
KENJI UENO
Birational geometry of algebraic threefolds 311

INTRODUCTION

Les Journées de Géométrie algébrique ont eu lieu du 2 au 7 juillet 1979 à la Faculté des Sciences d'Angers, dans le cadre des Journées SMF. Le thème choisi était "variétés de petite dimension". Ce volume reproduit le texte de la plupart des conférences, regroupées suivant quatre thèmes: fibrés vectoriels, périodes des variétés algébriques, surfaces, variétés de dimension trois.

Trois articles sont consacrés aux fibrés vectoriels: les deux premiers sont des exposés de synthèse, qui dressent l'historique du sujet: avant 1975 (Van de Ven) et de 1975 à aujourd'hui (Hartshorne). L'exposé de Barth, plus spécialisé, étudie une relation curieuse entre le fibré de Horrocks–Mumford et les surfaces de Kummer.

Le chapitre "périodes" contient l'article de Carlson et Griffiths, qui faisait l'objet de 3 exposés à Angers; il s'agit d'un vaste programme pour établir le théorème de Torelli pour les hypersurfaces, et plus généralement les variétés "extrémales". Les articles de Carlson et Carlson–Cattani–Kaplan développent très largement l'exposé de Carlson à Angers. Les autres articles, qui correspondent à des exposés oraux d'une heure, font le tour des problèmes que pose actuellement la théorie des périodes: conjecture de Hodge, injectivité ou surjectivité de l'application des périodes.

Le chapitre suivant regroupe des exposés assez divers sur les surfaces: automorphismes, groupe de Chow, etc. Enfin trois exposés sont consacrés aux variétés de dimension 3: ceux de Kawamata et Ueno font le point sur la classification, et celui de Reid ébauche une théorie des modèles canoniques en dimension 3.

Un certain nombre de conférences exposaient des articles parus ou à paraître dans des revues; leur liste (avec références) est donnée à la page suivante.

Les Journées ont bénéficié d'un soutien matériel important de la SMF et de l'Université d'Angers; nous les en remercions. Nous remercions également les conférenciers, les participants, et le personnel de la Faculté des Sciences, qui ont contribué au succès de ces Journées – tout comme la douceur de l'été angevin.

A. Beauville
G. Pourcin

LISTE DES EXPOSES NON PUBLIES DANS CE VOLUME

E. ARBARELLO, M. CORNALBA, P. GRIFFITHS: Three lectures on the Brill–Nœther problem
(référence: P. Griffiths, J. Harris: *Dimension of the variety of special divisors on an algebraic curve, Duke Math. Journal (1980)*)

S. BLOCH: On the Chow group of certain rational surfaces
(*Annales de l'ENS, à paraître*)

S. BLOCH: Mori's proof of Hartshorne conjecture
(S. Mori: *Projective manifolds with ample tangent bundle, Ann. of Math. 110 (1979)*)

F. CATANESE: Babbage's conjecture, contact of surfaces, symmetric determinantal varieties and applications
(*à paraître*)

H. CLEMENS: Double solids
(*à paraître*)

M. DESCHAMPS–R. MENEGAUX: Surfaces de type général dominées par une variété fixe
(*Note aux C.R. Acad. Sci. Paris, t.288 (1979)*)

I. DOLGACHEV: Where to embed algebraic varieties?
(*non publié*)

O. GABBER: Properties of étale cohomology
(*non publié*)

L. GRUSON: Courbes gauches
(L. Gruson, C. Peskine: *Genre des courbes de l'espace projective, Proceedings of the Tromsø conference, Springer Lecture Notes 687*)

U. PERSSON, H. PINKHAM: Degeneration of surfaces with trivial canonical bundle
(*à paraître aux Annals of Math.*)

R. PIENE: Riemann–Roch pour les variétés projectives de dimension 3
(R. Piene, F. Ronga: *A geometric approach to the arithmetic genus of a projective manifold of dimension 3, à paraître dans Topology*)

P. SWINNERTON-DYER: Hilbert modular surfaces
(*non publié*)

A.N. TODOROV: The period map is onto for Kähler K-3 surfaces
(*Compositio Mathematica, à paraître*)

E. VIEHWEG: Klassifikationstheorie algebraischer Varietäten der Dimension drei (*Compositio Mathematica 41* (1980))

LISTE DES PARTICIPANTS

AL AMRAMI (Rabbat)
AMBROGIO, E. (Turin)
ANDUJO, C. (Brandeis)
ANGERMULLER, G. (Erlangen)
ARBARELLO, E. (Rome)
AVRITZER, D. (Londres)
BARATTERO, R. (Gênes)
BARLOW (Warwick)
BARTH, W. (Erlangen)
BEAUVILLE, A. (Angers)
BECKER, J. (Buffalo)
BELTRAMETTI, M. (Gênes)
BENVENISTE, X. (Paris)
BERTHELOT, P. (Rennes)
BERTIN, J. (Toulouse)
BLOCH, S. (Chicago)
BORCHERS (Mannheim)
BOUTOT, J.F. (Strasbourg)
BRAND, R. (Leiden)
BRYLINSKI, J.L. (Paris)
BUCHWEIZ, R. (Hanovre)
CAMPILLO, A. (Madrid)
CANUTO, G. (Turin)
CARLSON, J. (Utah)
CATANESE, F. (Pisa)
CAVALIERE, M.P. (Gênes)
CERESA, G. (Utah)
CHAKIRIS, K. (New York)
CHERENAK, P. (Le Cap)
CHIPMAN, R. (M.I.T.)
CLEMENS, H. (Utah)
COLLINO, A. (Turin)
COLLIOT-THELENE (Paris)
CONRIERI-VERDI, L. (Florence)
CONTE, A. (Turin)
CONTESSA, M. (Rome)
CORAY, D. (Genève)
CORNALBA, M. (Pavie)
COX, D. (Rutgers)
CRANDER, B. (Columbia)
CUMINO, C. (Turin)
DECAUWERT, M. (Grenoble)
DESCHAMPS, M. (Paris)
DE SOUZA (Paris)
DIALLO, S. (Paris)
DLOUSSKY, G. (Marseille)
DOLGACHEV, I. (Ann Arbor)
DU BOIS, P. (Nantes)
DURET, J.L. (Angers)
ELENCWAJG, G. (Nice)
ELZEIN, F. (Paris)
ESNAULT, H. (Paris)
FATEMI, T. (Caen)
FERRAND, D. (Rennes)
FERRARESE (Turin)
FINAT (Madrid)
FLEXOR, M. (Paris)
FRANCIA, P. (Gênes)
GABBER, O. (Tel Aviv)
GONZALEZ, G. (Paris)
GREYSON, D. (Columbia)
GRIFFITHS, P. (Harvard)
GRUSON, L. (Lille)
HARBATER, D. (Harvard)
HARTSHORNE, R. (Berkeley)
HOSTMALINGEN, C. (Oslo)
ILLUSIE, L. (Orsay)
JARRAUD (Paris)
KAWAMATA, Y. (Mannheim)
KHALED, A. (Alger)
LAMARI, A. (Alger)
LANDMAN, A. (Brown)
LANGLANDS, R. (Princeton)
LANTERI, A. (Milan)
LE BARZ, P. (Nice)
LE DUNG TRANG (Paris)
LE POTIER, J. (Paris)
LIGOZAT, G. (Orsay)
LORENZANI, M. (Rome)
LUCAS, F. (Angers)
LUENGO, (Madrid)
MANARESI, M. (Bologne)
MANGOLINI, (Gênes)
MARINARI, M.G. (Gênes)
MATHIEU, P. (Louvain)
MAURER, J. (Bonn)

Mebkhoutz (Orléans)
Menegaux, N. (Paris)
Merindol, J.Y. (Angers)
Morales, M. (Grenoble)
Mitschi, C. (Oslo)
Miyaoka, Y. (Tolio)
Murre, J.P. (Leiden)
Nout Mehidi (Constantine)
Odetti, F. (Gênes)
Ogus, A. (Orsay)
Oort, F. (Utrecht)
Palleschi, M. (Milan)
Persson, U. (Columbia)
Peters, C. (Leiden)
Picco Botta, L. (Turin)
Piene, R. (Mittag-Leffler)
Pinkham, H. (Columbia)
Pourcin, G. (Angers)
Prabakhar Rao (Harvard)
Puts, P. (Leiden)
Raimondo, M. (Gênes)
Raynaud, M. (Orsay)
Recillas, S. (Mexico)
Recio, T. (Madrid)
Reid, M. (Warwick)
Rodino, N. (Reggio di C.)
Romagnoli, D. (Turin)
Rossi, M. (Gênes)
Saint-Loup (Paris)
Sabbah, C. (Paris)
Sansuc, J.J. (Paris)
Sanchez-Peregrino (Paris)
Schaub, D. (Angers)
Sernesi, E. (Ferrare)
Sheperd-Barron (Warwick)
Smith, R. (Georgia)
Soberon, S. (Oxford)
Soule (Paris)
Souville (Paris)
Stahl, C. (Westfield)
Steenbrink, J. (Leiden)
Swinnerton-Dyer (Cambridge)
Szpiro, L. (Paris)
Tamone, G. (Gênes)
Todorov, A. (Sofia)
Toet, A. (Leiden)
Treger, R. (Tel-Aviv)
Turrini, C. (Milan)
Treibich (Paris)
Ueno, K. (Kyoto)
Van de Ven, A. (Leiden)
Vermeulen, L. (Amsterdam)
Verra, A. (Turin)
Viehweg, E. (Mannheim)
Vogelaar, H. (Leiden)
Wagreich, P. (Chicago)
Wajnryb, B. (Bonn)
Welters, G. (Utrecht)
Wilson, P. (Cambridge)

Fibrés vectoriels

TWENTY YEARS OF CLASSIFYING ALGEBRAIC VECTOR BUNDLES

A. Van de Ven

Preface

At the Angers conference six lectures were given on algebraic vector bundles: two by Barth, two by Hartshorne and two by myself. My own lectures, which are reproduced in this paper, served as an historic introduction, covering the period 1955–1975. Consequently, this paper contains little that is new, and there is a considerable overlapping with some other reports on the same or related subjects, like [20] and in particular [34]. However, the full content of my lectures is not contained in either of these.

Though in principle I only sketch some aspects of what happened with respect to the classification of algebraic vector bundles in the years 1955–1975, I have indicated some later results where it seemed natural to do so. On the other hand, some subjects which came up before 1975 have not even been mentioned. This holds in particular for the classification of bundles by monads. Though monads were introduced by Horrocks as early as 1964, they did not gain a prominent place until the theory of stable bundles had started flowering.

I. Introduction

Vector bundles were first introduced in topology around 1935. In those years H. Hopf and Stiefel studied the existence of everywhere independent vector fields on compact differentiable manifolds and introduced the Stiefel–Whitney classes for the tangent bundle as obstruction classes. In connection with some other concrete problems, this time about embeddings, Whitney introduced real vector bundles in general and proved both the existence of an universal bundle and the main property of Stiefel–Whitney classes, now known as (Whitney) duality.

In 1946 Chern published his fundamental paper ([13]) on continuous and differentiable complex vector bundles. He proved that on a finite polyhedron every complex k-vector bundle can be obtained as inverse image of the universal (sub-)bundle on a Grassmann manifold. More precisely he showed that the functor attaching to a finite polyhedron X the set of continuous complex k-bundles on X can be represented by the homotopy classes of maps from X into the Grassmann variety $G_{n,k}$ of k-planes in $\mathbf{C}^n$, with n large enough. In this way the classification problem of continuous complex vector bundles is reduced to a problem of homotopy theory.

In the first years to follow several more concrete results on the classification of vector bundles were proved. Some of them are as easy as they are frequently used, like the fact that continuous complex k-bundles on a Riemann surface are completely determined by rank and first Chern class. Some other results are less easy, like the following one, due to Wu, that later on will play a role in these lectures.

PROPOSITION I.1 ([45], p. 68): *Let X be a finite polyhedron of dimension 4. For any two cohomology classes $c_1 \in H^2(X, \mathbf{Z})$ and $c_2 \in H^4(X, \mathbf{Z})$ there is one and only one continuous complex 2-vector bundle $\mathcal{V}$ with Chern classes $c_1(\mathcal{V}) = c_1$ and $c_2(\mathcal{V}) = c_2$.*

In the fifties and sixties enormous progress was made in homotopy theory, not in the last place since algebraic geometry inspired topologists to develop continuous K-theory. As a consequence, nowadays many questions concerning the topological classification of complex bundles can be satisfactorily answered. Nevertheless, when it comes to the precise classification of continuous vector bundles on a given space, relatively few cases have been treated. As far as complex projective spaces are concerned, I mention the following results.

(1) As a consequence of Proposition I.1, the 2-bundles on $\mathbf{P}_2$ are classified by their Chern classes, which may obtain any values.

(2) (compare [4], p. 131) Given $c_1, c_2 \in \mathbf{Z}$, there exists a 2-bundle on $\mathbf{P}_3$ with these Chern classes if and only if $c_1c_2 \equiv 0\ (2)$. Furthermore, if c_1 is odd, then there is one such bundle, and if c_1 is even, then there are exactly two of these bundles (every element of $H^{2k}(X, \mathbf{Z})$, $0 \leq k \leq n$, can be seen as an integer in a canonical way).

(3) (compare [21], p. 166) In order that there exists at least one continuous k-vector bundle on $\mathbf{P}_n$ with Chern classes $c_1, \ldots, c_k$, these classes must satisfy a finite number of congruences. This set of congruences depends on n and k, and can be calculated explicitly for any given n and k. Thus for a 2-bundle on $\mathbf{P}_3$ we get $c_1c_2 \equiv 0\ (2)$, for a 2-bundle on $\mathbf{P}_4$ in addition $(2c_1^2 - 3c_1 + c_2 + 1)c_2 \equiv 0\ (12)$ and for a 3-bundle on $\mathbf{P}_3$ we get $c_1c_2 - c_3 \equiv 0\ (2)$. In general these congruences give only a necessary condition, however in the case of 2-bundles on $\mathbf{P}_3$ and $\mathbf{P}_4$, but not on $\mathbf{P}_5$, the condition is also sufficient. For later use I observe the following consequence: on $\mathbf{P}_4$ there

are many continuous 2-bundles with $c_1^2 - 4c_2 > 0$ which are indecomposable because of Chern class reasons.

(4) (compare [41]) For $k \geq n$ there is, by way of the Chern classes, a 1–1 correspondence: k-bundles on $\mathbf{P}_n \leftrightarrow (c_1, \ldots, c_k)$, satisfying the aforementioned conditions. In particular, every bundle on $\mathbf{P}_2$ is completely determined by its Chern classes.

Once the concept of a complex manifold and that of a fibre bundle were introduced, it was of course clear how to define holomorphic bundles. Soon after Leray, Cartan and others developed sheaf theory in the late forties, it was realised that this theory was perfectly suited for the holomorphic case. So, for example, the finiteness theorems for the cohomology groups of coherent sheaves on compact complex spaces were proved, as well as the Hirzebruch Riemann–Roch theorem. This theorem could be applied to concrete problems thanks to vanishing theorems and Serre duality.

In algebraic geometry fibre bundles and sheaf theory were introduced a little bit later, around 1954. The classical reference here is of course Serre's paper FAC ([36]).

In 1954 Atiyah wrote ([1], p. 407): "Although much work has been done in the topological theory of fibre bundles, very little appears to be known on the complex analytic side." I don't think that Atiyah intended to belittle the developments I have just indicated. All he wanted to say was that, whereas in principle it was known how to classify continuous vector bundles (and other fibre bundles), and whereas this classification had been carried out in several cases, nothing of the like was true for holomorphic vector bundles (and other holomorphic fibre bundles). In fact an attempt had been made a few years earlier (compare [1], introduction and appendix) to prove a classification theorem for holomorphic vector bundles analogous to the one for continuous bundles, but such an approach can't possibly work. It is true that for every algebraic bundle V on an algebraic variety X there is $r_0 \in \mathbf{Z}$, such that $V(r)$ for $r > r_0$ is induced (as an algebraic bundle) from the universal quotient bundle by a suitable regular map. But this r does not only depend on X, but also on V. Still this fact can be used as a starting point for classifying certain types of bundles, though in a way quite different from the topological case.

So Atiyah took up the classification problem for holomorphic vector bundles in the first case which was open: 2-bundles on algebraic curves. I say: "the first case which was open", for the case of 1-bundles on algebraic varieties and complex manifolds was already dealt with. As everybody knows they are classified by means of the Picard variety and, on compact Kähler manifolds, the Lefschetz theorem on (1,1)-classes.

Before I start with my proper story, I first give some preliminaries.

II. Preliminaries

In general all considerations and theorems will hold in algebraic geometry over a closed field K. If not, then this is either stated explicitly or clear from the context.

I shall use the following (mostly standard) notations:

$\mathbf{P}_n(K)=\mathbf{P}_n$: n-dimensional projective space;
$G_{n,m}(K)=G_{n,m}$: Grassmann variety of m-planes in K^n;
$\mathcal{V}, \mathcal{W}, \mathcal{L}, \ldots$: vector bundles, or equivalently, their sheaf of sections;
$\mathcal{V}^*$: dual bundle of $\mathcal{V}$;
$Proj(\mathcal{V})$: projective bundle of $\mathcal{V}$ (not of $\mathcal{V}^*$), i.e. if $\mathcal{V}$ is a k-vector bundle, then $Proj(\mathcal{V})$ is the associated bundle with fibre $\mathbf{P}_{k-1}$;
$\Gamma(X,\mathcal{V})$: space of sections of the bundle $\mathcal{V}$ over X;
$\mathcal{O}_X$: structure sheaf of X, or the trivial line bundle on X;
$\mathcal{T}_X$: tangent bundle of X (X smooth);
$\mathcal{O}_X(r)$: line bundle $\mathcal{L}^r$, where $\mathcal{L}$ is a very ample line bundle on X, given by the situation (on $\mathbf{P}_n$ always the hyperplane bundle);
$\mathcal{V}(r)$: $\mathcal{V}\otimes\mathcal{O}(r)$;
$c_i(\mathcal{V})$: i-th Chern class of $\mathcal{V}$.

I shall also use the following simple and well-known facts

FACT II.1 ([16], p. 123): *If $\mathcal{L}_1$ and $\mathcal{L}_2$ are line bundles on a variety X such that $H^1(X, \mathcal{L}_1\otimes\mathcal{L}_2^*)=0$, then every exact sequence of bundles*

$$0 \longrightarrow \mathcal{L}_1 \longrightarrow \mathcal{V} \longrightarrow \mathcal{L}_2 \longrightarrow 0$$

splits. Since $H^1(\mathbf{P}_n, \mathcal{O}(r))=0$ for all r if $n\geq 2$ it follows that on $\mathbf{P}_n$ every extension of two line bundles splits.

FACT II.2: *Every continuous k-bundle with $k>n$ on an n-dimensional finite polyhedron is the direct sum of an n-dimensional bundle and a trivial bundle of rank $n-k$.*

FACT II.3: *Given an algebraic k-bundle $\mathcal{V}$ on an n-dimensional projective variety X, there exists an r_0, such that $\mathcal{V}(r)$ is very ample if $r>r_0$.*

FACT II.4: *Every k-bundle on an n-dimensional projective variety X has $(k-n)$-subbundles as soon as $k>n$.*

Given any vector bundle $\mathcal{V}$ on $\mathbf{P}_n$, we can derive other bundles from it by the following two procedures:

(i) tensoring with $\mathcal{O}(r)$ to obtain the bundles $\mathcal{V}(r)$,

(ii) taking the bundles $f^*(\mathcal{V})$, where $f:\mathbf{P}_n\to\mathbf{P}_n$ is any ramified covering.

By varying f we obtain from $\mathcal{V}$ families of bundles.

III. The first years

At the end of the introduction it was mentioned that the first paper dealing with the classification of holomorphic k-vector bundles with $k \geq 2$ was the paper [1] by Atiyah. In this paper Atiyah classifies 2-vector bundles on (smooth) curves of genus 0, 1 and 2. De facto he classifies $\mathbf{P}_1$-bundles over such curves, but that makes no difference here. He is of course well aware of the fact that the classification of algebraic $\mathbf{P}_1$-bundles over curves is equivalent with the classification problem for ruled surfaces.

In 1957 Grothendieck proved ([16], Théorème 2.1)

THEOREM III.1: *Every algebraic vector bundle on* $\mathbf{P}_1$ *is isomorphic to a direct sum of line bundles, which are uniquely determined up to isomorphisms and permutations.*

In a hidden form this result was already known in the 19-th century; for $k = 2$ it had been reproved by Atiyah in the paper cited above.

Grothendieck's theorem plays an important role in the classification of bundles as will become clear in section VI.

Thus already on $\mathbf{P}_1$ there is a difference between the topological and the algebraic case: in both cases every vector bundle splits into a sum of line bundles, but in the second case the factors are determined up to isomorphisms, whereas in the first case they are not.

About the same time Atiyah proved quite generally ([2], Theorem 3) that algebraic vector bundles on complete varieties have the Krull–Schmidt property: *if* $\mathscr{V} = \mathscr{V}_1 \oplus \cdots \oplus \mathscr{V}_\ell$ *and* $\mathscr{V} = \mathscr{W}_1 \oplus \cdots \oplus \mathscr{W}_m$, *with the* $\mathscr{V}_i$ *and* $\mathscr{W}_j$ *indecomposable (of some rank* ≥ 1*), then* $\ell = m$ *and* $\mathscr{V}_i \cong \mathscr{W}_i$, *provided that the* $\mathscr{W}_j$*s are properly ordered.*

What about curves of higher genus? Since a k-bundle on a variety of dimension n has always an $(k-n)$-subbundle if $k > n$, there exist for every vector bundle $\mathscr{V}$ of rank k over a curve C a series of subbundles

$$\mathscr{D}_1 \subset \mathscr{D}_2 \subset \cdots \subset \mathscr{D}_{k-1} \subset \mathscr{V}$$

such that $\mathscr{D}_i/\mathscr{D}_{i-1}$ has always rank 1. But such a series if far from unique and the successive extensions may not split even as the bundle splits.

So let C be a smooth curve of genus ≥ 1 over a closed field. Since $H^1(C, \mathcal{O}_C) \neq 0$, there are non-trivial extensions

$$0 \longrightarrow \mathcal{O}_C \longrightarrow \mathscr{V} \longrightarrow \mathcal{O}_C \longrightarrow 0.$$

Using projection in the bundle space of $\mathscr{V}$ you readily see that $\mathscr{V}$ is indecomposable. Thus on any smooth curves of genus ≥ 1 there are algebraically indecomposable 2-bundles. In fact, among all normal projective varieties X the projective line is singled out by the property that every algebraic vector bundle on X is a direct sum of line bundles. For smooth X

this is a consequence of the following result which I proved in 1959 ([37], p. 262).

THEOREM III.2: *Let X be a smooth algebraic variety, linearly normally embedded in* $\mathbf{P}_n$. *Then the restriction of the tangent bundle of* $\mathbf{P}_n$ *to X is a direct sum of line bundles if and only if X is isomorphic to* $\mathbf{P}_1$.

This theorem was later generalised by Simonis ([37], Corollary 3.3.4.) and Maruyama ([28]). The last author showed that on any n-dimensional smooth algebraic variety there are stable k-vector bundles for all $k \geq n$.

Though in general the classification of vector bundles on curves is complicated, the situation is still rather simple for elliptic curves. For this case Atiyah showed in 1957 ([3], p. 434)

THEOREM III.3: *Let X be an elliptic curve and let* $I(k, d)$ *be the set of indecomposable k-vector bundles* $\mathcal{V}$ *on X with* $deg(\wedge^k \mathcal{V}) = d$.

(i) *In* $I(k, 0)$ *there is one bundle* $\mathcal{E}_k$ *with* $\Gamma(X, \mathcal{E}_k) \neq 0$. *This bundle can be obtained by successive extensions*

$$0 \longrightarrow \mathcal{O}_X \longrightarrow \mathcal{E} \longrightarrow \mathcal{E}_{k-1} \longrightarrow 0.$$

(ii) *There are (non-canonical) 1–1 correspondences* $f_{k,d} : \mathcal{I}(k, 0) \rightarrow \mathcal{I}(k, d)$ *such that every bundle in* $\mathcal{I}(k, d)$ *can be written as* $f_{k,d}(\mathcal{E}_k) \otimes$ *(line bundle)*.

In fact, Atiyah's results are much more precise, taking also into account multiplicative properties coming from tensor products of bundles.

From the late fifties on there have been several major developments, all but one of which I can only mention:

(A) *Vector bundles on curves.* The case of 2-bundles has been treated extensively by Tyurin ([30], Chapter V, §7), and several people, above all Mumford, Seshadri and M. Narasimhan have developed a beautiful theory of stable bundles. These have good moduli spaces, of which many properties are now known (see e.g. [29]).

(B) *Vector bundles on affine varieties.* Much (but not all) of the work done in this direction has centered around a conjecture of Serre: every vector bundle over an affine space is (algebraically) trivial. Many special cases have been proved in the course of the years, but the general result has resisted until 1975, in which year it was proved by Quillen ([27]). This direction has close ties with developments in homological algebra, and Quillen's theorem is even more universally known in an algebraic form: every finitely generated projective module over a polynomial ring with respect to a closed field is a free module.

(C) *Homogeneous bundles.* These are defined as vector bundles on n-dimensional homogeneous manifolds, which arise from the tangent bundle by a representation of $GL(n)$. In this case classification is of course a matter of representation theory. The cohomology of homogeneous bundles has been studied extensively, for example already in Bott's fundamental paper [12].

(D) *Vector bundles of rank ≥ 2 on higher dimensional projective varieties.* I believe that people thinking about such bundles at the end of the fifties had several strategical questions in mind:

(a) Which continuous complex vector bundles (over a smooth algebraic variety) carry algebraic structures? Think of the resemblance with the Hodge conjecture and the existence of algebraic structures on topological manifolds. There is an obvious necessary condition: the Chern classes must be algebraic.

(b) If a continuous bundle has algebraic structures, do these form something nice? Think of the case of line bundles, where you get Pic^0.

(c) Are there perhaps also for $k \geq 2$ relations between k-bundles and the Chow group in codimension k?

But too few facts were known. What was needed was a study of some particular cases. Such a study was made for 2-bundles on algebraic surfaces, and in particular on $\mathbf{P}_2$, by Schwarzenberger ([32] and [33]).

IV. 2-Bundles on surfaces

Though on $\mathbf{P}_2$ there are no non-trivial extensions of line bundles by line bundles, it is not difficult to find some examples of algebraic 2-bundles which are indecomposable since their Chern classes don't split: the tangent bundle, or the normal bundle if $\mathbf{P}_2$ is embedded as a projected Veronese surface in $\mathbf{P}_4$. But it is less obvious how to obtain examples in a more or less systematic way.

Schwarzenberger invented two methods for producing many bundles.

(1) Let X and Y be smooth algebraic varieties, and let $f: Y \to X$ be a 2-fold ramified covering map. If $\mathscr{L}$ is a line bundle on Y, then $f_*(\mathscr{L})$ is (the sheaf of sections of) a 2-bundle $\mathscr{V}$ on X, whereas $R^i f_*(\mathscr{L}) = 0$ for $i \geq 1$. Application of Grothendieck–Riemann–Roch yields the Chern classes of $\mathscr{V}$. Taking for X the projective plane and for Y 2-fold coverings ramified along smooth curves of degree 2 and 4, Schwarzenberger found many indecomposable bundles on $\mathbf{P}_2$. In fact every 2-bundle on an algebraic surface can be obtained by this method, but not in a canonical way.

(2) Let X be again a smooth algebraic surface, $\mathscr{V}$ an algebraic 2-vector bundle on X and s a section in $\mathscr{V}$, which vanishes transversally in finitely many points $x_1, \ldots, x_k$ and nowhere else. Let $\tilde{X}$ be obtained from X by blowing up $x_1, \ldots, x_k$, and let $p: \tilde{X} \to X$ be the projection. Then $p^*(\mathscr{V})$ has

an induced section, vanishing simply on the exceptional curves $E_1, \ldots, E_k$ and nowhere else. In other words, the bundle $p^*(\mathcal{V})$ has a 1-subbundle and it is therefore an extension of two line bundles (independently of the fact whether $\mathcal{V}$ itself is such an extension or not). Furthermore $p_* p^*(\mathcal{V}) \cong \mathcal{V}$. Thus it is natural to try to find a priori extensions $\mathcal{W}$ of two line bundles on X, such that $p_*(\mathcal{W})$ is locally free and indecomposable. Schwarzenberger proved that $p_*(\mathcal{W})$ is locally free (of rank 2 of course) if and only if the restriction of $\mathcal{W}$ to all curves E_i is the trivial 2-bundle. You can also blow up more times in the same point; the condition remains that the restriction of $\mathcal{W}$ to all curves in all exceptional trees is trivial. In this way Schwarzenberger obtained many new indecomposable bundles on $\mathbf{P}_2$, and by varying points and extensions he obtained algebraic families of bundles. In particular Schwarzenberger constructed bundles with arbitrarily prescribed Chern classes. Combining this with Proposition I.1 and Remark II we find

THEOREM IV.1: *Every continuous d-bundle on* $\mathbf{P}_2(\mathbf{C})$ *has an algebraic structure.*

By the same method Schwarzenberger obtained also the following more general result.

THEOREM IV.2: *Let X be a smooth complex algebraic surface. A continuous vector bundle* $\mathcal{V}$ *over X carries an algebraic structure if and only if* $\mathcal{V}$ *has algebraic Chern classes, i.e. if and only if* $c_1(\mathcal{V})$ *is of type* (1,1).

Also method (2) is not immediately suited for the classification of bundles, simply because a bundle can be obtained in many different ways as the direct image of an extension on a blown-up surface. But it can be used to classify almost-decomposable bundles, i.e. bundles admitting endomorphisms different from scalar multiplication. However this classification is not much more than a mere listing, for no good moduli spaces exist.

It is an immediate consequence of Theorem 1 in [32] that on $\mathbf{P}_n$ the almost-decomposable 2-bundles are exactly the non-stable bundles. For a general discussion of stable bundles I refer, as always, to Hartshorne's lectures; but I give one possible definition of stability for 2-bundles on $\mathbf{P}_n$. It goes as follows. If $\mathcal{V}$ is any 2-bundle on $\mathbf{P}_n$, then $\Gamma(\mathbf{P}_n, \mathcal{V}(r)) = 0$ for $r \ll 0$ and $\Gamma(\mathbf{P}_n, \mathcal{V}(r)) \neq 0$ for $r \gg 0$. Thus there is one r_0 such that $\Gamma(\mathbf{P}_n, \mathcal{V}(r_0 - 1)) = 0$, but $\Gamma(\mathbf{P}_n, \mathcal{V}(r_0)) \neq 0$. Now $\mathcal{V}$ is called stable, semi-stable or unstable according to whether $c_1(\mathcal{V}(r_0)) > 0$, $= 0$ or < 0.

After the work of Schwarzenberger, what can be said about the questions (a), (b) and (c), raised at the end of section III? Clearly, for surfaces the answer to (a) is very satisfactory. As to (b) it had become rather clear that you can't expect to get a good moduli space for all algebraic structures on a given topological bundle. But the possibility that, say, the stable ones among them have a good moduli space, remained open. Finally, the new examples did not indicate any sensible answer to question (c).

Thus among the most obvious questions remaining were the following ones:

(1) classify stable 2-bundles on $\mathbf{P}_2$;

(2) study 2-bundles on $\mathbf{P}_n$, $n \geq 3$;

(3) study k-bundles on $\mathbf{P}_2$, $k \geq 3$.

The first of these problems was solved only much later and I refer to Hartshorne's lectures. As to the third question, I don't think that it was ever attacked in full generality (but compare [34], 2.3). However, the stable case was recently treated by Hulek ([25]).

Problem (2) forms the subject of the next section.

V. Vector bundles on $\mathbf{P}_n$, $n \geq 3$

Let us first look at 2-bundles on $\mathbf{P}_3$. Again it is easy enough to find at least one example of an indecomposable 2-bundle. Take homogeneous coordinates $(\xi_1, \ldots, \xi_4)$ in $\mathbf{P}_3$ and attach to each point $(p_1, \ldots, p_4)$ the hyperplane $p_2\xi_1 - p_1\xi_2 + p_3\xi_4 - p_4\xi_3 = 0$, passing through p. This gives a 2-subbundle of the tangent bundle to $\mathbf{P}_3$, which has Chern classes $c_1 = 2$, $c_2 = 2$ and therefore is indecomposable. It is called a collineation bundle, and is closely related to classical projective geometry ("Nullsysteme" and linear complexes). Instead of taking as coefficients $(p_2, -p_1, p_4, -p_3)$ you can take certain other linear forms in the p_i's, but all bundles thus obtained are equivalent under the action of $PGL(3)$.

However, this is only one example; we would like to have more systematic ways of obtaining 2-bundles on $\mathbf{P}_3$. Can Schwarzenberger's successful methods be generalised?

As to method (1), you can't expect much by taking ramified coverings of higher dimensional varieties. Namely, let $\mathcal{V}$ be a very ample 2-vector bundle on a smooth algebraic variety X. We consider the embedding of $Proj(\mathcal{V})$ in a $\mathbf{P}_N$, given by the tautological bundle $\mathcal{L}$ on $Proj(\mathcal{V}^*) = Proj(\mathcal{V})$. When we take a general quadric Q in $\mathbf{P}_N$, then $Y = Proj(\mathcal{V}) \cap Q$ is a smooth n-dimensional variety. And if $f: Y \to X$ denotes the projection, then $f_*(\mathcal{L} \mid Y) \cong \mathcal{V}$. In case that $\dim X = 2$, the map f is a 2-fold covering, hence the fact that in this case every 2-bundle can be obtained by method (1). But if $\dim X \geq 3$, then Y will always contain fibres of $Proj(\mathcal{V})$. So if we take instead of coverings "generalised coverings" we obtain also for $n \geq 3$ all bundles by method (1). In the case $n = 3$ these "generalised coverings" are desingularised genuine 2-fold coverings, ramified over a surface S with ordinary double points only. This means that the unique point above a double point of S is replaced by a $\mathbf{P}_1$.

The direct image of a line bundle will be a 2-bundle as soon as the restriction to all of these $\mathbf{P}_1$'s has degree 1.

For example, if we take for S a Kummer surface in $\mathbf{P}_3$, then the corresponding generalised covering becomes a $\mathbf{P}_1$-bundle over a certain rational surface. There is, up to tensoring with $f^*(\mathcal{O}(k))$, exactly one line bundle on the generalised covering satisfying the condition above; its direct image on $\mathbf{P}_3$ is a collineation bundle, tensored with some $\mathcal{O}(k)$.

It will certainly be worth-while to consider the relations between generalised ramified coverings and 2-bundles. Only recently some work has been done in this direction ([8]).

On the other hand, Schwarzenberger's second method has been very much in the centre of the developments. In the higher dimensional case you have to blow up not points but subvarieties of codimension 2. Using extensions of line bundles on such blown-up varieties a general theorem can be proved, which for $\mathbf{P}_n$ reads as follows.

THEOREM V.1: *Let X be a smooth subvariety of codimension 2 and degree g in $\mathbf{P}_n$, $n \geq 3$, and let $\mathcal{N}$ be the normal bundle of X in $\mathbf{P}_n$. Assume that $\wedge^2\mathcal{N} = \mathcal{O}_X(c)$. Then there is a 2-vector bundle $\mathcal{V}$ on $\mathbf{P}_n$ with the following properties*:

(i) *$\mathcal{V}$ has a section, vanishing precisely and transversally on X*;
(ii) *$c_1(\mathcal{V}) = c$ and $c_2(\mathcal{V}) = g$*;
(iii) *$\mathcal{V}$ splits if and only if X is a complete intersection.*

This result is already implicitly contained in Serre's lecture [25] and Horrocks paper [22]. Nowadays this theory is formulated both more generally and more precisely, compare [19], Remark 1.1.1. As to the generality it is sufficient to require that X is locally a complete intersection (it is indeed possible to speak about the normal bundle in that case, see [18], p. 245). Furthermore, the theorem can be proved without any blowing-up. Namely, suppose $\mathcal{V}$ is a 2-bundle on the smooth variety X, with a section having a subscheme Y of codimension 2 as set of zeros. Then there is an exact sequence of *sheaves*:

$$0 \longrightarrow \mathcal{O}_X \longrightarrow \mathcal{V} \longrightarrow \mathcal{J}(Y) \otimes_{\mathcal{O}_X} \wedge^2\mathcal{V} \longrightarrow 0.$$

So $\mathcal{V}$ occurs as a special type of extension. The proof of Theorem V.1 then consists of conversely constructing $\mathcal{V}$ as such an extension.

As was observed by Horrocks ([22], Theorem 3) Theorem V.1 is sufficient to realise on $\mathbf{P}_3$ all possible Chern classes c_1, c_2 with $c_1c_2 \equiv 0\,(2)$ as the Chern classes of a 2-bundle. You just have to take for Y a suitable union of disjoint smooth complete intersections.

However, as we have seen in the introduction, there are sometimes two different topological bundles with the same Chern classes. So Horrock's argument does not yet show that every continuous 2-bundle on $\mathbf{P}_3$ has an algebraic structure. But this is nevertheless true, as was shown by Atiyah and Rees in 1976 ([4], p. 131).

Up to now no attempt has been made to classify unstable 2-bundles on $\mathbf{P}_3$, nor did anybody study 2-bundles on other special threefolds. For the classification of stable bundles we refer again to Hartshorne's lectures.

As to 2-bundles on higher dimensional projective spaces it was clear around 1970 that, certainly, there is this nice Theorem V.1, but it is not of much help. For it is very difficult to find smooth subvarieties of codimension 2 with induced canonical class (that is the same as induced $\Lambda^2\mathcal{N}$ where $\mathcal{N}$ is the normal bundle), which are not complete intersections. In fact, only one type was ever found: tori, embedded as surfaces of degree 10 in $\mathbf{P}_4(\mathbf{C})$ (using the theory of θ-divisors you can show that on a large class of 2-dimensional tori there is a very ample divisor D with $D^2 = 10$, and hence $dim|D| = 4$; strangely enough, this never has been explained in detail anywhere). Up to automorphisms of $\mathbf{P}_4(\mathbf{C})$, all these tori give one and the same bundle. This is the famous Horrocks–Mumford bundle, which can also be constructed directly, as was done in [24], by using linear algebra. It can in fact be constructed as the "cohomology bundle" of a "monad": on $\mathbf{P}_4(\mathbf{C})$ there is a sequence of bundles and homomorphisms

$$\underbrace{\mathcal{O}(2)\oplus\ldots\oplus\mathcal{O}(2)}_{5\text{ times}} \xrightarrow{h_1} \Lambda^2\mathcal{T}_{\mathbf{P}_4}\oplus\Lambda^2\mathcal{T}_{\mathbf{P}_4} \xrightarrow{h_2} \underbrace{\mathcal{O}(3)\oplus\ldots\oplus\mathcal{O}(3)}_{5\text{ times}}$$

with h_1 injective, h_2 surjective and $h_2 \circ h_1 = 0$, and the Horrocks–Mumford bundle is nothing but kernel(h_2)/image(h_1) (compare [9], §4).

The standard procedures, applied to this bundle, give many indecomposable 2-bundles on $\mathbf{P}_4$ and hence many smooth surfaces, with induced canonical bundle which are not complete intersections. It is remarkable that, with exception of the forementioned tori, none of these surfaces has ever been described directly.

Not being able to produce more indecomposable 2-bundles on $\mathbf{P}_4$ starting from smooth surfaces, people have tried to use instead other local (but not global) complete intersections, in particular configurations of planes and quadrics, but it was all in vain. In the complex case, and more generally in all cases except that of characteristic 2, no other indecomposable 2-bundles have been found except for the Mumford–Horrocks bundle and its satellites. There is, however, an indecomposable 2-bundle on $\mathbf{P}_5$ in characteristic 2, due to Tango ([40], p. 206).

What about higher dimensional bundles? There are enough examples of indecomposable k-bundles with $k \geq n$, though classifying them is another matter, but compare [34], 2.3. For all n odd you can define on $\mathbf{P}_n$ an $(n-1)$-bundle in the same way as the collineation bundle was defined on $\mathbf{P}_3$ at the beginning of this section; they all are indecomposable. Also for n

even there are indecomposable $(n-1)$-bundles on $\mathbf{P}_n$ ([39]). But otherwise there are few examples. Using representation theory, Horrocks has constructed a 3-bundle on $\mathbf{P}_5$, which is indecomposable in all characteristics but 2.

In his Leiden thesis ([44]) Vogelaar has given a theorem analogous to Theorem V.1 for 3-bundles. This result enables him to construct 3-bundles on $\mathbf{P}_3$ with prescribed Chern classes c_1, c_2 and c_3, provided of course that the topological condition $c_1c_2 - c_3 \equiv 0\,(2)$ is satisfied.

Combining Vogelaar's result with (I.4), Fact II.2 and the results of Atiyah and Rees (loc. cit.) we find:

THEOREM V.2: *Every continuous vector bundle on* $\mathbf{P}_3(\mathbf{C})$ *admits an algebraic structure.*

The interest in the existence of indecomposable bundles of low rank on projective spaces was much stimulated by Barth's discovery (compare [6], [7]) that if Y is a smooth r-dimensional subvariety of $\mathbf{P}_n$, then $H^i(\mathbf{P}_n, \mathbf{Z}) \overset{\approx}{\to} H^i(X, \mathbf{Z})$ and $\pi_i(X) \overset{\approx}{\to} \pi_i(\mathbf{P}_n)$ for $i \leq 2r-n$. In other words, in a certain sense smooth varieties of small codimension look very much like complete intersections. This result, combined with Theorem V.1 implies in particular that for $n \geq 6$ the existence of indecomposable 2-bundles is equivalent to the existence of smooth subvarieties of codimension 2, which are not complete intersections.

The general feeling is that there will be very few indecomposable 2-bundles on $\mathbf{P}_n$, $n \geq 5$, and probably none. In fact, Hartshorne conjectured in [17] (p. 1017) that every smooth subvariety of codimension $<\frac{1}{3}n$ in $\mathbf{P}_n$ is a complete intersection. This implies that on $\mathbf{P}_n$, $n \geq 7$, every 2-bundle is a direct sum of two line bundles.

Though these conjectures in general remain unproved, there are however some partial results.

Let us first consider unstable 2-bundles $\mathcal{V}$ on $\mathbf{P}_n$, $n \geq 3$. By definition there is a $r_0 \in \mathbf{Z}$, with $c_1(\mathcal{V}(r_0)) < 0$, such that $\mathcal{V}(r_0)$ has a section, whereas $\mathcal{V}(r_0-1)$ has not. This section has a zero scheme Y of dimension $n-2$. Now Y can't be smooth. For if a 2-vector bundle $\mathcal{V}$ on a smooth X has a section with a smooth zero set Y of codimension 2, then $\mathcal{V} \mid Y$ is isomorphic to the normal bundle $\mathcal{N}$ of Y in X. But if $X = \mathbf{P}_n$, then $\mathcal{N}$, being a quotient of $\mathcal{T}_{\mathbf{P}} \mid Y$, is generated by its sections. Hence $c_1(\mathcal{N}) \geq 0$, whereas $c_1(\mathcal{V} \mid Y) < 0$ if $\mathcal{V}$ is unstable.

Nevertheless, there are unstable 2-bundles on $\mathbf{P}_3$; the zero scheme of the "first section" is a complicated curve.

In [15] Grauert and Schneider tried to prove that (in the complex case) this no longer is possible for $n \geq 4$. However, there is a gap in their proof. If this gap can be filled, then there are important consequences. I mention one of them. It can be proved that a 2-bundle on $\mathbf{P}_n$ with $c_1^2 - 4c_2 > 0$ is always unstable. But according to what is said in the introduction about continuous 2-bundles on $\mathbf{P}_4$, there exist many indecomposable continuous bundles with $c_1^2 - 4c_2 > 0$. It would follow that on $\mathbf{P}_4(\mathbf{C})$ there are continuous 2-bundles (trivially with algebraic Chern classes) which have no algebraic structure. But for the moment this remains unproved and there still is no example of any continuous vector bundle on $\mathbf{P}_n$, which has no algebraic structure.

And then there is another type of theorems concerning the decomposability of bundles on projective spaces. These theorems will be treated in the last section.

VI. The tower theorems

Already long before Schwarzenberger, people were well aware of the importance of the jumping lines of a vector bundle on $\mathbf{P}_n$, but this author was the first to use them in a more systematic way.

Let $\mathcal{V}$ be a 2-vector bundle on $\mathbf{P}_n$ with $c_1(\mathcal{V})$ either 0 or -1. Since for every 2-bundle $\mathcal{W}$ there is a $r \in \mathbf{Z}$ such that $c_1(\mathcal{W}(r)) = 0$ or -1, this is no restriction as far as the decomposability of $\mathcal{V}$ is concerned. Let us focus on the case $c_1 = 0$; the case $c_1 = -1$ is completely analogous.

If L is any line in $\mathbf{P}_n$, then we have by Theorem III.1:

$$\mathcal{V} \mid L \cong \mathcal{O}_L(a) \oplus \mathcal{O}_L(-a),\ a \geq 0.$$

We denote by ℓ the point in $G_{n+1,2}$ corresponding to L, and set $a = a(\ell)$. Furthermore we define

$$S_a(\mathcal{V}) = S_a = \{\ell \in G_{n+1,2},\ a(\ell) = a\}.$$

On the product $\mathbf{P}_n \times G_{n+1,2}$ we consider the flag manifold F, i.e. the smooth algebraic variety, formed by the pairs (x,ℓ) with $x \in L$. Let $p: F \to \mathbf{P}_n$ and $q: F \to G_{n+1,2}$ be the projections. Then we have

$$S_a = \{\ell \in G_{n+1,2}\colon dim\ \Gamma(q^{-1}(\ell), p^*(\mathcal{V}) \mid q^{-1}(\ell)) = a + 1\}.$$

From standard theorems concerning direct images by flat maps (for $q: F \to G_{n+1,2}$ is a $\mathbf{P}_1$-bundle) it follows that there is a sequence $a_1 < a_2 < \cdots < a_k$ (k depending on $\mathcal{V}$), such that $S_{a_1} \supset S_{a_2} \supset \cdots \supset S_{a_k}$, with $\bigcup_{i=1}^{k} S_{a_i} = G_{n+1,2}$ and S_{a_1} Zariski-open in $G_{n+1,2}$, next S_{a_2} Zariski-open in $\bar{S}_{a_2} = G - S_{a_1}$, next S_{a_3} Zariski-open in $\bar{S}_{a_3} = G - S_{a_1} - S_{a_2}, \ldots$, finally S_{a_k} Zariski-closed in $G_{n+1,2}$. So on an open set $a(\ell)$ is constant, whereas $a(\ell)$ jumps successively on smaller and smaller locally closed subsets of the complement. Naturally, the lines L with $\ell \notin S_{a_1}$ are called the jumping lines of $\mathcal{V}$. We set $a_1 = m(\mathcal{V}) = m$ and $a_k = M(\mathcal{V}) = M$.

Among many other things the jumping lines have been used by Barth and myself to prove

THEOREM VI.1 (Babylonian tower theorem) ([10] Theorem I): *Be given for all* $n \geq 0$ *a linear embedding* $f_n : \mathbf{P}_n \to \mathbf{P}_{n+1}$, *and on every* $\mathbf{P}_n$ *an algebraic 2-vector bundle* $\mathcal{V}_n$, *with* $\overset{*}{f}_n(\mathcal{V}_{n+1}) \cong \mathcal{V}_n$. *Then all bundles* $\mathcal{V}_n$ *are decomposable.*

This theorem is an immediate consequence of

THEOREM VI.2 ([10] Theorem II): *There exists a function* $F : \mathbf{Z} \times \mathbf{Z} \to \mathbf{Z}$ *with the following property: every 2-bundle* $\mathcal{V}$ *on* $\mathbf{P}_n$ *with* $c_1(\mathcal{V}) = 0$ *or* -1, *with* $a(\mathcal{V}) = a$ *and* $c_2(\mathcal{V}) = c_2$ *splits as soon as* $n > F(a,c_2)$.

REMARK: In [16] such an F is explicitly given as a polynomial.

I shall sketch our proof of this theorem. The idea goes back to [43].

For every point $x \in \mathbf{P}_n$ we set $m(x) = \min_{x \in L} a(\ell)$ and $M(x) = \max_{x \in L} a(\ell)$. We assume $M(x) \geq 1$, the case $M(x) = 0$ for some point x being only a minor nuisance. Consider points $\ell \in S_{\max(x)} \cup q(p^{-1}(x))$, i.e. points ℓ, such that L passes through x and $a(\ell)$ is as big as possible for such lines. The restriction $\mathcal{V} | L$ has a unique 1-subbundle of degree $M(x)$; this subbundle determines a unique point in the fibre of *Proj*($\mathcal{V}$) over x. Thus we get a regular map from $S_{\max(x)} \cap q(p^{-1}(x))$ into $\mathbf{P}_1$. Now $S_{\max(x)} \cap q(p^{-1}(x))$ is a subvariety of the projective $(n-1)$ space $q(p^{-1}(x))$; as soon as it has codimension $\leq \frac{1}{2}(n-1)$ in this projective space, the forementioned map must be constant. If this holds for all points $x \in \mathbf{P}_n$, then we get a section in *Proj*($\mathcal{V}$), which turns out to be regular, as soon as $n \geq 5$. But a regular section in *Proj*($\mathcal{V}$) means a 1-subbundle of $\mathcal{V}$, which makes $\mathcal{V}$ decomposable. The codimension condition is proved to be satisfied if S_M has codimension $\leq \frac{1}{2}(n-1)$ in G. Now *Proj*($p^*(\mathcal{V})$) can be seen as a family of Hirzebruch surfaces over $G_{n+1,2}$. Locally, this family is induced from the versal family by a map into the versal base space. It follows from the situation in the versal family of such Hirzebruch surfaces that the codimension condition is satisfied if $M \leq m + \frac{1}{4}(n-1)$. Therefore we shall be ready as soon as we have shown: there exists a universal function $N : \mathbf{Z} \times \mathbf{Z} \to \mathbf{Z}$ with the property that for every 2-bundle on a projective space (with $c_1 = 0$) the inequality

$$M(\mathcal{V}) \leq N(m(\mathcal{V}), c_2)$$

holds. This is first of all proved for a 2-bundle on $\mathbf{P}_2$, and then for a bundle on $\mathbf{P}_n$ by applying the result for $\mathbf{P}_2$ two times: once to a plane passing through a line L with $a(\ell) = m$ and once to a plane passing through a line L with $a(\ell) = M$, in such way that both planes have a line in common.

Theorem VI.2 implies a theorem for submanifolds, which was proved independently by Hartshorne ([17]), and which was eventually generalised by Barth and myself to subvarieties in $\mathbf{P}_n$ of any codimension ([7], p. 409):

THEOREM VI.3: *If X is a smooth subvariety on $\mathbf{P}_n$ of dimension d and degree g, then X is a complete intersection as soon as $d \geq g(g-1)$ (and even $d \geq 4g-7$, if $d = n-2$).*

Later these results were again extended in several directions.

(a) Infinitely extendable vector bundles on Grassmann varieties.

The situation is as follows. For a field K the natural embeddings

$$K^d \subset K^{d+1} \subset \cdots \subset K^n \subset \cdots$$

induce embeddings of Grassman varieties

$$G_{d,d} \subset G_{d+1,d} \subset \cdots \subset G_{n,d}(K) \subset \cdots.$$

Which are the vector bundles existing on the whole tower? More precisely, one would like to find all sequences $\{\mathcal{V}_n\}_{n \geq d}$, where $\mathcal{V}_n$ is a k-bundle on $G_{n,d}$ with the property that $\mathcal{V}_n \mid G_{n-1,d} \cong \mathcal{V}_{n-1}$ for all $n \geq d$.

An example of such an infinite bundle tower is obtained by taking for $\mathcal{V}_n$ the universal subbundle. Other examples are provided by the bundles which are derived from the universal bundle by linear algebra, that is by taking duals, direct sums, tensor products, symmetric products and exterior products. (These include all sums of line bundles). Hartshorne in fact conjectures that there are no others. This was proved by Barth and myself in a special case ([10], Theorem 1):

THEOREM VI.4: *Consider the embeddings of (complex) Grassmannians*

$$G_{2,2} \subset G_{3,2} \subset \cdots \subset G_{n,2} \subset \cdots$$

induced by the natural embeddings

$$\mathbf{C}^2 \subset \mathbf{C}^3 \subset \cdots \subset \mathbf{C}^n \subset \cdots.$$

If on every $G_{n,2}$ there is a 2-bundle $\mathcal{V}_n$ with $\mathcal{V}_n \mid G_{n-1,d} \cong \mathcal{V}_{n-1}$, then either all bundles $\mathcal{V}_n$ are decomposable or for all n they are isomorphic to the universal subbundle, tensored with a line bundle.

As a corollary we obtained a tower theorem for smooth subvarieties of codimension 2 on Grassmann varieties ([11], Theorem 3).

We also proved a "finite form" of Theorem VI.4 ([11], Theorem 4). This has consequences for the representability of 2-codimensional homology classes by smooth subvarieties, a question which was tackled from the topological side by Hartshorne, Rees and E. Thomas.

(b) Sato ([31]) and Tyurin ([42]) generalised Theorem VI.1 to bundles of any rank. Sato also proved a slight extension of Theorem VI.4.

(c) Recently, Elencwajg and Forster ([14]) generalised an important step in the proof of Theorem VI.2. They proved that if for a normalised k-bundle $\mathcal{V}$ on $\mathbf{P}_n$ (so $c_1(\mathcal{V})$ is one of the numbers $0, 1, \ldots, k-1$) the general splitting type and the second Chern class are given, then in terms of these you can find effective bounds for the higher Chern classes, for all cohomology groups and for the splitting types that may occur. For stable bundles this implies that "everything can be bounded in terms of c_1 and c_2". This shows some resemblance with a recent theorem of Yau: if on a compact Kähler manifold with ample canonical bundle c_1 and c_2 (of the tangent bundle) vanish, then all Chern classes vanish.

At the end of the period we have considered it was clear that, as far as the existence of indecomposable bundles of low rank on projective spaces is concerned, new ideas had to be awaited. On the other side it was realised that very nice results could and should be proved concerning stable 2-bundles on $\mathbf{P}_2$ and $\mathbf{P}_3$. In this way vector bundles could take their place in the rapid development of Mumford's geometric invariant theory. Thus started with the Grauert-Mülich theorem ([34], p. 18) and Hulsbergen's thesis ([26]) the period which will be treated in Hartshorne's lectures.

REFERENCES

[1] M.F. Atiyah: Complex fibre bundles and ruled surfaces. Proc. London Math. Soc. (3) *5* (1955) 407–434.

[2] M.F. Atiyah: On the Krull–Schmidt theorem with applications to sheaves. Bull. Soc. Math. France *84* (1957) 307–317.

[3] M.F. Atiyah: Vector bundles over an elliptic curve. Proc. London Math. Soc. *7* (1957) 414–452.

[4] M.F. Atiyah and E. Rees: Vector bundles on projective 3-space. Inv. Math. *35* (1976) 131–153.

[5] W. Barth: Der Abstand von einer komplex-projektiven Mannigfaltigkeit im komplex-projektiven Raum. Math. Ann. *187* (1970) 150–162.

[6] W. Barth: Larsen's theorem on the homotopy groups of projective manifolds of small embedding codimension. In: Algebraic Geometry. Proc. Symp. Pure Math. Am. Math. Soc. *29* (1974) 307–313.

[7] W. Barth: Submanifolds of low codimension in projective space. Proc. Int. Congr. Math. Vancouver, vol. *I* (1974) 409–413.

[8] W. Barth: Counting singularities of quadratic forms on vector bundles. To appear.

[9] W. Barth: Kummer surfaces associated with the Mumford-Horrocks bundle. This volume.

[10] W. Barth and A. Van de Ven: A decomposability criterion for algebraic 2-bundles on projective spaces. Inv. Math. *25* (1974) 91–106.

[11] W. BARTH and A. VAN DE VEN: On the geometry in codimension 2 of Grassmann manifolds. In: Classification of algebraic varieties and compact complex manifolds. Lecture Notes in Math. *412*, Springer, Berlin, Heidelberg, New York, (1974) 1–35.

[12] R. BOTT: Homogeneous vector bundles. Ann Math. *66* (1957) 203–248.

[13] S.-S. CHERN: Characteristic classes of hermitian manifolds. Ann. Math. *47* (1946) 85–121.

[14] G. ELENCWAJG and O. FORSTER: Bounding cohomology groups of vector bundles on $\mathbf{P}_n$. Math. Ann. *246* (1980) 251–270.

[15] H. GRAUERT and M. SCHNEIDER: Komplexe Unterräume und holomorphe Vektorbündel von Rang zwei. Math. Ann. *230* (1977) 75–90.

[16] A. GROTHENDIECK: Sur la classification des fibrés holomorphes sur la sphère de Riemann. Am. J. Math. *79* (1957) 121–138.

[17] R. HARTSHORNE: Varieties of small codimension in projective space. Bull. Am. Math. Soc. *80* (1974) 1017–1032.

[18] R. HARTSHORNE: Algebraic Geometry. Graduate Texts in Math. *52*, Springer, Berlin, Heidelberg, New York (1977).

[19] R. HARTSHORNE: Stable vector bundles of rank 2 on $\mathbf{P}_3$. Math. Ann. *238* (1978) 229–280.

[20] R. HARTSHORNE: Algebraic vector bundles on projective spaces: a problem list. To appear.

[21] F. HIRZEBRUCH: Topological methods in algebraic geometry. Third edition. Translated by R.L.E. Schwarzenberger. Springer, Berlin, Heidelberg, New York (1966).

[22] G. HORROCKS: A construction for locally free sheaves. Topology *7* (1968) 117–120.

[23] G. HORROCKS: Examples of rank-three vector bundles on five-dimensional projective space. J. London Math. Soc. (2) *18* (1978) 15–27.

[24] G. HORROCKS and D. MUMFORD: A rank-2 vector bundle on $\mathbf{P}^4$ with 15,000 symmetries. Topology *12* (1973) 63–81.

[25] K. HULEK: On the classification of stable k-bundles on $\mathbf{P}_2$. To appear.

[26] W. HULSBERGEN: Vector bundles on the complex projective plane. Thesis Leiden (1976).

[27] T.Y. LAM: Serre's conjecture. Lecture Notes in Mathematics *635*. Springer, Berlin, Heidelberg, New York (1978).

[28] M. MARUYAMA: On a family of algebraic vector bundles. In: Number theory, algebraic geometry and commutative algebra (in honor of Y. Akizuki). Kinokunija, Tokio (1973).

[29] M.S. NARASIMHAN and C.S. SESHADRI: Stable and unitary vector bundles on a compact Riemann surface. Ann. Math. (2) *82* (1965) 540–567.

[30] I.R. SAFAREVIČ et al.: Algebraic surfaces. Proc. Steklov Inst. Math. *75* (1965), translated by Am. Math. Soc. (1967).

[31] E. SATO: On the decomposability of infinitely extendable vector bundles on projective spaces and Grassmann varieties. J. Math. Kyoto Univ. *17* (1977) 127–150.

[32] R.L.E. SCHWARZENBERGER: Vector bundles on algebraic surfaces. Proc. London Math. Soc. (3) *11* (1961) 601–622.

[33] R.L.E. SCHWARZENBERGER: Vector bundles on the projective plane. Proc. London Math. Soc. (3) *11* (1961) 623–640.

[34] M. SCHNEIDER: Holomorphic vector bundles on $\mathbf{P}_n$. Sém. Bourbaki no 530 (1978).

[35] J.-P. SERRE: Sur les modules projectifs. In: Sém. Dubreuil–Pisot 1960/61.

[36] J.-P. SERRE: Faisceaux algébriques cohérents. Ann. Math. *61* (1955) 197–278.

[37] J. SIMONIS: A class of indecomposable algebraic vector bundles. Math. Ann. *192* (1971) 262–278.

[38] R. SWITZER: Complex 2-plane bundles over complex projective space. To appear in Math. Zeitschrift.

[39] H. TANGO: An example of an indecomposable vector bundle of rank $n-1$ on $\mathbf{P}_n$. J. Math. Kyoto Univ. *16* (1976) 137–141.

[40] H. TANGO: On morphisms from projective space $\mathbf{P}_n$ to the Grassmann variety $Gr(n, d)$. J. Math. Kyoto Univ. *16* (1976) 201–207.

[41] A. THOMAS: Almost complex structures on projective spaces. Trans. Am. Math. Soc. *193* (1974) 123–132.

[42] A.N. TYURIN: Finite dimensional vector bundles over infinite varieties. Math. U.S.S.R. Isvestija *10* (1976) 1187–1204.

[43] A. VAN DE VEN: On uniform vector bundles. Math. Ann. *195* (1972) 245–248.

[44] J. VOGELAAR: Constructing vector bundles from codimension-2 subvarieties. Thesis Leiden (1978).

[45] W.-S. WU and G. REEB: Sur les espaces fibrés et les variétés feuilletées. Act. Scient. et Industr. 1183. Hermann, Paris (1952).

Mathematisch Instituut
Rijksuniversiteit Leiden
Wassenaarseweg 80, Leiden (Netherlands)

FOUR YEARS OF ALGEBRAIC VECTOR BUNDLES: 1975–1979

Robin Hartshorne

During the last four years, work on algebraic vector bundles has focused on three areas: existence of varieties of moduli; classification of bundles of low rank on low-dimensional varieties; and the relations with instantons and mathematical physics. I will describe briefly each of these areas, but refer to the Bourbaki seminar talk [34], the forthcoming lecture notes [32], and the problem list [21] for further details and references.

§1. Variety of moduli

As Van de Ven explained in his lectures, there does not exist a good variety of moduli for the set of all algebraic vector bundles of given rank r and given Chern classes $c_1, \dots, c_r$ on a fixed nonsingular projective algebraic variety X. There are two reasons for this. One is that (except for a few special cases, such as curves of genus 0, 1) there are too many such bundles. They cannot all be obtained in a family parametrized by a scheme of finite type over the base field k. For example, one can show [18, 2.5] that on $\mathbf{P}^3$, for every choice of integers c_1, c_2 satisfying $c_1c_2 \equiv 0 \pmod 2$, there exist families of mutually nonisomorphic rank 2 bundles parametrized by varieties of arbitrarily high dimension.

The other reason is that the topology on the set of isomorphism classes of vector bundles dictated by algebraic families is bad. There exist families parametrized by the t-line where the bundles $\mathscr{E}_t$ are all isomorphic to each other for $t \neq 0$, while the bundle $\mathscr{E}_0$ is different. Or, there exist families defined for $t \neq 0$ which can be completed in many different ways to a family defined for all t.

Therefore, in order to obtain a reasonable variety of moduli – that is, a quasiprojective algebraic scheme whose points should parametrize isomorphism classes of bundles – one must restrict the class of bundles under consideration. For this Mumford [29] introduced the notion of a *stable* bundle. It is a condition which allows one to apply the methods of geometric invariant theory developed in his book [30].

Mumford's definition, for a vector bundle $\mathscr{E}$ on a nonsingular curve C, is this: for every nontrivial subbundle $\mathscr{F} \subseteq \mathscr{E}$,

$$\deg \mathscr{F}/\operatorname{rank} \mathscr{F} < \deg \mathscr{E}/\operatorname{rank} \mathscr{E}.$$

Using this definition, Mumford was able to prove the existence of a coarse moduli scheme for stable vector bundles on curves in characteristic 0. These moduli schemes were studied in the complex case by Narasimhan and Seshadri [31], and later Seshadri [35] was able to extend the existence proof to arbitrary characteristic.

On varieties of higher dimension it is not so evident what definition of stability one should use. One definition, due to Mumford and Takemoto [37], is a direct generalization of the above. Let X be a nonsingular projective variety with a given very ample divisor H. A vector bundle $\mathscr{E}$ is *stable* if for every coherent subsheaf $\mathscr{F} \subseteq \mathscr{E}$ (not necessarily locally free) with rank $\mathscr{F} <$ rank $\mathscr{E}$,

$$\deg c_1(\mathscr{F})/\operatorname{rank} \mathscr{F} < \deg c_1(\mathscr{E})/\operatorname{rank} \mathscr{E}.$$

Here the *degree* of the first Chern class c_1 is measured with respect to the given very ample divisor H.

Another more subtle definition is used by Gieseker [16] and Maruyama [26]. They found that on a nonsingular projective variety X it was more natural to consider all coherent torsion-free sheaves, not just the locally free ones which correspond to vector bundles. For any torsion-free coherent sheaf $\mathscr{F}$, define $P(\mathscr{F})$ to be ($1/\operatorname{rank} \mathscr{F}$) times the Hilbert polynomial of $\mathscr{F}$ with respect to the given projective embedding of X. Then a sheaf $\mathscr{E}$ is said to be *stable* if for every subsheaf $\mathscr{F} \subseteq \mathscr{E}$, of rank $<$ rank $\mathscr{E}$, $P(\mathscr{F}) < P(\mathscr{E})$, where inequality for polynomials means the lexicographic ordering on the coefficients. Using this definition, Maruyama [27] was able to prove, using techniques developed by Gieseker in some special cases, the existence of variety of moduli for stable torsion-free coherent sheaves of all ranks with given Hilbert polynomial on a nonsingular projective variety X, in any characteristic, except for the question of boundedness. That is, Maruyama showed these sheaves are parametrized by the union of an increasing family of quasiprojective schemes. If one could prove they form a bounded family, then the moduli space is in fact quasiprojective. Maruyama [28] also proved boundedness in some special cases: rank 2 in all characteristics, and rank 3, 4 in characteristic 0.

Another definition of stability, or rather, of *unstability* has been proposed in characteristic 0 by Bogomolov (see [9] and [33]). A bundle $\mathscr{E}$ of rank r is *unstable* if there exists a representation ρ of $GL(r)$ of determinant 1 such that the associated bundle $\mathscr{E}^{\rho}$ has a nonzero section which vanishes at some point of X. For bundles on curves and projective spaces, this definition agrees with those above (where "unstable" is taken to mean "not semistable", and *semistable* is the notion obtained by replacing $<$ by $\leq$ in the definitions). Note that the definition of Bogomolov does not depend on a given projective embedding of X. The methods of Bogomolov seem very powerful, and have not yet been fully exploited.

It remains to discuss the problem of boundedness. On a nonsingular projective variety X, do the stable vector bundles of given rank r and Chern classes $c_1, \ldots, c_r$ form a bounded family – i.e. can they be parametrized by a scheme of finite type over k? Or, to put the question in the generality of torsion-free coherent sheaves, do the stable sheaves of given rank with given Hilbert polynomial form a bounded family?

In characteristic 0, this problem has recently been solved in the affirmative by Forster, Hirschowitz, and Schneider [15], based on earlier work of Grauert and Mülich [17], Spindler [36], and Elencwajg and Forster [13]. The key point is to show that the restriction of a stable bundle to a general curve cut out by hyperplane sections in X has a filtration by subbundles whose degrees are not too far apart. This technique is a generalization of the method described by Van de Ven of restricting a bundle on projective space to a general line and studying its decomposition type there.

It is interesting to note that in Maruyama's work on the boundedness problem, an important step was an inductive result saying that the restriction of a semistable bundle on X to a general hyperplane section is also semistable. He could only prove this under restrictive conditions on the rank, and indeed a general theorem of this type is still unknown, even for vector bundles on projective spaces. However it was possible to solve the boundedness problem without this inductive step.

The boundedness problem in characteristic $p > 0$ remains open.

§2. Bundles of low rank on projective spaces

A big effort has gone into the study of rank 2 bundles on projective spaces, that being the first nontrivial case. Knowing that a variety of moduli exists for stable bundles, one can ask for a description of the bundles and their variety of moduli. As Van de Ven mentioned in his lectures, aside from the rank 2 bundle on $\mathbf{P}^4$ discovered by Horrocks and Mumford [23], no other rank 2 bundles are known (in characteristic 0) on any $\mathbf{P}^n$ with $n \geq 4$. However, much is now known about rank 2 bundles on $\mathbf{P}^2$ and $\mathbf{P}^3$.

On $\mathbf{P}^2$, an old result of Horrocks [22] shows that a bundle $\mathscr{E}$ is uniquely determined if one knows the sum of cohomology groups $\bigoplus_{l \in \mathbf{Z}} H^1(\mathbf{P}^2, \mathscr{E}(l))$, considered as a finite-length graded module over the homogeneous coordinate ring $S = k[x_0, x_1, x_2]$ of $\mathbf{P}^2$. A remarkable discovery of Barth [6] was that in the case of a stable rank 2 bundle $\mathscr{E}$ with $c_1 = 0$, the bundle is actually determined by knowing the middle pieces of this module, namely $H^1(\mathscr{E}(-2))$ and $H^1(\mathscr{E}(-1))$, together with the linear mappings $x_i : H^1(\mathscr{E}(-2)) \to H^1(\mathscr{E}(-1))$, $i = 0, 1, 2$, and the Serre duality isomorphism $H^1(\mathscr{E}(-1)) \cong H^1(\mathscr{E}(-2))'$. In this way the study of the bundles is reduced to linear algebra. Barth determines exactly which such linear algebra data can arise from stable bundles. Then using these methods, he is able to show that the variety of moduli of stable rank 2 bundles on $\mathbf{P}^2$ with $c_1 = 0$ and given c_2 (necessarily $c_2 \geq 2$) is an irreducible nonsingular rational variety of dimension $4c_2 - 3$.

This method was extended by Hulek [24] to give a similar result for the case of stable rank 2 bundles on $\mathbf{P}^2$ with $c_1 = -1$. Independently Ellingsrud and Strømme [14] have shown by a different method that this variety of moduli in the case $c_1 = -1$ is irreducible nonsingular and rational.

Another question which arises concerning these varieties of moduli is the existence of a universal family. In general, the methods of geometric invariant theory used to prove the existence of the variety of moduli guarantee only a *coarse* moduli scheme. That means that the points of the moduli space are in one-to-one correspondence with the isomorphism classes of bundles, and that for any family of bundles parametrized by a scheme T, there is a natural corresponding morphism of T to the moduli scheme M. However, in general there is no *universal family* of bundles parametrized by M, i.e. a family over M which would pull back to any given family parametrized by a scheme T. So one can ask when such a universal family exists. Maruyama [27] has some results in this direction, and Le Potier [25] has determined, in the case of stable rank 2 bundles over $\mathbf{P}^2_{\mathbf{C}}$, for exactly which pairs c_1, c_2 the moduli space has a universal family.

On $\mathbf{P}^3$ the situation is much more complicated. For simplicity I will discuss only the case of rank 2 bundles with even first Chern class, which we can normalize by tensoring with a line bundle so that $c_1 = 0$. Then it is known that stable bundles exist for all values of $c_2 > 0$. The moduli spaces for $c_2 = 1, 2$ are irreducible and nonsingular of dimension 5, 13, respectively. For $c_2 \geq 3$ there are at least two separate connected components of the moduli space, corresponding to the topological mod 2 invariant α of rank 2 bundles with $c_1 = 0$ on $\mathbf{P}^3$. The dimension suggested by deformation theory for the moduli space is $8c_2 - 3$, however for $c_2 \geq 5$ it is known that there may be irreducible components of larger dimension. The problem of determining all the irreducible components of the moduli space of stable rank 2 bundles on $\mathbf{P}^3$ with given c_1, c_2 thus remains open. See Barth [5], Barth and Hulek [8], and Hartshorne [18] for discussion of these questions.

An invariant of a stable rank 2 bundle $\mathscr{E}$ on $\mathbf{P}^3$ with $c_1 = 0$ has been introduced by Barth and Elencwajg [7]. Let L be a general line in $\mathbf{P}^3$, let $p: X \to \mathbf{P}^3$ be the blowing-up of L, and let $q: X \to \mathbf{P}^1$ be the morphism determined by the pencil of planes passing through L. Then the sheaf $R^1 q_* p^* \mathscr{E}(-1)$ is locally free of rank c_2 on $\mathbf{P}^1$. Therefore it is isomorphic to a direct sum of line bundles $\bigoplus_{i=1}^{c_2} \mathcal{O}(k_i)$, with $k_i \in \mathbf{Z}$. Barth and Elencwajg show that the set of integers $\{k_i\}$, which I will call the *spectrum* of $\mathscr{E}$, is *symmetric*, i.e. $\{-k_i\} = \{k_i\}$, and *connected*, i.e., there are no gaps among the integers which occur. Thus if $c_2 = 3$, for example, there are just two possibilities for the spectrum: $0, 0, 0$ and $-1, 0, 1$. The first corresponds to the bundles with α – invariant 0, the second to those with α – invariant 1. I believe the spectrum will play an important role in the eventual determination of the irreducible components of the variety of moduli.

One special case is of particular interest, namely the bundles whose spectrum consists entirely of 0's. These bundles are also characterized by the property $H^1(\mathscr{E}(-2))=0$, and so are called *instanton* bundles, because they include all the bundles arising from the instantons of mathematical physics described in §3 below. It seems reasonable to conjecture, following Tyurin, that the instanton bundles with given c_2 form an irreducible nonsingular rational variety of dimension $8c_2-3$.

About bundles of rank ≥ 3 very little is so far known. Elencwajg [12] has begun a study of bundles of higher rank on $\mathbf{P}^2$, and has found examples of bundles of rank 4 which are *uniform* in the sense that their splitting type is the same on all lines, but which are not homogeneous under the action of PGL(2). Vogelaar [38] has shown on $\mathbf{P}^3$ that there exist rank 3 bundles with all possible Chern classes c_1, c_2, c_3 allowed by topology, i.e. $c_1c_2 \equiv c_3$ (mod 2). But it is not yet known exactly which c_1, c_2, c_3 can occur for *stable* rank 3 bundles on $\mathbf{P}^3$, although Elencwajg and Forster [13] have given a bound on c_3 in terms of c_1 and c_2.

§3. Instantons

In the last few years there have been several striking cases of differential equations arising in mathematical physics revealing a hidden connection with algebraic geometry. This confluence of ideas provides stimulation and new insight to both areas. The differential equation which concerns us here is the Euclidean classical Yang-Mills equation, whose self-dual solutions are called *instantons* by the physicists, and which should give some insight into the quantized Yang-Mills equation on Minkowski space corresponding to the "real world". For some years Penrose has been advocating the interpretation of the physics of real Euclidean or Minkowski space in terms of complex manifolds via his theory of twistors. Atiyah and Ward [1] showed by this method that the instantons could be interpreted as holomorphic (hence algebraic) vector bundles on $\mathbf{P}^3_{\mathbf{C}}$ with an added reality condition. Thus the problem of finding all instantons was reduced to the classification problem for these particular vector bundles.

This problem has now been solved, or more accurately, reduced to a problem in linear and quadratic algebra, by the efforts of Horrocks, Barth, Atiyah, Hitchin, Drinfeld and Manin.

Let us discuss the rank 2 case for simplicity. Horrocks showed that any vector bundle $\mathscr{E}$ on $\mathbf{P}^3$ could be represented by a *monad*, which is a sequence

$$\mathscr{A} \xrightarrow{\alpha} \mathscr{B} \xrightarrow{\beta} \mathscr{C}$$

of vector bundles, each a direct sum of line bundles, with α injective, β surjective, $\beta\alpha = 0$, and $\mathscr{E} = \ker \beta/\mathrm{im}\, \alpha$. Then Barth [8] observed that in the special case of stable bundles $\mathscr{E}$ with $c_1 = 0$ and $H^1(\mathscr{E}(-2)) = 0$, the monad could be written in the form

$$\mathcal{O}(-1)^{c_2} \xrightarrow{\alpha} \mathcal{O}^{2c_2+2} \xrightarrow{\beta} \mathcal{O}(1)^{c_2}.$$

Thus the bundle $\mathscr{E}$ is completely determined by the maps α,β which are matrices of linear forms in the homogenous coordinates x_0, x_1, x_2, x_3 of $\mathbf{P}^3$, satisfying certain linear and quadratic conditions.

Now in general a stable rank 2 bundle $\mathscr{E}$ on $\mathbf{P}^3$ with $c_1 = 0$ need not satisfy the condition $H^1(\mathscr{E}(-2)) = 0$ (see §2 above). However, using the extra reality conditions on the bundles arising from instantons, Atiyah and Hitchin [3], and independently Drinfeld and Manin [11], were able to prove this vanishing theorem for those bundles.

Thus the problem of classifying instantons was "reduced to linear algebra". The problem is not yet solved however. One knows that the variety of moduli of the instantons with instanton number k (corresponding to bundles with second Chern class $c_2 = k$) is a real analytic manifold of dimension $8k - 3$. But one does not yet know whether this manifold is connected (except for $k = 1, 2$) nor what its topology is like.

For further references about instantons and their corresponding bundles, see [2], [4], [10], [19], [20].

REFERENCES

[1] M.F. Atiyah and R.S. Ward: Instantons and algebraic geometry. Comm. Math. physics *55* (1977) 117–124.

[2] M.F. Atiyah, N.J. Hitchin and I.M. Singer: Self-duality in four-dimensional Riemannian geometry. Proc. R. Soc. Lond. A *362* (1978) 425–46.

[3] M.F. Atiyah, N.J. Hitchin, V.G. Drinfeld and Ju. I. Manin: Construction of instantons. Physics Letters *65A* (1978) 185–187.

[4] M.F. Atiyah: Geometry of Yang-Mills fields. (Fermi lectures).

[5] W. Barth: Some properties of stable rank–2 vector bundles on $\mathbf{P}_n$. Math. Ann. *226* (1977) 125–150.

[6] W. Barth: Moduli of vector bundles on the projective plane. Invent. math. *42* (1977) 63–91.

[7] W. Barth and G. Elencwajg: Concernant la cohomologie des fibrés algébriques stables sur $\mathbf{P}_n(\mathbf{C})$, in Variétés analytiques compactes, (Nice 1977) Lecture notes in Math. *683* Springer (1978) 1–24.

[8] W. Barth and K. Hulek: Monads and moduli of vector bundles, manusc. math. *25* (1978) 323–347.

[9] M. Deschamps: Courbes de genre géométrique borné sur une surface de type générale (d'après F.A. Bogomolov), Sem. Bourbaki *519* (1978).

[10] A. Douady and J.L. Verdier: Seminar notes on vector bundles and Yang-Mills fields, Astérisque *71–72* (1980).

[11] V.G. Drinfeld and Ju. I. Manin: Instantons and sheaves on $\mathbf{CP}^3$.

[12] G. ELENCWAJG: Des fibrés uniformes non homogènes, Math. Ann. *239* (1979) 185–192.
[13] G. ELENCWAJG and O. FORSTER: Bounding cohomology groups of vector bundles on $\mathbf{P}^n$, Math. Ann. *246* (1980) 251–270.
[14] G. ELLINGSRUD and S.A. STRØMME: On the moduli space for stable rank-2 vector bundles on $\mathbf{P}^2$. (to appear).
[15] O. FORSTER, A. HIRSCHOWITZ and M. SCHNEIDER: Type de scindage généralisé pour les fibrés stables, Nice conference 1979, (to appear).
[16] D. GIESEKER: On the moduli of vector bundles on an algebraic surface, Annals of Math. *106* (1977) 45–60.
[17] H. GRAUERT and G. MÜLICH: Vektorbündel vom Rang 2 über dem n-dimensionalen komplex-projektiven Raum, Manusc. math. *16* (1975) 75–100.
[18] R. HARTSHORNE: Stable vector bundles of rank 2 on $\mathbf{P}^3$, Math. Ann., *238* (1978) 229–280.
[19] R. HARTSHORNE: Stable vector bundles and instantons, Commun. math. Physics *59* (1978) 1–15.
[20] R. HARTSHORNE: Algebraic vector bundles on projective spaces, with applications to the Yang-Mills equation, in Complex Manifold Techniques in Theoretical Physics, Pitman Publ. Ltd. (1979) 35–44.
[21] R. HARTSHORNE: Algebraic vector bundles on projective spaces: a problem list, Topology *18* (1979) 117–128.
[22] G. HORROCKS: Vector bundles on the punctured spectrum of a local ring, Proc. Lond. Math. Soc. (3) *14* (1964) 689–713.
[23] G. HORROCKS and D. MUMFORD: A rank 2 vector bundle on $\mathbf{P}^4$ with 15,000 symmetries., Topology *12* (1973) 63–81.
[24] K.W. HULEK: Stable rank-2 bundles on $\mathbf{P}_2$ with c_1 odd, Math. Ann. *242* (1979) 241–266.
[25] J. LE POTIER: Fibrés stables de rang 2 sur $\mathbf{P}_2(\mathbf{C})$, Math. Ann. *241* (1979) 217–256.
[26] M. MARUYAMA: Moduli of stable sheaves, I., J. Math. Kyoto Univ. *17* (1977) 91–126.
[27] M. MARUYAMA: Moduli of stable sheaves, II., J. Math. Kyoto Univ. *18* (1978) 557–614.
[28] M. MARUYAMA: Boundedness of semi-stable sheaves of small ranks, (to appear).
[29] D. MUMFORD: Projective invariants of projective structures and applications, Proc. I.C.M. (1962) 526–530.
[30] D. MUMFORD: Geometric invariant theory, Ergebnisse der Math *34*, Springer Verlag (1965)
[31] M.S. NARASIMHAN and C.S. SESHADRI: Stable and unitary vector bundles on a compact Riemann Surface, Annals of Math *82* (1965) 540–567.
[32] C. OKONEK, M. SCHNEIDER and H. SPINDLER: Vector bundles on complex projective spaces, Birkhäuser Boston, Inc. (to appear).
[33] M. RAYNAUD: Fibrés vectoriels instables – applications aux surfaces (d'après Bogomolov) Sem. Geom. Alg., Orsay (1977–8) exp. no. 3.
[34] M. SCHNEIDER: Holomorphic vector bundles on $\mathbf{P}_n$, Séminaire Bourbaki *530* (1978–9) 23 pp.
[35] C.S. SESHADRI: Quotient spaces modulo reductive algebraic groups and applications to moduli of vector bundles on algebraic curves, Actes Congrès intern. math. (ICM) (1970), I, 479–482.
[36] H. SPINDLER: Der Satz von Grauert-Mülich für beliebige semistabile holomorphe Vektorbündel über den n-dimensionalen komplex-projektiven Raum, Math. Ann. *243* (1979) 131–141.
[37] F. TAKEMOTO: Stable vector bundles on algebraic surfaces, Nagoya Math. J. *47* (1972) 29–48.
[38] J.A. VOGELAAR: Constructing vector bundles from codimension-two subvarieties, Proefschrift, Univ. Leiden (1978).

Department of Mathematics,
University of California,
Berkeley, CA, 94720
(USA)

KUMMER SURFACES ASSOCIATED WITH THE HORROCKS-MUMFORD BUNDLE

Wolf Barth

0. Introduction

The aim of the talk was to point out this relation between classical geometry of Kummer surfaces and the rank-2 vector bundle F on $\mathbf{P}_4$ constructed by Horrocks and Mumford in [4]: If $x \in \mathbf{P}_4$ is a general point, then the jumping lines for F through x form a variety of dimension one, they are the generators of a cone with top x over a curve $R_x \subset \mathbf{P}_3$. This curve has degree 8 and is a curve of contact of two Kummer surfaces. Both surfaces have all their nodes on the curve R_x.

This statement has to be taken with a grain of salt, because I only believe that for general x the curve R_x is nondegenerate. Unfortunately, at the moment I cannot prove this. A better understanding of the geometry of R_x – although it will be interesting for its own sake – will probably lead to a construction of more rank-2 vector bundles on $\mathbf{P}_4$. This construction, described in sections 6–8 below, starts with a curve of contact of two surfaces in $\mathbf{P}_3$ and not with a surface in $\mathbf{P}_4$. At the moment, however, no example of a curve with the properties necessary to apply the construction is known. The first example should be the curve R_x above.

I was led to this construction by the definition of "canonical subspaces" given by Grauert and Mülich [2, §4]. Independently, in some of its details earlier than by myself, it was also discovered by Gruson and Peskine when studying Halphen's work on space curves [7]. I want to thank L. Gruson for informing me about their progress on occasion of the Angers conference.

As to the contents: Sections 1 and 2 recall the classical theory of Kummer surfaces, sections 3–5 give an elementary account of the linear algebra behind the Horrocks-Mumford bundle, and sections 6–8 contain the construction of bundles on $\mathbf{P}_4$ mentioned above.

Notation: If E is a complex vector space of dimension n, then $\mathbf{P}(E) = \mathbf{P}_{n-1}(E)$ denotes the projective space of lines in E. Hoping that no confusion arises, a non zero vector $e \in E$ and the corresponding point $\mathbf{C} \cdot e \in \mathbf{P}(E)$ are usually denoted by the same symbol.

Because of $\dim \Lambda^n E = 1$, one may identify $\Lambda^n E = \mathbf{C}$. Then $\Lambda^p E$ and $\Lambda^{n-p} E$ are dual to each other under the exterior product. Such an identification $\Lambda^n E = \mathbf{C}$ will be assumed very often without mentioning it explicitly.

If F is a vector bundle and x a point of the base space, then F_x is the fibre at x (not the stalk of the sheaf $\mathcal{O}(F)$).

1. Quadratic complexes and Kummer surfaces

Kummer was the first one to discover the relation between a quadratic complex and a quartic surface with 16 nodes, which now is called Kummer's surface. This classical relation [6, p. 97 f; 5, p. 51 f; 3, p. 762] is responsible for the appearance of such surfaces in connection with the bundle of Horrocks-Mumford. So it seems appropriate to recall here the relevant theory.

Fix a 4-dimensional complex vector space E and put

$$\mathbf{P}_3 = \mathbf{P}(E), \quad \mathbf{P}_5 = \mathbf{P}(\Lambda^2 E).$$

The decomposable tensors $w = x \wedge y \in \Lambda^2 E$, corresponding to lines $\mathbf{P}_1(\mathbf{C}x + \mathbf{C}y) \subset \mathbf{P}_3$, lie on the quadric hypersurface $G = \mathrm{Grass}(1,3) \subset \mathbf{P}_5$ defined by Plücker's relation

$$Q_0(w) = w \wedge w = 0.$$

A *quadratic complex* is the intersection $G \cap \{Q = 0\}$, where $Q \in \Gamma(\mathcal{O}_{\mathbf{P}_5}(2)) = S^2(\Lambda^2 E)$ is linearly independent of Q_0. This quadratic complex is in general a nonsingular threefold.

The *Kummer surface* $S_Q \subset \mathbf{P}_3$ is defined as follows: Koszul's complex on $\mathbf{P}_3$ induces an exact sequence

$$0 \longrightarrow T_{\mathbf{P}_3}(-2) \longrightarrow \Lambda^2 E \otimes \mathcal{O}_{\mathbf{P}_3} \longrightarrow \Omega_{\mathbf{P}_3}(2) \longrightarrow 0. \tag{1}$$

At $y \in \mathbf{P}_3$, the image of $T_{\mathbf{P}_3}(-2)_y$ in $\Lambda^2 E$ is just $y \wedge E \subset \Lambda^2 E$. The quadratic form Q on $\Lambda^2 E$ restricts to a quadratic form on $T_{\mathbf{P}}(-2)$. Now S_Q is the discriminant surface of this form,

$$S_Q = \{y \in \mathbf{P}_3 \colon Q \mid y \wedge E \text{ is singular}\}.$$

An explicit description of S_Q is this: $y \in S_Q$ if and only if there is some $y' \in E$, $y' \notin \mathbf{C}y$, such that for all $e \in E$

$$Q(y \wedge y', y \wedge e) = 0. \tag{2}$$

Condition (2) means there is some $y'' \in E$ such that for all $w \in \Lambda^2 E$

$$Q(y \wedge y', w) = y \wedge y'' \wedge w. \tag{3}$$

In general, S_Q is a quartic surface, nonsingular except for 16 nodes. These nodes are exactly the points y where the rank of $Q \mid y \wedge E$ drops one step further: $y \in \mathbf{P}_3$ is a node of S_Q if and only if $\mathrm{rank}(Q \mid y \wedge E) = 1$.

2. Special pencils of quadratic complexes

Consider a pencil

$$\alpha Q_1 + \beta Q_2, \quad (\alpha:\beta) \in \mathbf{P}_1$$

of quadratic complexes (Q_o, Q_1, Q_2 must be linearly independent). We call it *special* if

$$\text{rank } Q_2 = 1,$$

i.e. if there is some $q \in \Lambda^2 E$ such that for all $w \in \Lambda^2 E$

$$Q_2(w) = (q \wedge w)^2,$$

and Q_2 is the square of the linear complex $q \wedge w = 0$.

THEOREM: *The Kummer surfaces $S_{\alpha:\beta} = S_{\alpha Q_1 + \beta Q_2}$ form a (linear) pencil. Their common intersection is a curve $R \subset \mathbf{P}_3$ of degree eight. Any two Kummer surfaces in this pencil touch each other along R. This curve R contains the 16 nodes of each $S_{\alpha:\beta}$.*

PROOF: Consider the exact sequence (1). The tensor $q \in \Lambda^2 E$ is mapped onto a section $s \in H^0(\Omega(2))$, which can be viewed as a linear form on $T(-2)$. In fact, for $y \wedge e \in y \wedge E = T(-2)_y$ one has

$$s(y \wedge e) = q \wedge y \wedge e.$$

If we denote by Q_1 also the restriction of Q_1 to $T(-2)$, then the equation for $S_{\alpha:\beta}$ is

$$\det(\alpha Q_1 + \beta s^2) = 0.$$

To understand this equation, trivialize $T(-2)$ locally on $\mathbf{P}_3$ and write

$$Q_1(x) = (a_{ij}(x))_{1 \le i,j \le 3}$$
$$s(x) = (s_1(x), s_2(x), s_3(x)).$$

Let $a^{ij}(x)$ be the i–j-minor of $Q_1(x)$. Then

$$\det(\alpha Q_1 + \beta s^2) =$$
$$= \alpha^3 \det Q_1 + \alpha^2 \beta \sum_{i,j} a^{ij} s_i s_j.$$

So after dividing by α^2, the surface $S_{\alpha:\beta}$ depends linearly on $(\alpha:\beta)$.

Also, every node $x_0 \in S_{(1:0)}$ belongs to $S_{\alpha:\beta}$, because rank $(a_{ij}(x)) = 1$ and the rank of $(a_{ij}(x_0) + s_i(x_0)s_j(x_0))$ cannot exceed two.

We still have to show that $S_{(\alpha:\beta)}$ touches $S_{(1:0)}$ outside of the nodes of this surface. Since rank $(a^{ij}(x)) = 1$ on the regular part of $S_{(1:0)}$, we can write $a^{ij}(x) = \alpha^i(x)\alpha^j(x)$ locally on $S_{(1:0)}$ outside of its nodes. There

$$\sum a^{ij}(x)s_i(x)s_j(x) = \left(\sum \alpha^i(x)s_i(x)\right)^2$$

indeed is a square. □

For general choice of q, the octic curve R is nonsingular of genus 5. It was studied in classical geometry [5, sections 37, 40, 93].

R is the biregular image of a curve $R' \subset \mathbf{P}_5$. This R' is the complete intersection of the linear complex $\mathbf{P}_4 : \{w \wedge q = 0\}$ with the three quadric hypersurfaces

$$Q_0 = Q_1 = Q_1 Q_0^{-1} Q_1 = 0.$$

R' is a canonical octic curve in $\mathbf{P}_4$, but the map $R' \to R$ is *not* obtained by projecting $\mathbf{P}_4$ onto $\mathbf{P}_3$. So R is *linearly normal.*

3. Some linear algebra

Fix once for all a basis $e_0, \ldots, e_4$ of $V = \mathbf{C}^5$. Following Horrocks and Mumford [4, p. 69], define maps f^+ and $f^- : V \to \Lambda^2 V$ by

$$f^+\left(\sum v_i e_i\right) = \sum v_i e_{i+2} \wedge e_{i+3}$$

$$f^-\left(\sum v_i e_i\right) = \sum v_i e_{i+1} \wedge e_{i+4}.$$

Here $i = 0, 1, \ldots, 4$ modulo 5. Furthermore, for $x \in V$ define $p^+(x)$ and $p^-(x)$: $V \to \Lambda^2 V \wedge x \subset \Lambda^3 V$ by

$$p^\pm(x)v = f^\pm(v) \wedge x.$$

Now $\Lambda^2 V \wedge x = \Lambda^2 E$ with $E = V/\mathbf{C}x$, a space of dimension 4. Consider the quadratic form Q_0 from section 1 as a self-duality on $\Lambda^2 V \wedge x$. It is defined by the exterior product, so it is compatible with the duality $(\Lambda^2 V)^* = \Lambda^3 V$ via

$$\Lambda^2 V \xrightarrow{\wedge x} \Lambda^2 V \wedge x \subset \Lambda^3 V.$$

In particular there are maps

$$(f^+)^t, (f^-)^t : \Lambda^3 V \longrightarrow V^*$$

$$p^+(x)^t, p^-(x)^t : \Lambda^2 V \wedge x \longrightarrow V^*$$

satisfying

$$\langle (f^\pm)^t w, v \rangle = w \wedge f^\pm v$$

for all $v \in V$, $w \in \Lambda^3 V$ and

$$p^\pm(x)^t(z \wedge x) = (f^\pm)^t(z \wedge x)$$

for all $z \in \Lambda^2 V$.
Now explicit computation shows for all $v = \Sigma\, v_i e_i$ and $x = \Sigma\, x_j e_j \in V$

$$p^+(x)v = \sum_i x_i v_i \cdot e_i \wedge e_{i+2} \wedge e_{i+3} + (x_{i+1}v_{i+2} + x_{i+4}v_{i+3})\, e_i \wedge e_{i+1} \wedge e_{i+4}, \tag{4}$$

$$p^-(x)v = \sum_i (-x_{i+2}v_{i+4} - x_{i+3}v_{i+1})\, e_i \wedge e_{i+2} \wedge e_{i+3} + x_i v_i\, e_i \wedge e_{i+1} \wedge e_{i+4}. \tag{5}$$

Also, for all $u = \Sigma\, u_i e_i \in V$

$$\begin{aligned}\langle p^+(x)^t p^+(x)v, u\rangle &= p^+(x)v \wedge p^+(x)u \\ &= f^+(v) \wedge f^+(u) \wedge x = \sum (x_{i+1}v_{i+2} + x_{i+4}v_{i+3})u_i \\ &= \sum x_i(v_{i+1}u_{i-1} + v_{i-1}u_{i+1}),\end{aligned} \tag{6}$$

$$\begin{aligned}\langle p^-(x)^t p^-(x)v, u\rangle &= f^-(v) \wedge f^-(u) \wedge x \\ &= -\sum x_i(v_{i+2}u_{i-2} + v_{i-2}u_{i+2}),\end{aligned} \tag{7}$$

$$\begin{aligned}\langle p^+(x)^t p^-(x)v, u\rangle &= f^-(v) \wedge f^+(u) \wedge x \\ &= \sum x_i v_i u_i \\ &= \langle p^-(x)^t p^+(x)v, u\rangle = f^+(v) \wedge f^-(u) \wedge x.\end{aligned} \tag{8}$$

Define tensors

$$z^+ = z^+(x) = \sum x_{i+2}x_{i+3}e_{i+2} \wedge e_{i+3} \in \Lambda^2 V$$

$$z^- = z^-(x) = \sum x_{i+1}x_{i+4}e_{i+1} \wedge e_{i+4} \in \Lambda^2 V$$

$$w^+ = w^+(x) = \sum x_i x_{i+2} x_{i+3} e_i \wedge e_{i+2} \wedge e_{i+3} \in \Lambda^3 V$$

$$w^- = w^-(x) = \sum x_i x_{i+1} x_{i+4} e_i \wedge e_{i+1} \wedge e_{i+4} \in \Lambda^3 V$$

Then explicit computation shows

$$z^+ \wedge x = w^+ + 2w^-,$$

$$z^- \wedge x = w^- - 2w^+.$$

Inverting these equations we find that

$$w^+ = u^+ \wedge x \text{ where } u^+ = \tfrac{1}{5}(z^+ - 2z^-)$$

$$w^- = u^- \wedge x \text{ where } u^- = \tfrac{1}{5}(2z^+ + z^-),$$

so w^+ and w^- are contained in $\Lambda^2 V \wedge x \subset \Lambda^3 V$.

These tensors will be needed later, because they annihilate the image of f^+, resp. f^-:

$$w^+ \wedge f^+(v) = w^- \wedge f^-(v) = 0 \text{ for all } v \in V. \tag{9}$$

Finally, one easily checks

$$\begin{aligned} -u^+ \wedge w^+ &= 2x_o x_1 x_2 x_3 x_4 = u^- \wedge w^- \\ u^+ \wedge w^- &= x_0 x_1 x_2 x_3 x_4 = u^- \wedge w^+. \end{aligned} \tag{10}$$

The following lemma holds for "general" $x \in V$. Unfortunately I do not know for which. My only way of proving the assertion is to prove it in one point x and to observe that the statement is Zariski-open.

LEMMA 1: *For general $x \in \mathbf{P}_4$, the vectors $p^+(x)v$ and $p^-(x)v$ are linearly independent if $v \neq 0$.*

PROOF: Choose $x = (0, -1, 1, 1, 1)$. By (4) and (5) one must check the independence of

$$\begin{array}{cccccccccc} 0 & -v_1 & v_2 & v_3 & v_4 & -v_2+v_3 & v_3 & v_4-v_0 & v_0+v_1 & v_2 \\ -v_4-v_1 & -v_0-v_2 & -v_1 & v_4 & v_3-v_0 & 0 & -v_1 & v_2 & v_3 & v_4. \end{array}$$

If these rows are dependent, then either $v_1 = v_2 = v_3 = v_4 = v_0 = 0$, which is impossible, or

$$v_3 = v_2, \quad v_4 = -v_1$$

and the system becomes

$$\begin{array}{cccccccc} -v_1 & v_2 & v_2 & -v_1 & v_2 & -v_0-v_1 & v_0+v_1 & v_2 \\ -v_0-v_2 & -v_1 & -v_1 & v_2-v_0 & -v_1 & v_2 & v_2 & -v_1. \end{array}$$

One finds $v_1 = v_2 = 0$ and finally $v_0 = 0$. □

COROLLARY: *For general $x \in V$ the maps*

$$p^{\pm}(x) : V \longrightarrow \Lambda^2 V \wedge x$$

are injective and

$$p^+(x)V = \{w \in \Lambda^2 V \wedge x : w \wedge u^+ = 0\}$$

$$p^-(x)V = \{w \in \Lambda^2 V \wedge x : w \wedge u^- = 0\}.$$

4. Description of the Horrocks-Mumford bundle and of its jumping lines

Denote by F the Horrocks-Mumford bundle [4] on $\mathbf{P}_4$, normalized such that

$$c_1(F) = -1, \quad c_2(F) = 4.$$

Horrocks and Mumford describe F as the cohomology bundle of a monad

$$V \otimes \mathcal{O}_{\mathbf{P}_4}(-1) \xrightarrow{p} 2(\Lambda^2 T_{\mathbf{P}_4})(-3) \xrightarrow{q} V^* \otimes \mathcal{O}_{\mathbf{P}_4}. \tag{11}$$

When constructing F in this way, the essential difficulty consists in finding maps p and q satisfying

i) p is injective (fibre-wise),

ii) q is surjective,

iii) $q \circ p = 0$.

This is reduced to linear algebra by using the middle part of Koszul's complex on $\mathbf{P}_4$

$$\begin{array}{ccccccc} V\otimes\mathcal{O}(-2) \xrightarrow{\wedge s} & \Lambda^2 V\otimes\mathcal{O}(-1) & \xrightarrow{\wedge s} & \Lambda^3 V\otimes\mathcal{O} & \xrightarrow{\wedge s} \Lambda^4 V\otimes\mathcal{O}(1) \\ & {\scriptstyle p_0}\searrow & & \nearrow{\scriptstyle q_0} & \\ & & (\Lambda^2 T)(-3) & & \\ & & \nearrow \quad \searrow & & \\ & & 0 \qquad\quad 0 & & \end{array} \tag{12}$$

Here $s: \mathcal{O}_{\mathbf{P}}(-1) \to V \otimes \mathcal{O}_{\mathbf{P}}$ is the tautological subbundle ($s(x) = x$). The exterior product provides the sequence (11) with a self-duality (with values in $\mathcal{O}_{\mathbf{P}}(-1)$). It is compatible with the self-duality on $(\Lambda^2 T)(-3)$: If φ in $\Lambda^2 V \otimes \mathcal{O}(-1)$ and ψ in $(\Lambda^2 T)(-3)$ are local sections, then

$$\langle \varphi, q_0 \psi \rangle = \varphi \wedge q_0 \psi = p_0 \varphi \wedge \psi.$$

This self-duality is extended to $2 \cdot (12) = \mathbf{C}^2 \otimes (12)$ by tensoring with the standard symplectic form on $\mathbf{C}^2$.

Now p and q are defined as compositions

$$p: \quad V \otimes \mathcal{O}(-1) \xrightarrow{f} 2\Lambda^2 V \otimes \mathcal{O}(-1) \xrightarrow{p_0} 2(\Lambda^2 T)(-3),$$

$$q: \quad 2(\Lambda^2 T)(-3) \xrightarrow{q_0} 2\Lambda^3 V \otimes \mathcal{O} \xrightarrow{g} V^* \otimes \mathcal{O}.$$

Here f: $V \to 2\Lambda^2 V$ is just the pair (f^+, f^-) from section 3, and $g = f^t$ is the transposed map. i.e., $g = ((f^-)^t, -(f^+)^t)$. Then $p = (p^+, p^-)$ and $q = p^t = ((p^-)^t, -(p^+)^t)$ in the notation from the preceding section.

The conditions i)–iii) above are reformulated in terms of f^+, f^-:

i) For all non zero vectors $x, v \in V$, one has
$$(f^+(v) \wedge x, \quad f^-(v) \wedge x) \neq (0, 0).$$
ii) is dual, hence equivalent, to i),
iii) for all $x, u, v \in V$ one has
$$f^-(u) \wedge f^+(v) \wedge x = f^+(u) \wedge f^-(v) \wedge x.$$

Now assertion iii) has been verified already (8) and i) has a surprisingly elegant proof [4, p. 69]. If $f^+(v) \wedge x = f^-(v) \wedge x = 0$, then $f^+(x) = x^+ \wedge x$ and $f^-(x) = x^- \wedge x$, so $f^+(v) \wedge f^-(v) = 0$. But

$$f^+(v) \wedge f^-(v) = \sum v_i^2 e_i \wedge e_{i+1} \wedge e_{i+2} \wedge e_{i+4}.$$

To describe the *jumping lines* of F, define for every line $L \subset \mathbf{P}_4$ the integer $k = k(L) \geq 0$ by

$$F \mid L = \mathcal{O}_L(-1-k) \oplus \mathcal{O}_L(k),$$

or equivalently

$$k(L) = h^0(F \mid L) - 1.$$

Then use the display

$$\begin{array}{ccccccccc}
 & & & & 0 & & 0 & & \\
 & & & & \downarrow & & \downarrow & & \\
0 & \longrightarrow & V \otimes \mathcal{O}(-1) & \longrightarrow & \mathcal{C}^*(-1) & \longrightarrow & F & \longrightarrow & 0 \\
 & & \| & & \downarrow & & \downarrow & & \\
0 & \longrightarrow & V \otimes \mathcal{O}(-1) & \xrightarrow{p} & 2(\Lambda^2 T)(-3) & \longrightarrow & \mathcal{C} & \longrightarrow & 0 \\
 & & & & {\scriptstyle q}\downarrow & & \downarrow & & \\
 & & & & V \otimes \mathcal{O} & = & V \otimes \mathcal{O} & & \\
 & & & & \downarrow & & \downarrow & & \\
 & & & & 0 & & 0 & &
\end{array}$$

of the monad (11) to verify that

$$h^0(F \mid L) = h^0(\mathcal{C}^*(-1) \mid L)$$
$$= \dim \operatorname{Ker} (q : H^0(2(\Lambda^2 T)(-3) \mid L) \longrightarrow V^*).$$

So it is necessary to identify $H^0((\Lambda^2 T)(-3) \mid L)$.

Lemma 2: *Let the line L be given by $x, y \in V$. Then $q_0(H^0((\Lambda^2 T)(-3) \mid L)) \subset \Lambda^3 V$ is just the three-dimensional subspace $x \wedge y \wedge V$.*

Proof: Since $T \mid L = \mathcal{O}_L(2) \oplus 3\mathcal{O}_L(1)$, one has

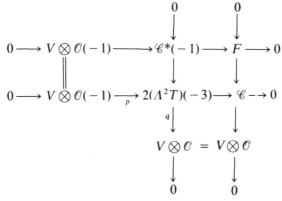

$$(\Lambda^2 T)(-3)\,|\,L = 3\mathcal{O}_L \oplus 3\mathcal{O}_L(-1),$$

and $q_0 H^0((\Lambda^2 T)(-3)\,|\,L) \subset \Lambda^3 V$ will be a subspace of dimension three. On the other hand, this subspace is contained in

$$x \wedge y \wedge V = (x \wedge \Lambda^2 V) \cap (y \wedge \Lambda^2 V),$$

because $x \wedge \Lambda^2 V$ is the image of q_0 over x and $y \wedge \Lambda^2 V$ is that of q_0 over y. □

This description shows that

$$\begin{aligned} h^0(F\,|\,L) &= \dim \operatorname{Ker}(g : 2x \wedge y \wedge V \longrightarrow V^*) \\ &= 6 - \operatorname{rank}(f^t : 2x \wedge y \wedge V \longrightarrow V^*). \end{aligned}$$

The rank of the map

$$2V \xrightarrow{x \wedge y} 2x \wedge y \wedge V \xrightarrow{f^t} V^*$$

equals the rank of its transpose

$$V \xrightarrow{f} 2\Lambda^2 V \xrightarrow{x \wedge y} 2\Lambda^4 V.$$

This proves the

PROPOSITION 1: *$k(L)$ equals the dimension of the kernel of the map*

$$\begin{aligned} &V \longrightarrow 2\Lambda^4 V \\ &v \longrightarrow (f^+(v) \wedge x \wedge y,\quad f^-(v) \wedge x \wedge y). \end{aligned}$$

In particular, the line L determined by $x, y \in V$ is a jumping line if and only if for some nonzero $v \in V$ one has

$$f^+(v) \wedge x \wedge y = f^-(v) \wedge x \wedge y = 0.$$

To evaluate this criterion, fix some nonzero $x \in V$, and ask: which lines $L = \mathbf{C}x + \mathbf{C}y$ are jumping lines?

LEMMA 3: *Given $x \in V$, $x \neq 0$, the following conditions on $v \in V$ are equivalent:*

(a) *There is some $y \in V$, $y \notin \mathbf{C}x$, such that*

$$f^+(v) \wedge x \wedge y = f^-(v) \wedge x \wedge y = 0$$

(b) $f^+(v) \wedge f^+(v) \wedge x = f^-(v) \wedge f^-(v) \wedge x = f^+(v) \wedge f^-(v) \wedge x = 0.$

PROOF OF (a) $\Rightarrow$ (b): Condition (a) implies

$$\begin{aligned} f^+(v) &= x \wedge x^+ + y \wedge y^+ \\ f^-(v) &= x \wedge x^- + y \wedge y^- \end{aligned}$$

for some $x^{\pm}, y^{\pm} \in V$. Then (b) is obvious.

PROOF OF (b)⇒(a): The vanishing of $f^+(v) \wedge f^+(v) \wedge x$ and $f^-(v) \wedge f^-(v) \wedge x$ implies

$$f^+(v) = a^+ \wedge b^+ + x \wedge x^+$$

$$f^-(v) = a^- \wedge b^- + x \wedge x^-$$

for some $a^{\pm}, b^{\pm}, x^{\pm} \in V$. Then

$$f^+(v) \wedge f^-(v) \wedge x = a^+ \wedge b^+ \wedge a^- \wedge b^- \wedge x$$

vanishes if and only if a^+, b^+, a^-, b^- are linearly dependent modulo x. So there is some $y \in V$, $y \notin \mathbf{C}x$ with

$$f^+(v) = y \wedge y^+ + x \wedge x_1$$
$$f^-(v) = y \wedge y^- + x \wedge x_2$$

and

$$f^+(v) \wedge x \wedge y = f^-(v) \wedge x \wedge y = 0. \qquad \square$$

Lemma 3 describes a correspondence between the variety of jumping lines through x and the following variety $R'_x \subset \mathbf{P}_4(V)$:

$$R'_x = Q^+ \cap Q^- \cap Q^0$$

with three quadrics Q^+, Q^-, Q^0 of equation (cf. section 3)

$$f^+(v) \wedge f^+(v) \wedge x = \sum x_i v_{i+2} v_{i+3},$$

$$-f^-(v) \wedge f^-(v) \wedge x = \sum x_i v_{i+1} v_{i+4},$$

$$f^+(v) \wedge f^-(v) \wedge x = \sum x_i v_i^2.$$

PROPOSITION 2: *For general $x \in \mathbf{P}_4$, there is a surjective rational morphism of R'_x onto R_x, the variety of jumping lines through x.*

PROOF: The assertion is that the line $\mathbf{C}x + \mathbf{C}y$ from lemma 3(a) is uniquely determined by the nonzero vector $v \in V$ from lemma 3(b). But this is true, unless the tensors $a^+ \wedge b^+$ and $a^- \wedge b^-$ are linearly dependent modulo x, or equivalently, $f^+(v) \wedge x$ and $f^-(v) \wedge x$ are linearly dependent. Now use lemma 1. $\square$

I believe–but up to now failed to prove it–that for general x, the set R'_x is a nonsingular curve (of degree 8, genus 5, and canonically embedded in

$\mathbf{P}_4$), and that the map $R'_x \to R_x$ is biregular. All I know is that (in general) R'_x is a *curve* of degree ≥ 6. Consider e.g. the point

$$x = (0, 0, 0, 1, 1).$$

The equations become

$$Q^+(x): \quad v_0 v_1 + v_1 v_2 = 0$$
$$Q^-(x): \quad v_2 v_4 + v_0 v_3 = 0$$
$$Q^0(x): \quad v_3^2 + v_4^2 = 0.$$

So $R'(x)$ is the union of the six lines

$$v_4 = \pm i v_3,$$
$$v_1(v_0 + v_2) = v_3(v_0 \pm i v_2) = 0.$$

COROLLARY: *The set of jumping lines for F forms in* Grass $(1, 4)$ *a subvariety of codimension* 2.

5. Kummer surfaces associated with F

Fix some point $x \in \mathbf{P}_4$, sufficiently general so that lemma 1 holds for it. It will be shown here, that the projective line $\mathbf{P}(F_x)$ parametrizes a pencil of quadratic complexes in $\mathbf{P}_5(\Lambda^2 T_x)$. This pencil is special in the sense of section 2. So it determines a pencil of Kummer surfaces in $\mathbf{P}_3(T_x)$ touching each other along a curve R_x. Notice that $\mathbf{P}_3(T_x)$ can be identified with the space of lines through x. It should turn out that the jumping lines through x are exactly those lines lying over R_x.

The two-dimensional vector space F_x is the cohomology of

$$V \xrightarrow{p(x)} 2\Lambda^2 T_x \xrightarrow{p(x)^t} V^*.$$

The condition $p(x)^t p(x) = 0$ means that $p(x)\,V$ is a totally isotropic subspace of symplectic space $2\Lambda^2 T_x$. Every one-dimensional subspace of F_x is determined by a six-dimensional subspace $I \subset 2\Lambda^2 T_x$ which contains the image of $p(x)$ and is contained in the kernel of $p(x)^t$.

LEMMA 4: *These subspaces I are effectively parametrized by*

$$I_{\alpha:\beta} = p(x)V + \mathbf{C}(\alpha w^-, \beta w^+), \quad (\alpha : \beta) \in \mathbf{P}_1$$

PROOF: (a) Each $I_{\alpha:\beta}$ is totally isotropic: We have to show

$$\beta f^+(v) \wedge w^+ = \alpha f^-(v) \wedge w^-.$$

But this is a trivial consequence of (9).
(b) dim $I_{\alpha:\beta} = 6$ and all $I_{\alpha:\beta}$ are different: this follows from the claim

$$p(x)V \cap \{\mathbf{C}(w^-, 0) + \mathbf{C}(0, w^+)\} = 0$$

in $2\Lambda^3 V$. To prove the claim assume that for some $v \in V$

$$p^+(x)v = \alpha w^-, \quad p^-(x)v = \beta w^+.$$

Then

$$\alpha u^+ \wedge w^- = w^+ \wedge f^+(v) = 0$$
$$\beta u^- \wedge w^+ = w^- \wedge f^-(v) = 0$$

shows $\alpha = \beta = 0$ because of (10). □

Now let

$$P = (P^+_{\alpha:\beta}, P^-_{\alpha:\beta}): \quad I_{\alpha:\beta} = V \oplus \mathbf{C} \to 2\Lambda^2 T_x$$

be the map embedding $I_{\alpha:\beta}$. Explicitly

$$P^+_{\alpha:\beta}(v, c) = p^+(x)v + \alpha c w^-,$$

$$P^-_{\alpha:\beta}(v, c) = p^-(x)v + \beta c w^+.$$

Since $P^-_{\alpha:\beta}$: $V \oplus \mathbf{C} \to \Lambda^2 T_x$ is invertible for all $\beta \neq 0$, we may consider for these β the maps

$$P^+_{\alpha:\beta}(P^-_{\alpha:\beta})^{-1} : \Lambda^2 T_x \longrightarrow \Lambda^2 T_x$$

and the bilinear forms on $\Lambda^2 T_x$

$$Q_{\alpha:\beta}(t\,; t') = \beta t \wedge P^+_{\alpha:\beta}((P^-_{\alpha:\beta})^{-1} t').$$

PROPOSITION 3: *The forms $Q_{\alpha:\beta}$ are symmetric on $\Lambda^2 T_x$ and depend linearly on $\alpha:\beta$. The pencil of quadratic complexes on $\mathbf{P}_3(T_x)$ formed in this way is special in the sense of section 2. For general x and general $\alpha:\beta$ the quadratic complex $Q_{\alpha:\beta}$ is nonsingular.*

PROOF: We introduce coordinates $(v, c) = (v_0, \ldots, v_4, c)$ on $\Lambda^2 T_x$ via

$$(v, c) \longrightarrow p^-(v) + c w^+ = P^-_{\alpha:1}(v, c).$$

In these coordinates (use (9) and (10))

$$\begin{aligned} Q_{\alpha:\beta}(v, c; v', c') &= \beta(p^-(v) + c w^+) \wedge P^+_{\alpha:\beta}(v', c'/\beta) \\ &= \beta(p^-(v) + c w^+) \wedge \left(p^+(v') + \frac{\alpha}{\beta} c' w^-\right) \\ &= \beta f^-(v) \wedge f^+(v') \wedge x + \alpha c c' \;\; x_0 x_1 x_2 x_3 x_4. \end{aligned}$$

By (8) this expression is symmetric in v, c and v', c'. Obviously rank $Q_{1:0} = 1$.

Next, let

$$Q_0(v, c) = (p^-(v) + cw^+) \wedge (p^-(v) + cw^+)$$

be the equation defining the Grassmannian in $\mathbf{P}(\Lambda^2 T_x)$. The assertion is: the intersection

$$Q_0 = Q_{\alpha:\beta} = 0$$

is nonsingular for general $\alpha:\beta$. By Bertini's theorem, this will be the case except perhaps for points in the base locus $Q_0 = Q_{1:0} = Q_{0:1} = 0$. So assume that $p^-(v) \in \{Q_{1:0} = 0\}$ is a singularity of $Q_0 = Q_{0:1} = 0$. This implies that there are $\lambda, \mu \in \mathbf{C}^*$ such that for all $dv \in V$, $dc \in \mathbf{C}$

$$\lambda p^-(v) \wedge (p^-(dv) + dc\, w^+) + \mu p^+(v) \wedge p^-(dv) = 0.$$

For $dc = 0$ this becomes

$$(\lambda p^-(v) - \mu p^+(v)) \wedge p^-(V) = 0$$

or equivalently (corollary to lemma 1) for some $\nu \in \mathbf{C}$

$$\lambda p^-(v) - \mu p^+(v) + \nu w^- = 0.$$

However, $dv = 0$ put into the expression above shows (because of $u^+ \wedge w^- \neq 0$ by (10))

$$\nu u^+ \wedge w^- = \lambda u^+ \wedge p^-(v) = 0, \text{ so } \nu = 0.$$

But $p^+(v)$ and $p^-(v)$ cannot be linearly dependent (lemma 1). □

Now let $S_{\alpha:\beta} \subset \mathbf{P}_3(T_x)$ be the *Kummer surface* defined by the quadratic complex $Q_{\alpha:\beta}$. Since the complex is nonsingular for general $\alpha:\beta$, the surface $S_{\alpha:\beta}$ will be a quartic with 16 nodes [3, p. 771].

PROPOSITION 4: *The curve $R_x \subset \mathbf{P}_3(T_x)$ describing the jumping lines through x is contained in each surface $S_{\alpha:\beta}$.*

PROOF: Recall (lemma 3) that $z \wedge y$ is a jumping line if and only if for some $v \neq 0$

$$f^+(v) = x \wedge x^+ + y \wedge y^+,$$
$$f^-(v) = x \wedge x^- + y \wedge y^-.$$

So for all $w \in \Lambda^2 T_x$

$$Q_{\alpha:\beta}(x \wedge y \wedge y^-, w) = \beta w \wedge P^+_{\alpha:\beta}(v, 0)$$
$$= w \wedge y \wedge \beta y^+.$$

This is just the condition (3) for $y' = y^-$, $y'' = \beta y^+$. □

Obviously R_x would be the curve of contact of the surfaces $S_{\alpha:\beta}$, if one could prove that R_x is of degree 8 (and if one shows that the $S_{\alpha:\beta}$ are different for different parameters $\alpha:\beta$).

6. Ferrand's double curves

This section and the remaining two ones give an account on a construction of bundles on $\mathbf{P}_4$ – which at the moment still is quite theoretical. Let $R \subset \mathbf{P}_3$ be a nonsingular space curve. Every sub-linebundle $S \subset N_{R/\mathbf{P}_3}$ defines an ideal sheaf $\mathcal{J} \subset \mathcal{J}_R$ as the kernel of

$$\mathcal{J}_R \longrightarrow \mathcal{J}_R/\mathcal{J}_R^2 = N^*_{R/\mathbf{P}_3} \longrightarrow S^*.$$

(A function f vanishing on R belongs to $\mathcal{J}$ if its partial derivative in the direction determined by S vanishes.) The subscheme $C = (R, \mathcal{O}_\mathbf{P}/\mathcal{J})$ of $\mathbf{P}_3$ is locally a complete intersection.

There is the standard exact sequence

$$0 \longrightarrow S^* \longrightarrow \mathcal{O}_C \xrightarrow{restr} \mathcal{O}_R \longrightarrow 0, \tag{13}$$

which after application of $\mathcal{E}xt^2_{\mathcal{O}_\mathbf{P}}(-, \omega_\mathbf{P})$ leads to

$$0 \longrightarrow \omega_R \longrightarrow \omega_C \xrightarrow{r} S \otimes \omega_R \longrightarrow 0. \tag{14}$$

Now ω_C is an invertible $\mathcal{O}_C$-sheaf. Obviously r vanishes on $\mathcal{J}_R \cdot \omega_C$, so $S \otimes \omega_R = \omega_C \mid R$ and r is just the restriction from C to R.

Assume next that

$$S \otimes \omega_R = \mathcal{O}_R(\ell) \quad \text{for some} \quad \ell \in \mathbf{Z}. \tag{15}$$

Ferrand [1, proposition 2] has shown the equivalence of the following two conditions:

$$\omega_C = \mathcal{O}_C(\ell), \tag{16}$$

the morphism $H^1(\mathcal{J}_R(\ell)) \longrightarrow H^1(N^*_{R/\mathbf{P}_3}(\ell)) \longrightarrow H^1(S^*(\ell))$ vanishes. (17)

In the sequel, some variant of his condition will be used:

LEMMA 5: *If $\Gamma(S^*)$ vanishes, then* (16) *is equivalent with*:

$$\text{restr: } \Gamma(\mathcal{O}_C(\ell)) \longrightarrow \Gamma(\mathcal{O}_R(\ell)) \text{ is surjective.} \tag{18}$$

PROOF: Condition (16) can be stated as $\omega_C(-\ell) = \mathcal{O}_C$, and is equivalent to the surjectivity of

$$\text{restr: } \Gamma(\omega_C(-\ell)) \longrightarrow \Gamma(\mathcal{O}_R).$$

The sequence (14) tensored with $\mathcal{O}_\mathbf{P}(-\ell)$, in view of $\Gamma(\omega_R(-\ell) = \Gamma(S^*) = 0$, shows the equivalence of this surjectivity with $h^0(\omega_C(-\ell)) \neq 0$. By Serre-Grothendieck duality on C

$$h^0(\omega_C(-\ell)) = h^1(\mathcal{O}_C(\ell)).$$

Now tensor (13) with $\mathcal{O}_{\mathbf{P}}(\ell)$ to obtain

$$0 \longrightarrow \omega_R \longrightarrow \mathcal{O}_C(\ell) \longrightarrow \mathcal{O}_R(\ell) \longrightarrow 0. \qquad (19)$$

Because of

$$h^1(\omega_R) = 1,$$
$$h^1(\mathcal{O}_R(\ell)) = h^0(S^*) = 0,$$

we have $h^1(\mathcal{O}_C(\ell)) \neq 0$ if and only if condition iii) holds. □

7. A curve of contact

Assume that we are given a nonsingular space curve $R \subset \mathbf{P}_3$ with the following five properties:

Set-theoretically, R is the complete intersection of two surfaces $S_1, S_2 \subset \mathbf{P}_3$, both of degree n. (20)

R is the curve of contact of these surfaces, i.e., no point $z \in R$ is singular for both S_1 and S_2, and the intersection-multiplicity $i_R(S_1, S_2)$ equals 2. (21)

By (21) above, the tangent planes to S_1 and/or S_2 define a line subbundle $S \subset N_{R/\mathbf{P}_3}$. Put $Q = N/S$.

The exact sequence

$$0 \longrightarrow S \longrightarrow N_{R/\mathbf{P}_3} \longrightarrow Q \longrightarrow 0 \qquad (22)$$

splits.

$$Q = S(-1). \qquad (23)$$

R is linearly normal, i.e. the restriction map

$$\Gamma(\mathcal{O}_{\mathbf{P}}(1)) \longrightarrow \Gamma(\mathcal{O}_R(1)) \qquad (24)$$

is surjective.

I collect some consequences of these properties.

(a) It follows from (21) that $\deg R = n^2/2$. In particular n is even.

(b) The adjunction formula shows

$$\omega_R = \det N_{R/\mathbf{P}_3} \otimes \omega_{\mathbf{P}} = S \otimes Q \otimes \mathcal{O}_R(-4) =$$
$$\doteq S^{\otimes 2}(-5).$$

But ω_R can also be determined from (14), where now C is the complete intersection of two surfaces of degree n:

$$S \otimes \omega_R = \omega_C \mid R = \mathcal{O}_R(2n-4). \tag{25}$$

Combining (25) with the expression above, one finds

$$S^{\otimes 3} = \mathcal{O}_R(2n+1)$$
$$\omega_R^{\otimes 3} = \mathcal{O}_R(4n-13). \tag{26}$$

These equations show that

$n^2(2n+1)$ is divible by 12,

or equivalently

$2 \mid n$ and $3 \mid n(n+2)$.

(c) If S_1 has only ordinary double points (nodes), then there is still another expression for ω_R, namely in terms of the divisor D on R of nodes of S_1. In fact, let $\varphi: F \to S_1$ be the minimal resolution of singularities of S_1. Nodes do not affect the adjunction formula, so

$$\omega_F = \varphi^* \mathcal{O}_{\mathbf{P}_3}(n-4).$$

But if we denote by R also the proper transform of our curve on F, then

$$\mathcal{O}_F(2R) = \varphi^* \mathcal{O}_{\mathbf{P}_3}(n) - \varphi^* \text{ (nodes)}.$$

The adjunction formula gives

$$\omega_R^{\otimes 2} = \mathcal{O}_R(3n-8) - \mathcal{O}_R(D),$$

and the square of equation (26) implies

$$\mathcal{O}_R(3D) = \mathcal{O}_R(n+2).$$

In particular, for the number ν of nodes one finds $\nu = n^2(n+2)/6$. The first few possibilities would be

n	:	4	6	10	12
deg R	:	8	18	50	72
ν	:	16	48	200	336.

(d) We can provide R not only with the nonreduced structure C induced by S (which is the complete intersection), but also with the one induced by Q: let $\mathcal{J}' \subset \mathcal{J}_R$ be the kernel of

$$\mathcal{J}_R \longrightarrow \mathcal{J}_R/\mathcal{J}_R^2 = N_R^* \longrightarrow Q^*$$

and put $C' = (R, \mathcal{O}_{\mathbf{P}}/\mathcal{J}')$. In place of (13) and (14) we then have

$$0 \longrightarrow Q^* \longrightarrow \mathcal{O}_{C'} \longrightarrow \mathcal{O}_R \longrightarrow 0$$
$$0 \longrightarrow \omega_R \longrightarrow \omega_{C'} \longrightarrow Q \otimes \omega_R \longrightarrow 0$$

with $Q \otimes \omega_R = S \otimes \omega_R(-1) = \mathcal{O}_R(2n-5)$.

LEMMA 6: $\omega_{C'} = \mathcal{O}_{C'}(2n-5)$

PROOF: By lemma 5, we have to prove the surjectivity of

$$\text{restr}: \Gamma(\mathcal{O}_{C'}(2n-5)) \longrightarrow \Gamma(\mathcal{O}_R(2n-5)),$$

but it suffices to show that

$$\text{restr}: \Gamma(\mathcal{O}_{\mathbf{P}}(2n-5)) \longrightarrow \Gamma(\mathcal{O}_R(2n-5))$$

is surjective. We may factorize this restriction over C instead of C':

$$\Gamma(\mathcal{O}_{\mathbf{P}}(2n-5)) \longrightarrow \Gamma(\mathcal{O}_C(2n-5)) \longrightarrow \Gamma(\mathcal{O}_R(2n-5)).$$

C being a complete intersection, the first arrow is surjective, and it suffices to prove that the second one is an epimorphism too.

Now consider the exact sequence

$$\Gamma(\mathcal{O}_C(2n-5)) \longrightarrow \Gamma(\mathcal{O}_R(2n-5)) \longrightarrow H^1(S^*(2n-5))$$
$$\longrightarrow H^1(\mathcal{O}_C(2n-5)) \longrightarrow H^1(\mathcal{O}_R(2n-5)) \longrightarrow 0.$$

We know that

$$h^1(\mathcal{O}_R(2n-5)) = h^1(Q \otimes \omega_R) = h^0(Q^*) = 0,$$
$$h^1(\mathcal{O}_C(2n-5)) = h^0(\mathcal{O}_C(1)) = 4,$$
$$h^1(S^*(2n-5)) = h^0(\mathcal{O}_R(1)) = 4,$$

by (25). This proves the assertion. □

COROLLARY: *There is a rank-2 bundle E on $\mathbf{P}_3$ admitting a section $s \in \Gamma(E)$ which vanishes on C'. Its Chern classes are $c_1 = 2n-1$ and $c_2 = n^2$.*

PROOF: The existence of E is just an application of Serre's construction.

8. The construction on $\mathbf{P}_4$

Assume that there exists a space curve $R \subset \mathbf{P}_3$ with (20)–(24). Additionally, fix some point $x_0 \in \mathbf{P}_4$ and let $\sigma: \tilde{P} \to \mathbf{P}_4$ the monoidal transform in x_0. Denote by $q: \tilde{P} \to \mathbf{P}_3$ the canonical projection sending a point $y \neq x_0$ to the line joining it with x_0. Put

$$Y = q^{-1}R$$
$$X = \sigma(q^{-1}R).$$

X is the cone over R with vertex x_0.

THEOREM: *Given R, there exists a stable rank-2 bundle F on $\mathbf{P}_4$. This bundle admits a section $f \in \Gamma(F)$ vanishing with multiplicity two on X. Its Chern classes are*

$$c_1(F) = 2n-1, \quad c_2(F) = n^2.$$

PROOF: (a) *A non-reduced structure on Y*: Since $Y = q^{-1}R$, the splitting of $N_{R/\mathbf{P}_3}$ induces a splitting

$$N_{Y/\tilde{P}} = q^*S \oplus q^*Q, \quad Q = S \otimes \mathcal{O}_R(-1).$$

Now put

$$U = q^*S \otimes \sigma^*\mathcal{O}_{\mathbf{P}_4}(-1).$$

Fix some hyperplane $H \subset \mathbf{P}_4$ not containing x_0. An equation for H defines an injective sheaf map

$$u_1 : U \longrightarrow q^*S$$

vanishing on $\sigma^{-1}H$ only. Furthermore, since

$$\sigma^*\mathcal{O}_{\mathbf{P}_4}(-1) = q^*\mathcal{O}_{\mathbf{P}_3}(-1) \otimes \mathcal{O}_{\tilde{P}}(-\Sigma), \tag{27}$$

with $\Sigma \subset \tilde{P}$ the exceptional divisor, an equation for Σ induces a morphism

$$u_2 : U \longrightarrow q^*Q$$

vanishing on $\Sigma \cap Y$ only. Now $\sigma^{-1}H \cap \Sigma$ is empty, so the pair (u_1, u_2) makes U a subbundle of $N_{Y/\tilde{P}}$.

Define the nonreduced surface $Y' = (Y, \mathcal{O}_{\tilde{P}}/\mathcal{K})$ with the ideal

$$\mathcal{K} = \operatorname{Ker}(\mathcal{J}_Y \longrightarrow N_{Y/\tilde{P}} \longrightarrow U^*).$$

This surface Y' is locally a complete intersection.

(b) *Extending $\omega_{Y'}$, to a line bundle L on $\tilde{P}$*: Consider the exact sequences

$$0 \longrightarrow U^* \longrightarrow \mathcal{O}_{Y'} \longrightarrow \mathcal{O}_Y \longrightarrow 0 \tag{28}$$

$$0 \longrightarrow \omega_Y \longrightarrow \omega_{Y'} \longrightarrow U \otimes \omega_Y \longrightarrow 0$$

corresponding to (13) and (14). By the adjunction formula we have

$$\omega_Y = \det N_{Y/\tilde{P}} \otimes \omega_{\tilde{P}}$$

with

$$\det N_{Y/\tilde{P}} = q^*(S \otimes Q) = q^*(S^{\otimes 2} \otimes \mathcal{O}_R(-1))$$

$$\omega_{\tilde{P}} = \sigma^*\mathcal{O}_{\mathbf{P}_4}(-5) \otimes \mathcal{O}_{\tilde{P}}(3\Sigma).$$

So by (26) and (27)

$$U \otimes \omega_Y = q^*(S^{\otimes 3} \otimes \mathcal{O}_R(-1)) \otimes \sigma^*\mathcal{O}_{\mathbf{P}_4}(-6) \otimes \mathcal{O}_{\tilde{P}}(3\Sigma) = L \mid Y$$

with

$$L = q^*\mathcal{O}_{\mathbf{P}_3}(2n-3) \otimes \sigma^*\mathcal{O}_{\mathbf{P}_4}(-3).$$

Now the question is, whether $\omega_{Y'} = L \mid Y'$. Argueing as in the case of curves, and using $h^0(U^*) = 0$, this is equivalent with

$$h^2(L \mid Y') = h^0(\omega_{Y'} \otimes L^*) \neq 0.$$

To verify this, consider the exact sequence

$$0 \longrightarrow L \mid Y' \longrightarrow L \otimes \mathcal{O}_{Y'}(\sigma^{-1}H + \Sigma)$$
$$\longrightarrow L \otimes \mathcal{O}_{Y'}(\sigma^{-1}H + \Sigma) \mid R_1 \cup R_2 \longrightarrow 0.$$

Here R_1 and R_2 are the two disjoint curves

$$R_1 = Y' \cap \sigma^{-1}H, \quad R_2 = Y' \cap \Sigma.$$

Under q we have isomorphisms

$$R_1 \longrightarrow C', \quad R_2 \longrightarrow C.$$

The exact sequence induces a sequence in cohomology

$$H^1(L \otimes \mathcal{O}_{Y'}(\sigma^{-1}H + \Sigma))$$
$$\longrightarrow H^1(L \otimes \mathcal{O}_{Y'}(\sigma^{-1}H) \mid R_1) \oplus H^1(L \otimes \mathcal{O}_{Y'}(\Sigma) \mid R_2) \longrightarrow H^2(L \mid Y'). \quad (29)$$

The first terms are computed as follows: From (28) one finds

$$H^1(U^* \otimes L \otimes \mathcal{O}_Y(\sigma^{-1}H + \Sigma)) \longrightarrow H^1(L \otimes \mathcal{O}_{Y'}(\sigma^{-1}H + \Sigma))$$
$$\longrightarrow H^1(L \otimes \mathcal{O}_Y(\sigma^{-1}H + \Sigma)),$$

where

$$\begin{aligned} &h^1(U^* \otimes L \otimes \mathcal{O}_Y(\sigma^{-1}H + \Sigma)) \\ &= h^1(q^*(S^* \otimes \mathcal{O}_R(2n-3) \otimes \sigma^*\mathcal{O}_{\mathbf{P}_4}(-1) \otimes \mathcal{O}_Y(\Sigma)) \\ &= h^1(q^*(S^* \otimes \mathcal{O}_R(2n-4)) = h^1(\omega_R) = 1, \\ &h^1(L \otimes \mathcal{O}_Y(\sigma^{-1}H + \Sigma)) = h^1(q^*\mathcal{O}_R(2n-4) \otimes \sigma^*\mathcal{O}_{\mathbf{P}_4}(-1)) = 0, \end{aligned}$$

because q_* and R^1q_* of the bundle in question vanish. This shows that the first term in (29) has dimension ≤ 1.

Now with regard to the second term in (29):

$$\begin{aligned} h^1(L \otimes \mathcal{O}_{Y'}(\sigma^{-1}H) \mid R_1) &= h^1(\mathcal{O}_{C'}(2n-5)) \\ &= h^1(\omega_{C'}) = 1, \\ h^1(L \otimes \mathcal{O}_{Y'}(\Sigma) \mid R_2) &= h^1(\mathcal{O}_C(2n-4)) \\ &= h^1(\omega_C) = 1. \end{aligned}$$

So this term has dimension 2.
It follows from (29), that $h^2(L \mid Y') \geq 1$.

(c) *Construction of F*: From (b) one deduces that there is a rank-2 bundle G on $\tilde{P}$ admitting a section g vanishing on Y' and such that

$$\begin{aligned} \det G &= L \otimes \omega_{\tilde{P}}^* \\ &= q^*\mathcal{O}_{\mathbf{P}_3}(2n) \otimes \sigma^*\mathcal{O}_{\mathbf{P}_4}(-1), \end{aligned}$$

(of course one must check that H^2 of the dual of this line bundle vanishes). Now $G \mid \Sigma$ has a section vanishing precisely on the curve R_2, which under

$q: \Sigma \to \mathbf{P}_3$ becomes the complete intersection C.
This shows that

$$G \mid \Sigma = q^*(2\mathcal{O}_{\mathbf{P}_3}(n)).$$

If we replace G by

$$\tilde{F} = G \otimes \mathcal{O}_{\tilde{P}}(n\Sigma),$$

then $\tilde{F} \mid \Sigma = 2\mathcal{O}_\Sigma$. So $\tilde{F}$ descends to a bundle F on $\mathbf{P}_4$ with $\sigma^* F = \tilde{F}$.

By construction, $\tilde{F}$ admits a section $\tilde{f}$ vanishing precisely on $Y' \cup n\Sigma$. The section $f = \sigma_* \tilde{f}$ therefore vanishes with multiplicity 2 on the cone X.

(d) *Chern classes and stability of F*: Since $\deg X = \deg R = n^2/2$, we have $c_2(F) = n^2$. Also

$$\begin{aligned}\sigma^*(\det F) &= q^*\mathcal{O}_{\mathbf{P}_3}(2n) \otimes \sigma^*\mathcal{O}_{\mathbf{P}_4}(-1) \otimes \mathcal{O}_{\tilde{P}}(2n\Sigma) \\ &= \sigma^*\mathcal{O}_{\mathbf{P}_4}(2n-1)\end{aligned}$$

shows that $c_1(F) = 2n - 1$. Finally, F is stable, if for all lines $D \subset \mathbf{P}_4$, except for a set of codimension 2, we find

$$F \mid D = \mathcal{O}_{\mathbf{P}_1}(n) \oplus \mathcal{O}_{\mathbf{P}_1}(n-1).$$

But for every line $q^{-1}z$, $z \notin R$, in $\tilde{P}$ we have

$$G \mid q^{-1}z = \mathcal{O}_{\mathbf{P}_1} \oplus \mathcal{O}_{\mathbf{P}_1}(-1). \qquad \square$$

Notice that the lines $q^{-1}z$, $z \in R$ are exactly the jumping lines of F. In fact, there must be jumping lines through x by Van de Ven's theorem [8], because F is indecomposable. They must form at least a curve and are contained in the irreducible curve R.

REFERENCES

[1] D. FERRAND: Courbes gauches et fibrés de rang 2. C.R. Acad. Sci. Paris, *281* (1975) A 345–347.
[2] H. GRAUERT and G. MÜLICH: Vektorbündel vom Rang 2 über dem n-dimensionalen komplex-projektiven Raum. Manuscr. math. *16* (1975) 75–100.
[3] PH. GRIFFITHS, J. HARRIS: Principles of Algebraic Geometry. Wiley & Sons (1978).
[4] G. HORROCKS, D. MUMFORD: A rank 2 vector bundle on $\mathbf{P}^4$ with 15,000 symmetries. Topology *12* (1973) 63–81.
[5] R.W.H.T. HUDSON: Kummer's Quartic Surface, Cambridge Univ. Press 1905.
[6] C.M. JESSOP: A Treatise on The Line Complex. Cambridge Univ. Press 1903, reprinted by Chelsea Publ. Comp. N.Y. (1969).
[7] C. PESKINE: To appear.
[8] A. VAN de VEN: On Uniform vector bundles. Math. Ann. *195* (1972) 245–248.

Mathematisches Institut
der Universität Erlangen-Nürnberg
Bismarckstr. 1 1/2
D-8520 Erlangen
West-Germany

Périodes des variétés algébriques

INFINITESIMAL VARIATIONS OF HODGE STRUCTURE AND THE GLOBAL TORELLI PROBLEM*

James A. Carlson and Phillip A. Griffiths

1. Introduction

The aim of this report is to introduce the infinitesimal variation of Hodge structure as a tool for studying the global Torelli problem. Although our techniques have so far borne fruit only in special cases, we are encouraged by the fact that for the first time we are able to calculate the degree of the period mapping when the Hodge-theoretic classifying space is not hermitian symmetric, as it is in the standard examples of curves, K-3 surfaces, and cubic threefolds. Our specific goal is therefore to prove the following:

THEOREM: *The period mapping for cubic hypersurfaces of dimension* $3m$ *is of degree one onto its image.*

REMARKS: (1) The classifying space fails to be Hermitian symmetric as soon as m exceeds unity.

(2) Our proof, unlike the Pyatetski–Shapiro–Shafarevich argument for K-3 surfaces, is constructive: a defining polynomial for the hypersurface is produced from the infinitesimal variation of Hodge structure.

(3) The general problem of determining when the period map is of degree one onto its image will be called the weak global Torelli problem.

In rough outline, our approach is as follows. To each variety X is associated the object $\Phi(X)$ which describes the first order variation of the Hodge structure on its kth cohomology H^k as X moves with general moduli. If the Hodge filtration satisfies $F^m \neq 0$ and $F^p = 0$ for $p > m$, then Φ defines a natural homomorphism

$$\phi : S^2F^mH^k(X) \longrightarrow S^{2m-k}H^1(X, \Theta)^*,$$

where S^ℓ represents the ℓ-th symmetric power, where * denotes the dual vector space. In certain cases the variety X can be reconstructed from the kernel of the homomorphism ϕ. Now suppose that the natural map

$$p : \mathcal{M} \longrightarrow D/\Gamma$$

*Both authors supported in part by the National Science Foundation. First author also supported in part by an Alfred P. Sloan Fellowship.

from moduli to the period space is generically finite-to-one onto its image. If it is not generically injective, we may choose a regular value y with distinct preimages in the moduli space, represented by inequivalent varieties X_1 and X_2. But p gives a local isomorphism of a neighborhood of X_i in $\mathcal{M}$ with a neighborhood of y in $p(\mathcal{M})$, and this in turn induces an isomorphism between $\Phi(X_1)$ and $\Phi(X_2)$, unique if y is not a fixed point of Γ. Thus, if there is a natural construction of X from $\Phi(X)$, then X_1 and X_2 must be isomorphic, and so the period map has degree one onto its image.[1]

REMARK: We argue the existence of a regular value as follows: choose a smooth variety X which admits no automorphisms, let U be a small simply-connected open set around X in $\mathcal{M}$, let $\tilde{\mathcal{M}} \to \mathcal{M}$ be the universal cover and let U be an open set in $\mathcal{M}$ which projects homeomorphically to U. Then the period map lifts to

$$\tilde{p} : \tilde{U} \longrightarrow D;$$

and by the local Torelli theorem, its differential is of maximal rank. Consequently the only nonregular values of p on U are images of points in $\tilde{p}(\tilde{U})$ which are fixed by some element in the arithmetic group Γ.

Now suppose that all points of $\tilde{p}(\tilde{U})$ are fixed by an element $S \in \Gamma$. Then S commutes with the variation of Hodge structure defined by all smooth hypersurfaces of a fixed degree, hence commutes with the monodromy representation. But the monodromy representation is irreducible and contains a generalized reflection,

$$T(\gamma) = \gamma + \langle \gamma, \delta \rangle \delta,$$

where δ, the vanishing cycle, is fixed. Since S commutes with T, we have also

$$T(\gamma) = \gamma + \langle \gamma, S\delta \rangle S\delta,$$

which implies that δ is an eigenvector for S:

$$S\delta = \lambda \delta.$$

Schur's lemma then asserts that $S = \lambda I$, and the fact that S preserves the integral lattice implies that $\lambda = \pm 1$. Since $\pm I$ acts by the identity on all of D, the claim is demonstrated.

To see that these ideas are not totally farfetched, consider the case of curves. The infinitesimal variation determines the natural map

$$S^2H^0(K) \longrightarrow H^1(\Theta)^* \cong H^0(2K)$$

[1]The crucial idea of iterating the differential is due to Mark Green. The general theory of infinitesimal variations of Hodge structure is the subject of a paper under preparation by the two authors, Mark Green and Joe Harris.

given by cup-product. The kernel of ϕ therefore determines the linear system of quadrics passing through the canonical curve. By the Enriques–Petri theorem, the base locus of this system is the canonical model of X, provided that X is neither hyperelliptic, trigonal nor a smooth plane quartic. Since the functor $X \to \Phi(X)$ has an inverse up to isomorphism, weak global Torelli holds.

REMARKS: (1) The importance of the above argument is that it proves weak global Torelli without using the Jacobian. This is an essential feature of any method which addresses the problem in higher dimension.

(2) Note that when $m = k$, the kernel of ϕ always defines a linear system of quadrics in the projective space of the canonical model of X.

The example suggests that we study the kernel of ϕ in other situations. Because this study requires the detailed calculation of cup-products in sheaf cohomology, we have restricted our investigation to hypersurfaces in projective space, where explicit formulas can be found. In this case the answer is of a somewhat different nature than for curves: given mild degree restrictions, the kernel of ϕ determines a certain homogeneous component of the Jacobian ideal of the defining equation. If this component is not zero for trivial degree reasons, we can use Macaulay's theorem to construct the vector space of first partials, and from this the defining equation itself.

Now the identification of F^m with a space of homogeneous polynomials requires the use of additional data beyond the Hodge structure, namely the Poincaré residue homomorphism. Nonetheless, there are situations in which the choice of identification is irrelevant: this occurs precisely when F^m is given by residues of rational differential whose adjoint polynomial is linear. A simple calculation shows that cubics of dimension $n = 3\ell$ are the only hypersurfaces for which this happens: in this case $F^pH^n(X)$ is zero for $p \geq 2\ell$, and the residue map gives an isomorphism

$$H^0(\mathbf{P}_{3m+1}, \mathcal{O}(1)) \longrightarrow F^{2\ell}H^n(X).$$

To give the complete proof, then, we must (i) make precise our remarks about infinitesimal variations (ii) show how to compute cup-products on a hypersurface in terms of the adjoint polynomials which define cohomology classes through the residue map, (iii) show how to construct X from its Jacobian ideal. This will occupy us for most of the body of the paper, the real work being in (ii). Finally, we shall close with a few thoughts on the general case.

The authors gratefully acknowledge the helpful comments of Chris Peters, Paul Roberts, and Domingo Toledo. The authors also wish to thank Arnaud Beauville for the opportunity to present these results both at the *Journées de Géométrie Algébrique* in Angers, and, in written form, in the present volume.

2. Infinitesimal variations of Hodge structure

An *infinitesimal variation of Hodge structure* is a triple (H, T, Φ) consisting of (i) an S-polarized Hodge structure H of weight k, (ii) a complex vector space T, and (iii) a graded set of homomorphisms

$$\Phi_p : T \longrightarrow \mathrm{Hom}(\mathrm{Gr}_F^p, \mathrm{Gr}_F^{p-1}),$$

where we have set

$$\mathrm{Gr}_F^p = F^p/F^{p-1}.$$

The homomorphisms are subject to the compatibility relations

(iv) $\Phi_{p-1}(v)\Phi_p(w) = \Phi_{p-1}(w)\Phi_p(v)$

(v) $S(\Phi_p(v)\omega, \omega') + S(\omega, \Phi_{q+1}(v)\omega') = 0,$

where $\omega \in \mathrm{Gr}_F^p$, $\omega' \in \mathrm{Gr}_F^{q+1}$, $p + q = k$. For the sake of brevity, we shall usually write Φ in place of (H, T, Φ). A morphism

$$G : \Phi \longrightarrow \Phi'$$

is given by a morphism of polarized Hodge structures,

$$G_H : H \longrightarrow H',$$

and a morphism of complex vector spaces

$$G_T : T \longrightarrow T'$$

such that

$$\Phi'(G_T v)(G_H \omega) = G_H(\Phi(v)\omega)$$

for all choices of v and ω.

We remark that an infinitesimal variation of Hodge structure is what the term suggests: a description of the derivative homomorphism of a variation of Hodge structure $B \to D$. To see that this is reasonable, observe that Φ determines a variation in a neighborhood U of B centered at zero: choose homomorphisms $\tilde{\Phi}_p$ from F^p to F^{p-1} which induce Φ_p, choose coordinates t_i in U, and define a variable Hodge filtration by

$$F^p(t) = \left(1 + \sum \xi_i \tilde{\Phi}_p(\partial/\partial t_i)\right) F^p.$$

Let us now construct the associated homomorphism ϕ mentioned in the introduction. Given an r-tuple of vectors in T, we may define a homomorphism

$$\Phi^r(v) = \Phi_{p+r-1}(v_r) \circ \cdots \circ \Phi_p(v_1)$$

from Gr_F^p to Gr_F^{p+r}. The axioms of an infinitesimal variation imply that Φ^r is multilinear and symmetric in the components of v, so that we in fact obtain a homomorphism

$$\Phi^r : S^r T \longrightarrow \mathrm{Hom}(\mathrm{Gr}_F^p, \mathrm{Gr}_F^{p+r}).$$

Now suppose that $F^m \neq 0$ and $F^p = 0$ for $p > m$, so that Gr_F^m and Gr_F^{k-m} represent the "ends" of the Hodge filtration. Because the ends are paired perfectly by S, we may define, for each couple (ω, ω') in $F^m \times F^m$, a linear functional on $S^{2m-k}T$ by $\phi(\omega, \omega') : v \mapsto S(\Phi^{2m-k}(v)\omega, \omega')$. Applying the axioms of an infinitesimal variation once more, we find that the right hand side is symmetric and bilinear on $F^m \times F^m$. Sending (ω, ω') to the corresponding functional, we arrive at the desired homomorphism,

$$\phi : S^2 F^m \longrightarrow S^{2m-k}T^*.$$

Although an infinitesimal variation defines many other natural tensors, the one just constructed is the one we shall study here. Moreover we shall only usc part of the data given by Φ, namely its kernel viewed as a linear system of quadrics on $S^2 F^m$.

3. An algebraic cup-product formula

a. Statement of the formula

The purpose of this section is to give an algebraic formula for the cup-product on a smooth projective hypersurface. To state it, we first recall how to describe rational differentials on projective space: Fix a polynomial Q, homogeneous of degree d in $n+2$ complex variables $x_0, \ldots, x_{n+1}$, and assume that the projective variety it defines is smooth:

$$X \subset \mathbf{P}_{n+1}.$$

Fix a nonzero section of

$$\Omega^{n+1}_{\mathbf{P}_{n+1}}(n+2) \cong \mathcal{O}_{\mathbf{P}_{n+1}},$$

say

$$\Omega = \sum (-1)^i x_i dx_0 \ldots \widehat{dx_i} \ldots dx_{n+1}.$$

Then the expression

$$\Omega_A = \frac{A\Omega}{Q^{a+1}}$$

defines a rational $(n+1)$-form with X as polar locus, provided that the degree of A is chosen to make the quotient homogeneous of degree zero. The polynomial in the numerator is called an adjoint of level a, and Ω_A is said to have adjoint level a.

To each k-dimensional cohomology class on the complement of X a $(k-1)$-dimensional cohomology class is defined on X itself by the topological residue: Given a $(k-1)$-cycle γ on X, let $T(\gamma)$ be the k-cycle in

$\mathbf{P}_{n+1} - X$ defined by forming the boundary of an ϵ-tubular neighborhood of γ. This construction defines a map

$$T : H_{k-1}(X) \longrightarrow H_k(\mathbf{P}_{n+1} - X)$$

whose formal adjoint, up to a factor of $2\pi\mathrm{i}$, is the topological residue. Thus, if the class on the complement is represented by a differential form, we have the explicit relation

$$\int_\gamma \operatorname{res} \alpha = \frac{1}{2\pi\mathrm{i}} \int_{T(\gamma)} \alpha.$$

The residue map satisfies the following properties[1]:

(i) $\operatorname{res} \Gamma\Omega^{n+1}((n+1)X) = H^n(X, \mathbf{C})_o$, where the right-hand side denotes the subspace of primitive cohomology.

(ii) $\operatorname{res} \Gamma\Omega^{n+1}((a+1)X) = F^{n-a}H^n(X, \mathbf{C})_o$.

The compatibility between filtration by pole order and filtration by Hodge level can be strengthened further: Let Q_i denote the partial derivative of Q with respect to x_i, and let

$$J_Q = (Q_0, \ldots, Q_{n+1}),$$

the *Jacobian ideal*, be the homogeneous ideal which they generate. Then

(iii) the residue of a form Ω_A of adjoint level a has Hodge level $n - a + 1$ if and only if A lies in the Jacobian ideal.

This motivates, but does not prove, the following vanishing criterion:

THEOREM 2: *Let Ω_A and Ω_B be rational forms of complementary adjoint level $(a + b = n)$. Then the cup-product of their residues vanishes if and only if the ordinary product AB lies in the Jacobian ideal*:

$$\operatorname{res} \Omega_A \cdot \operatorname{res} \Omega_B = 0 \Leftrightarrow AB \in J_Q.$$

To prove this result, we first use the properties of the Hodge decomposition to show that

$$\operatorname{res} \Omega_A \cdot \operatorname{res} \Omega_B = (\operatorname{res} \Omega_A)^{n-a,a} (\operatorname{res} \Omega_B)^{n-b,b}.$$

Second, we seek explicit Čech cocycles which represent the indicated Hodge components, using

$$H^{p,q}(X) \cong H^q(X, \Omega^p).$$

Multiplication of these yields a cocycle in $H^n(X, \Omega^n)$. Third, we use the coboundary in the Poincaré residue sequence,

$$\delta : H^n(X, \Omega^n) \longrightarrow H^{n+1}(\mathbf{P}_{n+1}, \Omega^{n+1})$$

to transfer the product to projective space. Performing all of the Čech

---

[1] To keep notation reasonable, we write $\Omega^r(sX)$ for $\Omega^r_{\mathbf{P}_{n+1}}(sX)$.

calculations using the cover $\mathscr{U}$ defined by

$$U_j = \{Q_j \neq 0\},$$

we obtain the $(n+1)$-cocycle

$$\eta(A, B) = c_{ab} \frac{AB\Omega}{Q_0 \dots Q_{n+1}},$$

where c_{ab} is a nonzero rational constant depending only on a and b. To summarize, we claim the following:

THEOREM 3: $\delta(\text{res}\, \Omega_A \cdot \text{res}\, \Omega_B) = \eta(A, B)$.

The vanishing criterion now follows, since it turns out that $\eta(A, B)$ cobounds precisely when AB lies in the Jacobian ideal.

REMARK: It is evident that the smoothness of X is essential, since otherwise the U_j do not cover $\mathbf{P}_{n+1}$. We shall refer to $\mathscr{U}$ as the Jacobian cover of $\mathbf{P}_{n+1}$ relative to Q.

b. A Čechist residue formula

We shall now establish an explicit algebraic formula for the residue in Čech cohomology. To state it, we introduce the following notation: Given a vector field Z on $\mathbf{C}^{n+2}$, let $\text{i}(Z)$ denote the operation of contraction, and let $K_j = \text{i}(\partial/\partial x_j)$. Given a multi-index $J = (j_0, \dots, j_q)$ of size q, let

$$K_J = K_{j_q} \dots K_{j_0},$$

let

$$\Omega_J = K_J \Omega,$$

and let

$$Q_J = Q_{j_0} \dots Q_{j_q}.$$

PROPOSITION: *Let Ω_A be a form of adjoint level a. Then we have*

$$(\text{res}\, \Omega_A)^{n-a,a} = c_a \left\{ \frac{A\Omega_J}{Q_J} \right\}_{|J|=q},$$

where c_a is a nonzero rational constant depending only on a.

Note that the right-hand side makes sense, since

$$\frac{A\Omega_J}{Q_J} \in \Gamma(U_J \cap X, \Omega_X^{n-a}),$$

where

$$U_J = U_{j_0} \cap \dots \cap U_{j_a}.$$

In other words, the right-hand side is an a-cocycle of the cover $\mathscr{U} \cap X$ with values in Ω_X^{n-a}.

In outline, the proof of the proposition goes as follows: The cohomology of $\mathbf{P}_{n+1} - X$ can be calculated either by the global deRham complex

(i) $\Gamma\Omega^{\cdot}(*X)$

of rational forms with poles of an arbitrary order on X, or by the hypercohomology of the Čech–deRham complex

(ii) $\mathscr{C}^{\cdot}\Omega^{\cdot}(\log X)$

of cocycles with a logarithmic singularity. Whereas the first complex is the natural abode of the rational forms Ω_A, it is on the second complex that the algebraic Poincaré residue is defined in hypercohomology: If

$$\omega = \left\{\omega_I \frac{\mathrm{d}f_j}{f_j}\right\}$$

is a cocycle in (ii), then

$$\operatorname{res} \omega = \{\omega_I \mid f_j = 0\}.$$

Because both complexes compute the cohomology of the complement, there must be a Čech–deRham element with a logarithmic pole which represents Ω_A. Our problem, however, is to find an explicit representative. To do this, we imbed both resolutions in the Čech–deRham complex with arbitrary algebraic singularities,

(iii) $\mathscr{C}^{\cdot}\Omega^{\cdot}(*X)$.

Inside this object we shall construct an explicit homotopy from (i) to (ii), and from this the desired formula will follow.

We begin the real work with the definition of a partial homotopy operator: Cover $\mathbf{P}_{n+1}$ by the affine open sets

$$U_j = \{Q_j \neq 0\}$$

and consider the double complex

(iv) $\mathscr{C}^q(\mathscr{U}, \Omega^p(*X))$.

The cohomology of the associated single complex with respect to the total differential

$$D = \mathrm{d} + (-1)^p\delta$$

then computes the complex cohomology of $\mathbf{P}_{n+1} - X$. Define an operator

$$H_\ell : \mathscr{C}^q(\mathscr{U}, \Omega^p(\ell X)) \longrightarrow \mathscr{C}^q(\mathscr{U}, \Omega^{p-1}((\ell - 1)X))$$

by the formula

$$(H_\ell \omega)_{j_0 \ldots j_q} = \frac{1}{1-\ell} \cdot \frac{Q}{Q_{j_0}} K_{j_0} \omega_{j_0 \ldots j_q}.$$

LEMMA: *For $\ell \geqq 2$, H satisfies the identity*

$$\mathrm{d}H_\ell + H_{\ell+1}\,\mathrm{d} \equiv 1$$

modulo the group

$$\mathscr{C}^q(\mathscr{U}, \Omega^p(\ell - 1)X).$$

PROOF: The essential point is that the exterior derivative on cochains of pole order ℓ is represented, up to a multiplicative constant and modulo cochains of lower pole order, by multiplication against $\mathrm{d} \log Q$:

$$\mathrm{d}\left(\frac{\alpha}{Q^\ell}\right) \equiv -\ell \frac{\mathrm{d}Q}{Q} \wedge \frac{\alpha}{Q^\ell}.$$

The identity follows from this and the commutation relation

$$K_j\,\mathrm{d}Q + \mathrm{d}QK_j = Q_j.$$

The operator H gives an explicit means of reducing the pole order of a cocycle inside (iv):

LEMMA: *Let α be a Čech–deRham cochain of pole order $\ell \geqq 2$. Then*

$$\alpha \equiv DH\alpha + HD\alpha$$

modulo cochains of pole order $\ell - 1$. In particular, if α is a cocycle then

$$\tilde{\alpha} = (1 - DH)\alpha$$

is cohomologous to α and has pole order $\ell - 1$.

PROOF: Let α_q^p denote the component of α in $\mathscr{C}^q(U, \Omega^p(*X))$ and write

$$\alpha - DH\alpha = \sum (\alpha_p^q - \mathrm{d}H\alpha_q^p - (-1)^{p-1}\delta H\alpha_q^p)$$

$$\equiv \sum (H\,\mathrm{d}\alpha_q^p + (-1)^p \delta H \alpha_q^p).$$

Because H and δ commute up to an operator which reduces pole order, this becomes

$$\alpha - DH\alpha \equiv \sum (H\,\mathrm{d}\alpha_q^p + (-1)^p H\delta\alpha_q^p)$$

$$\equiv HD\alpha,$$

as desired.

REMARK: Let α^{ν}_{μ} denote the component of maximal Čech degree in the cocycle $\alpha : \alpha^{p}_{q} = 0$ if $q > \mu$. Then the component of maximal Čech degree for $\tilde{\alpha}$ is given by

$$\tilde{\alpha}^{\nu-1}_{\mu+1} = (-1)^{\nu}\delta H \alpha^{\nu}_{\mu}.$$

Let us now consider the cocycle

$$\tilde{\Omega}_A = (1 - DH)^a \Omega_A,$$

where Ω_A has a pole of order $a + 1$. The pole order of $\tilde{\Omega}_A$ is one, and since it is D-closed,

$$\mathrm{d}\tilde{\Omega}_A = \sum (-1)^p \delta(\Omega_A)^p_q$$

also has pole order one. It follows that $\tilde{\Omega}_A$ has logarithmic singularities, and consequently is prepared for application of the algebraic residue map.

To calculate the residue explicitly, we calculate $(\tilde{\Omega}_A)_a$, the component of maximal Čech degree, using the following result:

LEMMA:

(i) $\Omega_A \equiv (-1)^n \left\{\dfrac{A\Omega_i}{Q_i Q^a} \wedge \dfrac{\mathrm{d}Q}{Q}\right\}.$

(ii) *If* $I = (i_0, \ldots, i_q)$, *then*

$$H_{r+1}\left\{\frac{A\Omega_I}{Q_I Q^r} \wedge \frac{\mathrm{d}Q}{Q}\right\} = \frac{(-1)^{n-q+1}}{r}\left\{\frac{A\Omega_I}{Q_I Q^r}\right\}.$$

(iii) *Under the same assumptions,*

$$\delta\left\{\frac{A\Omega_I}{Q_I Q^r}\right\} \equiv (-1)^{q+1}\left\{\frac{A\Omega_J}{Q_J Q^{r-1}} \wedge \frac{\mathrm{d}Q}{Q}\right\},$$

where $J = (j_0, \ldots, j_{q+1})$.

PROOF: (i) Let $\mathrm{d}V = \mathrm{d}x_0 \ldots \mathrm{d}x_{n+1}$ be the Euclidean volume form, let

$$E = \sum x_i \frac{\partial}{\partial x_i}$$

be the Euler vector field, and observe that

$$\mathrm{i}(E)\,\mathrm{d}V = \Omega.$$

Take the trivial identity

$$\mathrm{d}Q \wedge \mathrm{d}V = 0$$

and contract with the Euler field to get

$$(\deg Q) Q\,\mathrm{d}V - \mathrm{d}Q \wedge \Omega = 0.$$

which we write as

$$dQ \wedge \Omega \equiv 0.$$

Contraction of this identity with $\partial/\partial x_i$ then yields

$$Q_i\Omega \equiv dQ \wedge \Omega_i,$$

or

$$\Omega \equiv (-1)^n \frac{\Omega_i}{Q_i} \wedge \frac{dQ}{Q},$$

and hence the result.

(ii) The essential point is that

$$K_{i_0}\Omega_I = 0$$

and

$$K_{i_0}\, dQ = Q_{i_0}.$$

Then

$$H_{r+1}\left\{\frac{A\Omega_I}{Q_I Q^r} \wedge \frac{dQ}{Q}\right\} = \frac{(-1)^{n-q}}{-r}\left\{\frac{Q}{Q_{i_0}} \cdot \frac{A\Omega_I}{Q_I Q^r} \wedge \frac{Q_{i_0}}{Q}\right\},$$

hence the result.

(iii) By definition of the coboundary,

$$\delta\left\{\frac{A\Omega_I}{Q_I Q^r}\right\} = \left\{\frac{A \sum (-1)^\ell Q_{J_\ell}\Omega_{J_\ell}}{Q_J Q^r}\right\},$$

where

$$J_\ell = (j_0, \ldots, \hat{j}_\ell, \ldots, j_{q+1}).$$

Now take the fundamental identity

$$dQ \wedge \Omega \equiv 0$$

and apply the iterated contraction operator K_J to obtain

$$(-1)^{q+1} \sum (-1)^\ell Q_{j_\ell}\Omega_{J_\ell} \equiv dQ \wedge \Omega_J.$$

Substitution into the formula above then yields the claimed result.

To complete the proof of the proposition, we set

$$\Omega_A(\mu) = ((1 - DH)^\mu \Omega_A)_\mu^{n+1-\mu}$$

and prove inductively that

$$(I_\mu) \qquad \Omega_A(\mu) \equiv \tilde{c}_\mu \left\{\frac{A\Omega_J}{Q_J Q^{a-\mu}} \wedge \frac{dQ}{Q}\right\}_{|J|=\mu}$$

where $\tilde{c}_\mu$ is a rational constant depending only on n and μ. By assertion (i) of the lemma, I_0 holds. Assume that I_μ holds, and then calculate

$$\Omega_A(\mu+1) = (-1)^{n+1-\mu}\delta H_{a-\mu+1}\Omega_A(\mu),$$

using assertions (ii) and (iii) of the lemma to obtain

$$\Omega_A(\mu+1) = \frac{(-1)^{\mu+1}}{a-\mu}\left\{\frac{A\Omega_J}{Q_J Q^{a-\mu-1}} \wedge \frac{dQ}{Q}\right\}_{|J|=\mu+1}.$$

Thus

$$\tilde{c}_{\mu+1} = \frac{(-1)^{\mu+1}}{a-\mu}\tilde{c}_\mu \quad (\mu \geqq 0)$$

for $\mu > 0$ and

$$\tilde{c}_0 = (-1)^n.$$

Therefore

$$(\tilde{\Omega}_A)_a^{n-a+1} = c_a \left\{\frac{A\Omega_J}{Q_J} \wedge \frac{dQ}{Q}\right\}_{|J|=a} \tag{$*$}$$

with

$$c_a = \frac{(-1)^{n+a(a+1)/2}}{a!}.$$

Substitution of ($*$) into the algebraic residue now yields the proposition.

Remark: The proof of the residue formula in Čech cohomology gives still another proof that the pole filtration is compatible with the Hodge filtration: A rational form of polar order $a+1$ unfolds in the Čech–deRham complex to a cocycle

$$\tilde{\omega}_0 + \cdots + \tilde{\omega}_a \in \bigoplus_{i \leqq a} \mathscr{C}^i(\mathscr{U}, \Omega^{n+1-i}(\log X))$$

whose residue lies in

$$\bigoplus_{i \leqq a} \mathscr{C}^i(\mathscr{U}|_X, \Omega_X^{n-i}).$$

But the Hodge filtration on Čech–deRham cohomology is induced by exactly this, the "filtration bête".

c. Computation of the cup-product

To complete the proof of the theorem, we must first calculate the cup-product

$$(\operatorname{res}\Omega_A)^{a,b}(\operatorname{res}\Omega_B)^{b,a},$$

where $a+b=n$, and then find its image in $H^{n+1}(\mathbf{P}_{n+1}, \Omega^{n+1})$. To begin, let

"$\wedge$" denote the natural product

$$\mathscr{C}^q(\mathscr{U}|_X, \Omega_X^p) \times \mathscr{C}^s(\mathscr{U}|_X, \Omega_X^r) \longrightarrow \mathscr{C}^{q+s}(\mathscr{U}|_X, \Omega_X^{p+r})$$

given by the "front q-face, back s-face" rule, followed by exterior multiplication of forms. Then the twisted product

$$\alpha_q^p \alpha_s^r = (-1)^{qr} \alpha_q^p \wedge \alpha_s^r$$

is skew-commutative and satisfies the Leibnitz rule [4]:

$$D(\alpha_q^p \alpha_s^r) = (D\alpha_q^p)\alpha_s^r + (-1)^{p+q} \alpha_q^p (D\alpha_s^r).$$

Because the twisted product reduces to exterior multiplication of forms, it represents the topological cup-product on the level of hypercohomology. Consequently the residue product above is represented by the cocycle

$$\{\psi_L\} : \tilde{c}_{ab} \left\{ \frac{AB\Omega_{Rs}\Omega_{sT}}{Q_R Q_s^2 Q_T} \right\} \in \mathscr{C}^n(\mathscr{U}|_X, \Omega_X^n),$$

where

$$\tilde{c}_{ab} = (-1)^{b^2} c_a c_b = \frac{(-1)^{a(a-1)/2 + b(b-1)/2 + b^2}}{a!b!}.$$

and where the multi-index L is partitioned according to

$$L = (r_0, \ldots, r_{a-1}, s, t_1, \ldots, t_b).$$

To calculate the coboundary of ψ in the Poincaré residue sequence, first lift to a cocycle

$$\tilde{\psi} \in \mathscr{C}^n(\mathscr{U}, \Omega^n(\log X)),$$

namely

$$\tilde{\psi}_L = \psi_L \wedge \frac{\mathrm{d}Q}{Q},$$

and then apply the following result to simplify the numerator:

LEMMA: *Let $v \in \{0, \ldots, n+1\}$ be the index complementary to L. Then*

$$\Omega_{Rs} \wedge \Omega_{sT} \wedge \mathrm{d}Q \equiv (-1)^{n+v} x_v Q_s \Omega.$$

PROOF: (1) Let E denote contraction with the Euler vector field, and then write the left-hand side as

$$(K_{Rs} E\, \mathrm{d}V) \wedge (K_{sT} E\, \mathrm{d}V) \wedge \mathrm{d}Q = (-1)^a (EK_{Rs}\, \mathrm{d}V) \wedge (EK_{sT}\, \mathrm{d}V) \wedge \mathrm{d}Q.$$

Next, use the Euler identity

$$E\, \mathrm{d}Q = (\deg Q)Q$$

and the fact that $E^2 = 0$ to write the right-hand side as

(i) $(-1)^a E\{(EK_{Rs}\,\mathrm{d}V) \wedge (K_{sT}\,\mathrm{d}V) \wedge \mathrm{d}Q\}$,

up to congruence.

(2) The ordered set $(0, \ldots, n+1)$ can be written either as

$$(0, \ldots, n+1) = RsT'vT''$$

or as

$$(0, \ldots, n+1) = R'vR''sT,$$

depending on the position of the complementary index v. We shall prove in the first case that

(ii) $(EK_{Rs}\,\mathrm{d}V) \wedge (K_{sT}\,\mathrm{d}V) \wedge \mathrm{d}Q = (-1)^{b+v} x_v Q_s\,\mathrm{d}V$;

the proof in the second is similar. To begin, note that[1]

$$K_{Rs}\,\mathrm{d}V = \mathrm{d}x_{T'vT''} = (-1)^{T'}\,\mathrm{d}x_v\,\mathrm{d}x_T,$$

hence

(iii) $EK_{Rs}\,\mathrm{d}V = (-1)^{T'}\{x_v\,\mathrm{d}x_T - \mathrm{d}x_v\,E\mathrm{d}x_T\}$.

Next, we find that

$$K_{sT}\,\mathrm{d}V = (-1)^{RT+R+T''}\,\mathrm{d}x_{Rv},$$

hence

(iv) $(K_{sT}\,\mathrm{d}V) \wedge \mathrm{d}Q = (-1)^{RT+R+T''}\left\{Q_s\,\mathrm{d}x_{Rvs} + \sum_{t\in T} Q_t\,\mathrm{d}x_{Rvt}\right\}$.

The only nonvanishing term in the product (i) is easily seen to arise from the product of the first term in (iii) with the first term in (iv), which is

(v) $(-1)^{RT'+R+T} x_v Q_s\,\mathrm{d}x_{TRvs}$.

Putting the $\mathrm{d}x_i$ in their natural order and using $v = R + T' + 1$, we obtain (ii), as desired.

(3) Substitution of (ii) into (i) then yields the asserted formula, since $E\,\mathrm{d}V = \Omega$.

Let us now return to the calculation of $\delta\psi$: By the lemma,

$$\tilde{\psi}_L \equiv (-1)^{n+v} c_{ab} \left\{\frac{ABx_v\Omega}{\prod_{j\neq v} Q_j}\right\} \frac{1}{Q},$$

so that

$$\delta\psi \equiv (-1)^n \tilde{c}_{ab} \left\{\sum \frac{ABx_v Q_v \Omega}{Q_0 \ldots Q_{n+1}}\right\} \frac{1}{Q}.$$

[1]By convention, $(-1)^J = (-1)^q$, where $J = (j_1, \ldots, j_q)$.

Because of the Euler identity, this reduces to

$$\delta\psi \equiv c_{ab} \frac{AB\Omega}{Q_0 \ldots Q_{n+1}},$$

where

$$c_{ab} = \frac{(-1)^{a(a+1)/2+b(b+1)/2+n+b^2}}{a!b!} \deg Q.$$

With a cocycle representative for the residue product in hand, the proof of theorem 3 is complete. To obtain the vanishing criterion of theorem 2, it therefore suffices to show that

$$\eta(A, B) \text{ cobounds} \Leftrightarrow AB \in J_Q.$$

To see this, consider the complexes $\mathscr{C}^{\cdot}_{\ell}(\mathscr{U}, \Omega^{n+1})$ where a typical cochain ϕ of dimension q has the form

$$\phi_J = \frac{R_J\Omega}{Q_J^{\ell}}.$$

Denote the cohomology groups of this complex by $H^{\cdot}_{\ell}(\mathbf{P}_{n+1}, \Omega^{n+1})$ and observe that

$$H^{\cdot}(\mathbf{P}_{n+1}, \Omega^{n+1}) = \varinjlim H^{\cdot}_{\ell}(\mathbf{P}_{n+1}, \Omega^{n+1}),$$

where the map from $\mathscr{C}^{\cdot}_{\ell}$ to $\mathscr{C}^{\cdot}_{\ell+1}$ is given by multiplication of the numerator in the cocycle against $\tilde{Q} = Q_0 \ldots Q_{n+1}$. Now the residue product is represented in $\mathscr{C}^{\cdot}_1$, and it cobounds in this complex if and only if

$$\delta \left\{ \frac{R_j\Omega}{\prod_{i \neq j} Q_i} \right\} = \eta(A, B),$$

so that

$$AB = \sum (-1)^j Q_j R_j \in J_Q.$$

Consequently

$$\eta(A, B) \text{ cobounds in } \mathscr{C}^{\cdot}_1 \Leftrightarrow AB \in J_Q.$$

To complete the argument, we must show that multiplication against $\tilde{Q}$ defines an injective map

$$H^{n+1}_{\ell}(\mathbf{P}_{n+1}, \Omega^{n+1}) \longrightarrow H^{n+1}_{\ell+1}(\mathbf{P}_{n+1}, \Omega^{n+1}),$$

hence an injective map

$$H^{n+1}_1(\mathbf{P}_{n+1}, \Omega^{n+1}) \longrightarrow H^{n+1}(\mathbf{P}_{n+1}, \Omega^{n+1}).$$

This we shall do in the next section using local duality theory.

d. The Grothendieck residue symbol and Macaulay's theorem

We begin by recalling the definition of the Grothendieck residue symbol. Let U be an open set in $\mathbf{C}^m$, let

$$f : U \longrightarrow \mathbf{C}^m$$

be holomorphic and such that $f^{-1}(0) = 0$, and let

$$\Gamma(\epsilon) = \{|f_j| = \epsilon : j = 1, \ldots, m\},$$

where f_j is the j-th coordinate of f. For any holomorphic function g on U, define the meromorphic differential

$$\omega_f(g) = \frac{g \, \mathrm{d}x_1 \ldots \mathrm{d}x_m}{f_1 \ldots f_m},$$

and set

$$\operatorname{Res}_0 \left\{ \begin{matrix} g \\ f_1 \ldots f_m \end{matrix} \right\} = \left(\frac{1}{2\pi \mathrm{i}} \right)^m \int_{\Gamma(\epsilon)} \omega_f(g).$$

This is the Grothendieck residue symbol, which we shall usually write as $\operatorname{Res}_0(g)$. It defines a $\mathbf{C}$-linear homomorphism on the local ring $\mathcal{O}$ of holomorphic functions at the origin which annihilates the ideal

$$I = (f_1, \ldots, f_m).$$

Consequently the residue descends to a homomorphism

$$\operatorname{Res}_0 : \mathcal{O}/I \longrightarrow \mathbf{C}$$

which defines in turn a pairing

$$\operatorname{Res}_0 : (\mathcal{O}/I) \times (\mathcal{O}/I) \longrightarrow \mathbf{C}$$

by

$$(g, h) \longmapsto \operatorname{Res}_0(gh).$$

THEOREM 4: *The residue pairing is perfect.*

(For proofs, see [7] or [9]).

Let us now study the residue pairing on the graded polynomial subring

$$V = \mathbf{C}[x_1, \ldots, x_m].$$

If the generators f_j are homogeneous of degree d_j, then the ideal $I \subset V$ is naturally graded, as is the quotient

$$\bar{V} = V/I.$$

If we decree that the complex numbers form an algebra of degree zero, then the residue symbol preserves the grading, up to the appropriate shift:

LEMMA: *The residue symbol on $\bar{V}$ is homogeneous of degree $-\sigma$, where*

$$\sigma(I) = \sum (d_j - 1).$$

PROOF: Let λ_t denote multiplication of all the variables x_j by t, let $g \in V$ be homogeneous of degree j, and observe that

$$\lambda_t^* \omega_f(g) = t^{j-\sigma} \omega_f(g).$$

The asserted homogeneity of $\omega_f(g)$ then yields the integral formula below:

$$t^{j-\sigma} \int_{\Gamma(\epsilon)} \omega_f(g) = \int_{\Gamma(\epsilon)} \lambda_t^* \omega_f(g)$$

$$= \int_{\lambda_t \Gamma(\epsilon)} \omega_f(g) - \int_{\Gamma(\epsilon)} \omega_f(g).$$

COROLLARY: *The Grothendieck residue pairing on $\bar{V} \times \bar{V}$ satisfies the following, where the superscript indicates the natural degree:*

(i) *$\bar{V}^i$ is orthogonal to $\bar{V}^j$ unless $i + j = \sigma$.*
(ii) *$\bar{V}^i$ and $\bar{V}^{\sigma-i}$ are perfectly paired.*
(iii) *$\bar{V}^i = 0$ if $i \notin [0, \sigma]$.*
(iv) dim $\bar{V}^\sigma = 1$.

As a first application of the corollary, we complete the proof of the vanishing criterion given by theorem 2. Indeed, it suffices to prove the following:

LEMMA: *The Grothendieck residue defines an isomorphism*

$$\mathrm{Res}_0 : H_\ell^{n+1}(\mathbf{P}_{n+1}, \Omega^{n+1}) \longrightarrow \mathbf{C}$$

which commutes with multiplication by $\tilde{Q} = Q_0 \ldots Q_{n+1}$.

PROOF: Map $\mathscr{C}_\ell^{n+1}(\mathscr{U}, \Omega^{n+1})$ to the complex numbers by the obvious formula,

$$\frac{R\Omega}{\prod Q_j^\ell} \longmapsto \mathrm{Res}_0 \left\{ \frac{R}{Q_0^\ell \ldots Q_{n+1}^\ell} \right\}.$$

Because a general coboundary has the form

$$\frac{\left(\sum (-1)^j R_j Q_j^\ell\right) \Omega}{\prod Q_j^\ell} = \delta \left\{ \frac{R_j \Omega}{\prod_{i \neq j} Q_i^\ell} \right\},$$

the set of all such can be identified with the group

$$\left\{\frac{R\Omega}{\Pi\, Q_j^\ell}\,\middle|\, R \in (Q_0^\ell, \ldots, Q_{n+1}^\ell)\right\}.$$

The residue therefore descends to a natural linear functional on H_ℓ^{n+1}. It is an isomorphism because it factors through the identification

$$H_\ell^{n+1} \longrightarrow \bar{V}^\sigma,$$

where

$$\bar{V} = V/(Q_0^\ell, \ldots, Q_{n+1}^\ell).$$

It commutes with multiplication by $\tilde{Q}$ because of the obvious formula

$$\int_{\Gamma_{\ell+1}} \frac{RQ_0 \ldots Q_{n+1}\Omega}{Q_0^{\ell+1} \ldots Q_{n+1}^{\ell+1}} = \int_{\Gamma_\ell} \frac{R\Omega}{Q_0^\ell \ldots Q_{n+1}^\ell},$$

where

$$\Gamma_\ell = \{|Q_j^\ell| = \epsilon\} \cong \{|Q_j| = \epsilon^{1/\ell}\}.$$

Q.E.D.

As a second application of the corollary, we shall reprove the classical Macaulay theorem. To state it, consider an ideal I in a ring V, together with a subset D of V, and define the *ideal quotient*

$$(I : D) = \{x \in V \mid xd \in I \text{ for all } d \in D\}.$$

The result is clearly an ideal of V containing I. Let V be the graded ring $\mathbf{C}[x_i]$, and let

$$F^p V = \bigoplus_{i \geqq p} V^i$$

be the filtration naturally associated to the grading. Assuming that I is generated by a regular sequence of homogeneous elements as before, we have

MACAULAYS THEOREM:

$$(I : F^p V) = I + F^{\sigma - p + 1} V.$$

Because the ideal quotient is itself graded, we have also the following:

COROLLARY: *If $t < \sigma - p + 1$, then*

$$(I : F^p V)^t = I^t.$$

PROOF: Macaulay's theorem is equivalent to the assertion

$$\text{annihilator}(F^p \bar{V}) = F^{\sigma - p + 1} \bar{V}.$$

But this follows from part (ii) of the corollary via the relation

$$\text{annihilator}(\bar{V}^i) = \bigoplus_{j \neq \sigma - i} \bar{V}^j.$$

Q.E.D.

REMARKS: (1) Returning to the cup-product formula, let λ be the residue of the fundamental class of $\mathbf{P}_{n+1}$, viewed in $H^{n+1}(\mathbf{P}_{n+1}, \Omega^{n+1})$. Then we have

$$\int_X \operatorname{res} \Omega_A \operatorname{res} \Omega_B = \frac{1}{\lambda} \operatorname{Res}_0 \eta(A, B).$$

(2) One can construct an explicit "fundamental class" in $\bar{V}$, that is, an element of $\bar{V}^\sigma$ which has residue one. Such a class is in fact given by the Jacobian determinant,

$$\mathcal{J} = \det\left(\frac{\partial f_i}{\partial x_j}\right).$$

To see this, write

$$\omega_f(\mathcal{J}) = \frac{\mathrm{d}f_1}{f_1} \wedge \cdots \wedge \frac{\mathrm{d}f_m}{f_m},$$

and observe that

$$\int_\Gamma \frac{\mathrm{d}f_1}{f_1} \wedge \cdots \wedge \frac{\mathrm{d}f_m}{f_m} = \int_{f(\Gamma)} \frac{\mathrm{d}x_1}{x_1} \wedge \cdots \wedge \frac{\mathrm{d}x_m}{x_m}.$$

Now $f(\Gamma)$ is an integer multiple of the fundamental class of

$$(S^1)^m \cong \{|x_i| = \epsilon\},$$

so that the integral on the right is $(2\pi \mathrm{i})^m$ times a positive integer. To evaluate the integer (which is just the multiplicity of the ideal I), deform $(f_1, \ldots, f_m)$ in the set of regular sequences to $(x_1^{d_1}, \ldots, x_m^{d_m})$. Continuity of the integral then yields

$$\operatorname{Res}_0 \omega_f(\mathcal{J}) = d_1 \ldots d_m,$$

so that

$$\frac{\mathcal{J}}{d_1 \ldots d_m}$$

represents the fundamental class.

4. Proof of the main theorem

a. Computation of the kernel of ϕ

We now return to the proof of the Torelli theorem, the first step of which is to calculate the kernel of

$$\phi : S^2 F^m \longrightarrow S^{2m-n} T^*$$

for a fixed hypersurface X in $\mathbf{P}_{n+1}$, where $T = H^1(X, \Theta)$. In this case the homomorphism

$$H^1(X, \Theta) \longrightarrow \mathrm{Hom}(F^p/F^{p+1}, F^{p-1}/F^p)$$

is induced by the Gauss–Manin connection, which we can compute as follows: Let

$$Q + tR$$

be a pencil of hypersurfaces X_t with $X_0 = X$, so that differentiation with respect to t represents a tangent vector in $H^1(X, \Theta)$, which we shall denote also by $\partial/\partial R$. Fix a rational form Ω_A on $\mathbf{P}_{n+1} - X$, and let ω_A be its Poincaré residue. To extend ω_A to a family of cohomology classes on $\{X_t\}$, it is enough to extend Ω_A to a family of rational forms on $\{\mathbf{P}_{n+1} - X_t\}$, so we set

$$\Omega_A(t) = \frac{A\Omega}{(Q + tR)^{a+1}}$$

$$\omega_A(t) = \mathrm{res}\ \Omega_A(t).$$

Formal differentiation of the rational forms gives

$$\text{(i)} \qquad \frac{\partial}{\partial t} \Omega_A(t) = -(a+1)\Omega_{RA}(t),$$

and application of the Poincaré residue yields

$$\text{(ii)} \qquad \frac{\partial}{\partial t} \omega_A(t) = -(a+1)\omega_{RA}(t),$$

which we may write symbolically as

$$\frac{\partial}{\partial R} \omega_A = -(a+1)\omega_{RA}.$$

The passage from (i) to (ii) is justified by passage of differentiation under the integral sign in the residue formula. Because the result of differentiation on the level of rational forms is independent of the extension of Ω_A to $\Omega_A(t)$, modulo forms of lower pole order, the result of differentiation on the level of residues is well-defined up to cohomology classes of higher Hodge level. In conclusion, we find that

$$\Phi\left(\frac{\partial}{\partial R}\right) \omega_A = c\omega_{RA}$$

for a suitable integral constant c, so that iteration yields

$$\Phi^s \left(\frac{\partial}{\partial R_1} \otimes \cdots \otimes \frac{\partial}{\partial R_s}\right) \omega_A = c\omega_{R_1 \ldots R_s A}.$$

We shall now compute the linear functional ϕ on S^2F^m defined by

$$\phi(\omega, \omega'): v \longmapsto S(\Phi^{2m-n}(v)\omega, \omega').$$

Combining the formula for the iterated Gauss–Manin connection with the cup-product formula, we find that

$$\phi(\omega_A, \omega_B): \frac{\partial}{\partial R_1} \otimes \cdots \otimes \frac{\partial}{\partial R_{2m-n}} \longmapsto c \operatorname{Res}_0 \eta(R_1 \ldots R_{2m-n}AB),$$

where $c \neq 0$ is fixed. Since the Grothendieck residue vanishes if and only if the argument of η lies in the Jacobian ideal, we see that

$$\phi(\omega_A, \omega_B) \text{ annihilates } \frac{\partial}{\partial R_1} \otimes \cdots \otimes \frac{\partial}{\partial R_{2m-n}}$$

if and only if $R_1 \ldots R_{2m-n}AB \in J_Q$. Since the products $R_1 \ldots R_{2m-n}$ generate V^s, where $s = (2m-n)q$, we see that

$$\phi(\omega_A, \omega_B) = 0 \Leftrightarrow AB \in (J_Q : V^s).$$

More generally, let

$$\nu = \sum \nu_{ij}\omega_{A_i} \otimes \omega_{B_j}$$

be an arbitrary element of S^2F^m. Then

$$\phi(\nu) = 0 \Leftrightarrow \sum \nu_{ij}A_iB_j \in (J_Q : V^s).$$

This provides a description of the kernel of ϕ: Define an isomorphism from F^m to an appropriate space of adjoint polynomials,

$$\lambda : F^m \longrightarrow V^t$$

by

$$\lambda^{-1}(A) = \operatorname{res}(\Omega_A)$$

and consider the composition

$$\tilde{\mu} : S^2F^m \xrightarrow{S^2\lambda} S^2V^t \xrightarrow{\mu} V^{2t}$$

where the right-hand map is multiplication of polynomials. Then we have

$$\operatorname{kernel}(\phi) = \tilde{\mu}^{-1}((J_Q : V^s)^{2t}).$$

We now refine this assertion using Macaulay's theorem.

PROPOSITION:

$$\operatorname{kernel}(\phi) = \tilde{\mu}^{-1}(J_Q^{2t}).$$

PROOF: By Macaulay's theorem

$$(J_Q : V^s)^{2t} = (J_Q^{2t} + F^{\sigma-s+1} V)^{2t}.$$

It therefore suffices to show that

$$\sigma - s + 1 > 2t. \tag{*}$$

Now the degree of the adjoints is

$$t = (a+1)q - (n+2),$$

where $q = \deg Q$ and a is the least positive number for which $t \geqq 0$. The highest level in the Hodge filtration is therefore $m = n - a$, and $2m - n = n - 2a$ derivatives are required to map

$$H^{n-a,a} \longrightarrow H^{a,n-a}.$$

Therefore $s = (n-2a)q$. Putting all of this together, we see that (*) is equivalent to $n + 3 > 0$, and so must hold automatically.

b. Construction of X from its Jacobian ideal

Let us now suppose that the vector space of polynomials J_Q^{2t} is known. Then by Macaulay's theorem we can reconstruct J_Q^{q-1}, hence J_Q itself:

LEMMA: *If $\ell > n/2$, then*

$$J_Q^{q-1} = (J_Q^{2t} : V^{2t-q+1}).$$

PROOF: By Macaulay's theorem it suffices to show that

$$\sigma - (2t - q + 1) + 1 > q - 1.$$

But this is equivalent to

$$a < \frac{n}{2} + \frac{n+3}{q},$$

which is certainly satisfied if $a < n/2$. Since $\ell = n - a$, the proof is complete.

By the lemma, we may assume J_Q^{q-1} is known as a vector space of polynomials. To construct Q from this datum, we appeal to the following result:

LEMMA: *Let Q be a generic polynomial of degree at least three, and let x_i be a basis for V^1. Then there is a unique basis $F_0, \ldots, F_{n+1}$ for J_Q^{q-1}, up to a common nonzero multiple, such that*

$$\frac{\partial F_i}{\partial x_j} = \frac{\partial F_j}{\partial x_i}.$$

Since $(\partial Q/\partial x_i)$ is a basis for J_Q^{q-1} satisfying the given condition, the lemma implies that

$$\sum x_j F_j = \lambda \sum x_j \frac{\partial Q}{\partial x_j} = \lambda' Q$$

whenever Q is generic:

PROPOSITION: *If the degree exceeds two, then*

$$Q \longmapsto J_Q^{q-1}$$

is generically injective as a map from a projective space to a Grassmannian.

REMARK: The map $Q \mapsto J_Q^{q-1}$ is probably injective so long as $Q = 0$ does not possess singularities which are "large in relation to the degree".

We conclude this section with a proof of the lemma. To begin, let (F_j) be an arbitrary basis for J_Q^{q-1}, and write

$$F_i = \sum A_{i\alpha} \frac{\partial Q}{\partial x_\alpha}.$$

The condition on the partials then yields

$$\sum A_{i\alpha} \frac{\partial^2 Q}{\partial x_j \partial x_\alpha} = \sum A_{j\beta} \frac{\partial^2 Q}{\partial x_i \partial x_\beta},$$

which we may write as

$$AH = H^t A, \tag{$*$}$$

where

$$H = \left(\frac{\partial^2 Q}{\partial x_\alpha \partial x_\beta}\right)$$

is the Hessian matrix. The equality $(*)$ must hold for all $x \in \mathbf{C}^{n+2}$; we claim that for Q generic, this is possible only if A is a multiple of the identity which is what the lemma asserts.

Consider now the special case when

$$Q = \frac{1}{q(q-1)} \sum x_i^q$$

is the Fermat. Then the Hessian matrix is

$$H_{ij} = \delta_{ij} x_i^{q-2},$$

so that the equations $(*)$ become

$$A_{ij} x_i^{q-2} = A_{ji} x_i^{q-2}.$$

These imply that, for $q > 2$, $A_{ij} = 0$ if $i \neq j$, so that $A = (\mu_i)$ is a diagonal matrix. Thus, any basis of J_Q^{q-1} satisfying the hypothesis of the lemma has the form $F_j = \mu_j \partial Q/\partial x_j$, and

$$\sum x_j F_j = \lambda \left(\sum \mu_j x_j^q\right)$$

is projectively equivalent to Q.

Now let

$$Q(s) = Q + sQ'$$

represent a deformation of Q, and let

$$A(s) = A + sA' + \frac{s^2}{2} A'' + \cdots$$

be a family of matrices satisfying (*) for all s near zero. Differentiating this relation and evaluating at $s = 0$, we obtain the additional condition

$$A'H + AH' = H''A + H'A',$$

where H' is the hessian of Q'. Using what we have already found concerning A, this yields

$$A'_{ij} x_j^{q-2} + \mu_i H'_{ij} = H'_{ij}\mu_j + x_i^{q-2} A'_{ji}.$$

Suppose now that Q' is chosen so that H'_{ij} does not lie in the vector space spanned by x_i^{q-2} and x_j^{q-2}. Then we have

$$A'_{ij} x_j^{q-2} = x_i^{q-2} A'_{ji}$$

and

$$\mu_i H'_{ij} = H'_{ij}\mu_j.$$

The first equation implies that A' is diagonal, and the second implies that $\mu_i = \mu_j$. Consequently

$$A = \mu I + sA' + \frac{s^2}{2} A'' + \cdots$$

satisfies the desired conclusion to first order. Substitution back into the original equation (*) yields

$$\left(A' + \frac{s}{2} A'' + \cdots\right) H(s) = H(s) \left({}^tA' + \frac{s}{2} {}^tA'' + \cdots\right).$$

Arguing as before, we find that A' is also a multiple of the identity, say $\mu' I$. Iteration of the argument then yields

$$A(s) = \mu(s) I$$

for the generic deformation of Q. We conclude that the only solution to (*), for generic Q, is a multiple of the identity, as desired.

c. Construction of X from its local variation

It now remains to assemble the pieces: The kernel of the local variation of Hodge structure determines a piece of the Jacobian ideal by the formula

$$J_Q^{2t} = \tilde{\mu}(\text{kernel}(\phi)).$$

If this piece is nonzero, then Q is generically determined by the results of the preceding section. Unfortunately, to construct Q, we need not only the local variation Φ, but also the isomorphism

$$\lambda : F^m \longrightarrow V^t$$

provided by the residue map. Put another way, Q is determined by the pair (Φ, λ), where λ imposes a "polynomial structure" on F^ℓ.

The proof of the Torelli theorem therefore reduces to the problem of defining λ in an *a priori* manner. In general we do not know how to do this. However, note that the set of all isomorphisms from F^m to V^t is homogeneous under the action of $\mathrm{GL}(V^t)$. Moreover, it is not λ itself, but rather its $\mathrm{GL}(V^1)$-orbit which is of significance: any isomorphism $\mathrm{GL}(V^1)$-equivalent to λ produces, via the Euler formula, a variety $\mathrm{PGL}(V^1)$-equivalent to $Q = 0$. Thus, if $t = 1$, any isomorphism will do, and hence a generic X is determined, up to $\mathrm{PGL}(V^1)$-equivalence, by its local variation of Hodge structure.

Fortunately there is a class of hypersurfaces for which $t = 1$ and $J_Q^{2t} \neq 0$: these are exactly the cubics of dimension 3ℓ: It suffices to observe that

$$\Omega_A = \frac{A\Omega}{Q^{\ell+1}}$$

is homogeneous of degree zero if A is homogeneous of degree one. The proof of the main theorem is now complete.

Remarks: (1) The crucial feature of the cubic case is that the projective space inside which X sits is given naturally in terms of the Hodge structure: $V^1 \cong F^{2m}$, so that

$$X \subset \mathbf{P}((F^{2m})^*).$$

(2) To prove weak global Torelli in general, one must deduce the polynomial structure on F^m, up to $\mathrm{GL}(V^1)$-equivalence, from the local variation of Hodge structure. One possible approach is suggested by the factorization

$$\tilde{\mu} = \mu \circ S^2\lambda$$

which precedes the proposition of the last section. The factorization implies that

$$\text{kernel}(\phi) \supset (S^2\lambda)^{-1}(\text{kernel}(\mu)).$$

Consequently the kernel of ϕ contains a "fixed part" which depends on Q only through the polynomial structure of F^m. It is conceivable that this fixed part, which comes from a piece of the multiplication map and determines the Veronese imbedding of $\mathbf{P}(\hat{V}^t)$ in $\mathbf{P}(\hat{V}^{2t})$, could be defined Hodge-theoretically.

REFERENCES

[1] H.C. CLEMENS: Double Solids, to appear in Advances in Math.

[2] M. CORNALBA and P. GRIFFITHS: Some transcendental aspects of algebraic geometry. Proc. Sympos. Pure Math. *29*, Amer. Math. Soc., Providence, RI, (1975) 3–110.

[3] PIERRE DELIGNE: Travaux de Griffiths, Séminaire Bourbaki, 1969/70, No. 376.

[4] W. GREUB, S. HALPERIN and R. VANSTONE: Connections, Curvature and Cohomology, vol. 1, Academic Press, NY, 1972.

[5] PHILLIP A. GRIFFITHS: Complex Analysis and Algebraic Geometry, Bull. Am. Math. Soc., vol. 1, no. 4 (1979) 595–626.

[6] PHILLIP A. GRIFFITHS: On the periods of certain rational integrals I, II, Annals of Math. *90* (1969) 460–541.

[7] P. GRIFFITHS and J. HARRIS: Principles of Algebraic Geometry, John Wiley & Sons, 1978.

[8] A. GROTHENDIECK: Théorèmes de dualité pour les faisceaux algébriques cohérents, Séminaire Bourbaki, No. 49 (1957).

[9] R. HARTSHORNE: Residues and Duality, Springer-Verlag, Berlin–Heidelberg–NY, 1966.

[10] NICHOLAS KATZ: Algebraic solutions of differential equations (p-curvature and the Hodge filtration), Inv. Math. *18* (1972) 1–118.

[11] W. SCHMID: Variations of Hodge structures – the singularities of the period map, Inv. Math. *22* (1973) 211–319.

P.A. Griffiths
Dept. of Math.
Harvard University
Cambridge, Mass. 02138 (USA)

J.A. Carlson
Dept. of Math.
University of Utah
Salt Lake City, Utah 84112 (USA)

MIXED HODGE STRUCTURES AND COMPACTIFICATIONS OF SIEGEL'S SPACE (Preliminary Report)

James A. Carlson*, Eduardo H. Cattani**
and Aroldo G. Kaplan**

Introduction

Let $\mathcal{H}_g$ be the Siegel upper half-space of order g, and let $\Gamma = \mathrm{Sp}(g, \mathbf{Z})$ be the Siegel modular group. The quotient $\mathcal{H}_g/\Gamma$ is a normal analytic variety, which we shall call *Siegel's space.* It admits certain natural compactifications; in particular, the minimal compactification of Satake [12] and the various toroidal compactifications of Mumford [10] (see also [1], [2], [13], [14]). The aim of this report is to describe a compactification which is natural from the point of view of mixed Hodge theory: The boundary components will be distinguished quotients of classifying spaces for mixed Hodge structures. This compactification is also good from the standpoint of curve degenerations: the limit point at the boundary detects the extension-theoretic part (cf. [3]) of the limiting mixed Hodge structure, and this in turn carries geometric information about the central fiber of the degeneration.

Although the compactification presented here for Siegel's space turns out to agree with Mumford's smooth compactification for arithmetic quotients of Hermitian symmetric spaces, we have, nevertheless, included a direct construction in the hope that this will clarify some of the subtler geometric aspects of the toroidal techniques, while at the same time lay a foundation for the study of Hodge-theoretic compactifications for classifying spaces of Hodge structures of higher weight. To this end, the emphasis of this report is on descriptions and motivations, rather than on technical details. These will be supplied in a subsequent paper where we shall also discuss some important aspects of the compactification which are not treated here: analytic structure and extension of period mappings, among others.

The first two sections are of an introductory nature: after recalling some basic facts about degenerations of Hodge structures, Schmid's nilpotent orbit theorem and the limiting mixed Hodge structure, we analyze

*Partially supported by NSF Grant and an Alfred P. Sloan Memorial Fellowship.
**Partially supported by NSF Grant.

an example to show the geometric significance of the extension data and of the information given by the Satake compactification. In §3, we give a detailed description of the boundary components and pre-components, while in §4, we show that the latter are a C^∞-product of a Satake boundary component and a nilpotent group. Finally in section 5 we sketch the construction of the compactification, showing the simplicial data on which the compactification depends and indicating how the topology is defined.

1. Nilpotent orbits

In this section we explain how an abstract degeneration of Hodge structures in the Siegel space defines a nilpotent orbit of mixed Hodge structures with a distinguished semigroup (integral cone) of polarizations. We view these orbits as the natural "limit objects" with which one should try to compactify the Siegel modular space. As indicated in the next section, this is reasonable on geometric as well as formal grounds. In fact, we will show that an irreducible nodal rational curve Y is determined up to a group of order two by the nilpotent orbit associated to a degeneration with Y as central fiber.

We begin by establishing the basic notation. Let

$$\phi = \begin{pmatrix} 0 & I \\ -I & 0 \end{pmatrix}$$

be the standard symplectic form on $H_{\mathbf{Z}} = \mathbf{Z}^{2g}$, let Γ be the group of integral symplectic matrices, and let $\mathcal{H}_g$ be the Siegel upper half-space. Let Δ, Δ^* and $\mathcal{H}$ denote the unit complex disk, punctured disk, and complex upper half plane, respectively, and let

$$t : z \longmapsto \exp(2\pi i z)$$

represent the universal covering projection from $\mathcal{H}$ to Δ^*. Set $P = \Delta^n$, $P^* = \Delta^{*n}$, and $\tilde{P} = \mathcal{H}^n$. Then a *local degeneration* is given by a holomorphic map

$$f : P^* \longrightarrow \mathcal{H}_g/\Gamma,$$

which is required to lift along the natural projections to

$$\tilde{f} : \tilde{P} \longrightarrow \mathcal{H}_g.$$

Let $(\gamma, x) \mapsto \gamma x$ denote the action of $\pi_1(P^*)$ on $\tilde{P}$ and observe that $\tilde{f}(\gamma x)$ is another lifting of f. Consequently there is an element $\mu(\gamma) \in \Gamma$ such that

$$\tilde{f}(\gamma x) = \mu(\gamma)\tilde{f}(x).$$

This defines the *monodromy representation*

$$\mu : \pi_1(P^*) \longrightarrow \Gamma.$$

According to the monodromy theorem, the eigenvalues of $\mu(\gamma)$ are roots of unity. Consequently there is a minimal unramified cover $P^* \to P^*$ such that each $\mu(\gamma)$ for the induced representation is *unipotent*: its eigenvalues are all equal to unity.

For unipotent degenerations the standard power series defines the logarithm of a typical monodromy transformation T:

$$N = \log T = \sum \frac{(-1)^{j+1}}{j}(T-I)^j.$$

In the Siegel case, the monodromy theorem implies that $(T-I)^2 = 0$, so that $N = T - I$ is integral and $N^2 = 0$. Another special feature of the Siegel case, as we shall see later, is that

$$NN' = 0 \qquad (*)$$

for any two monodromy elements. Now observe that $\phi(Tx, Ty) = \phi(x, y)$ implies that

$$\phi(Nx, y) = -\phi(x, Ny),$$

so that N is in the symplectic Lie algebra. Consequently the bilinear form

$$\phi_N(x, y) = \phi(x, Ny)$$

is symmetric, and its null space is precisely the kernel of N. Asymptotic estimates from Hodge theory then show that ϕ_N is positive:

$$\phi_N \geqq 0.$$

(We always mean positive semidefinite, unless otherwise stated. Proofs of the positivity are given in [8], [9] and [15].)

Let us now define the canonical semigroup associated to a local degeneration. Let ξ_j be the loop in the j-th punctured disk which turns once counterclockwise around the origin, with orientation defined by the complex structure. Then

$$\gamma_Z(P^*) = \{\xi_1^{k_1} \ldots \xi_n^{k_n} \mid k_j \geqq 0\}$$

$$\gamma_Z^+(P^*) = \{\xi_1^{k_1} \ldots \xi_n^{k_n} \mid k_j > 0\}$$

define commutative semigroups, natural with respect to the homomorphisms induced by holomorphic maps. Thus we may define sets

$$\nu_Z(f) = \log \mu\gamma_Z(P^*)$$

$$\nu_Z^+(f) = \log \mu\gamma_Z^+(P^*),$$

and according to the Campbell–Hausdorff formula,

$$\log TT' = \log T + \log T' + [\log T, \log T'] + \cdots,$$

so that

$$\log TT' = \log T + \log T'$$

in view of the relation (∗). The sets just defined are therefore closed under addition, hence give natural semigroups. We remark that $\nu_Z(f)$ is the convex integral hull of $N_1, \ldots, N_n$, where

$$N_j = \log T_j$$

$$T_j = \mu(\xi_j).$$

Consequently $\nu_Z(f)$ has the structure of an integral polyhedral cone, with the N_j generating its edges.

Next, we define the weight filtration associated to f: set

$$W_2(N) = H_Z$$

$$W_1(N) = \text{kernel}(N)$$

$$W_0(N) = \text{image}(N)$$

for any $N \in \nu_Z(f)$. Since N lies in the symplectic Lie algebra, this filtration is self-dual with respect to ϕ: W_0 is the ϕ-orthogonal complement of W_1, and W_1 is the nullspace of ϕ_N.

(1.1) PROPOSITION: (1) *For any two* $N, N' \in \nu_Z^+(f)$, *we have*

$$W_\ell(N) = W_\ell(N').$$

(2) *If* $N^+ \in \nu_Z^+(f)$ *and* $N \in \nu_Z(f)$, *then*

$$W_1(N^+) \subseteq W_1(N)$$

and

$$W_0(N^+) \supseteq W_0(N).$$

As a consequence of the proposition, $\nu_Z^+(f)$ defines an unambiguous weight filtration.

To give the proof, set

$$N_\lambda = \sum \lambda_j N_j$$

where λ_j is a non-negative integer.

(1.2) LEMMA: (1) $\text{kernel}(N_\lambda) = \bigcap_{\lambda_j > 0} \text{kernel}(N_j)$

(2) $\text{image}(N_\lambda) = \sum_{\lambda_j > 0} \text{image}(N_j)$.

PROOF: Certainly the intersection of the kernels is contained in the kernel of N_λ. To prove the reverse inclusion, suppose that $N_\lambda x = 0$, and observe that

$$0 = \phi(x, N_\lambda x) = \sum_{\lambda_j > 0} \lambda_j \phi(x, N_j x).$$

But the right hand side is a sum of non-negative terms, so $\phi(x, N_j x)$ must vanish for each j. Consequently x lies in the kernel of each N_j, as desired. The second assertion now follows from the first by duality.

(1.3) COROLLARY: (1) $N_i N_j = 0$
(2) $\nu_{\mathbf{Z}}(f)$ *is generated by the* N_j
(3) $NN' = 0$.

PROOF: (1) By the lemma,

$$\text{image}(N_i + N_j) \subseteq \text{kernel}(N_i + N_j) \subseteq \text{kernel}(N_i).$$

Therefore $N_i(N_i + N_j) = 0$, which implies $N_i N_j = 0$.

(2) The Campbell–Hausdorff formula gives, by virtue of (1), the relation

$$\log T_1^{\lambda_1} \ldots T_n^{\lambda_n} = \lambda_1 N_1 + \cdots + \lambda_n N_n.$$

(3) This follows from (1) and (2).

The proposition now follows from the lemma and the second assertion of the corollary.

Let us denote the common weight filtration defined from $\nu_{\mathbf{Z}}^+$ by $W_*(f)$, and let us define the lattice of type (1, 1) elements

$$L^1(f) = W_2(f)/W_1(f).$$

Then the correspondence

$$N \longmapsto \phi_N$$

maps $\nu_{\mathbf{Z}}(f)$ injectively to a semigroup of positive forms on $L^1(f)$, which we shall denote by $\sigma_{\mathbf{Z}}(f)$. It is an integral polyhedral cone and $\sigma_{\mathbf{Z}}^+(f)$ is the infinitely generated sub-cone of definite elements.

Finally, we examine the "asymptotic Hodge structure": Let $F^1(z)$ be the filtration on $H_{\mathbf{C}}$ defined by $\tilde{f}(z) \in \mathcal{H}_g$. Then Schmid's nilpotent orbit theorem asserts that

$$F^1_\infty = \lim_{\mathrm{Im}(z_j) \to \infty} \exp(-N_z) F^1(z)$$

defines a filtration on $H_{\mathbf{C}}$, provided that the real parts of the z_j remain within suitable bounds as the limit is taken. Note, however, that the limit does depend on the parametrization: multiplication of t_j in the j-th punctured disk by $\exp(2\pi i \alpha_j)$ results in multiplication of the limit filtration by $\exp \alpha_j N_j$. Consequently the object naturally defined by the degeneration is not a single filtration, but rather a nilpotent orbit of such:

$$\mathcal{N}(f) = \{(\exp N) F^1_\infty \mid N \in \nu_{\mathbf{C}}\},$$

where in general

$$\nu_\Lambda = \{\Lambda\text{-module spanned by the semigroup } \nu\}.$$

(1.4) REMARK: The nilpotent orbit theorem states further that the two holomorphic maps $\tilde{f}$ and

$$\tilde{\nu}(z) = \exp(N_z)F^1_\infty$$

are asymptotic in the sense that for any Sp$(g, \mathbf{R})$ – invariant distance on $\mathcal{H}_g$, we have

$$d(\tilde{\nu}(z), \tilde{f}(z)) = 0\left(\sum |t_j|\right),$$

where

$$t_j = \exp(2\pi i z_j).$$

The fact that arbitrary degenerations lift to maps asymptotic to nilpotent orbits is further motivation for the use of the latter as boundary points for $\mathcal{H}_g$.

We claim next that the degeneration f defines a nilpotent orbit of mixed Hodge structures:

(1.5) PROPOSITION: *Let f be a local unipotent degeneration of Hodge structures in the Siegel space. Then the triple $(W_*(f), F^1, \Lambda)$ defines a polarized mixed Hodge structure for any F^1 in $\mathcal{N}(f)$ and any $\Lambda \in \sigma^+_{\mathbf{Z}}(f)$.*

PROOF: Because $\exp(N_\alpha)$ acts by the identity on W_2/W_1 and on W_1, it suffices to show that F^1 defines a mixed Hodge structure for some $F^1 \in \mathcal{N}(f)$. To see this, observe that

$$t \longmapsto f(t_1^{\lambda_1}, \ldots, t_n^{\lambda_n})$$

defines a one-variable degeneration with monodromy transformation $\exp(N_\lambda)$. Application of the SL_2-orbit theorem to this degeneration shows that F^1_∞ defines a mixed Hodge structure, polarized with respect to $\phi_{N_\lambda} \in \sigma^+_{\mathbf{Z}}$. Since any element of $\sigma^+_{\mathbf{Z}}$ arises from a one-variable degeneration in this way, the proposition is proved.

In order to give the geometric applications, let us translate this result into the language of one-motifs. A *one-motif* is a homomorphism

$$u : L \longrightarrow J$$

of a lattice into an extension of an abelian variety by a complex multiplicative group $(\mathbf{C}^*)^r$. According to a theorem of Deligne, there is an equivalence of categories between mixed Hodge structures of level one and one-motifs (cf. [6, III] section 10). Thus, we know from general principles that the nilpotent orbit of mixed Hodge structures constructed above is naturally equivalent to a nilpotent orbit of one-motifs. To describe it, recall that to any mixed Hodge structure H are associated canonical lattices and generalized Jacobians, defined by

$$L^pH = (\mathrm{Gr}^W_{2p})^{p,p}_{\mathbf{Z}}$$

$$J^pH = W_{2p-1,\mathbf{C}}/(F^pW_{2p-1} + W_{2p-1,\mathbf{Z}}).$$

The mixed Hodge structure then gives a homomorphism

$$u : L^pH \longrightarrow J^pH$$

which carries the structure of a one-motif if it has level one (cf. [3], section 2).

In the case of degenerations in the Siegel space, the above construction applies, with $p = 1$, to each structure in the nilpotent orbit, giving a one-motif

$$u_f : L^1(f) \longrightarrow J^1(f).$$

The domain and range are independent of the structure chosen from the orbit because N_α acts by zero on both W_1 and $H_{\mathbf{Z}}/W_1$. On the other hand, it defines a non-zero homomorphism from $H_{\mathbf{Z}}/W_1$ to $W_0 \otimes \mathbf{C}$, hence defines a one-motif

$$N_\alpha : L^1(f) \longrightarrow J^1(f).$$

Observe that it takes values in the multiplicative group of the extension, namely

$$J^1W_0 = \frac{W_{0,\mathbf{C}}}{W_{0,\mathbf{Z}}} \cong (\mathbf{C}^*)^r,$$

where $r = \mathrm{rank}\ W_{0,\mathbf{Z}}$. In conclusion, *a degeneration of Hodge structures in the Siegel space defines a canonical orbit of one-motifs* $\mathcal{U}(f)$ *which is homogeneous under the effective action* $u \mapsto u + N$, *where* $N \in \nu_{\mathbf{C}}$.

(1.6) REMARK: The complete formal object $\mathcal{U}$ defined by a degeneration consists of

(i) A family of one-motifs

$$u : L^1 \longrightarrow J^1.$$

(ii) A polyhedral cone $\sigma_{\mathbf{Z}}$ of positive integral forms on L^1.

(iii) A perfect pairing

$$\phi : L^1 \times L^0 \longrightarrow \mathbf{Z}$$

where $L^0 = W_0H_{\mathbf{Z}}$.

The perfect pairing is that defined by cup-product; it defines an isomorphism of L^0 with the dual of L^1, hence defines an isomorphism between the space of bilinear forms on L^1 and the space of homomorphisms from L^1 to

L^0. In this way σ itself defines the canonical action of ν upon the one-motifs of the nilpotent orbit.

2. Curve degenerations

2a. The polyhedral cone

Consider now a local degeneration of curves

$$f:\mathscr{X}\longrightarrow P\times P',$$

smooth above $P^*\times P'$ and with Y as central fiber. Any such degeneration is the pullback of the versal deformation of Y, up to a possible preliminary base extension. Consequently all polyhedral cones $\sigma(f)$ arise as subobjects of the "versal cone" $\sigma(Y)$. Our aim in this section is to give a direct description of the versal cone when Y is stable.

To this end, recall that Y has a canonical Mayer–Vietoris sequence,

$$0\longrightarrow H_1(\tilde{Y})\xrightarrow{\pi_*}H_1(Y)\xrightarrow{\partial}H_0(\tilde{\Sigma})$$
$$\xrightarrow{\delta_*}H_0(\tilde{Y})\oplus H_0(\Sigma)\longrightarrow H_0(Y)\longrightarrow 0.$$

Here $\pi:\tilde{Y}\to Y$ is the normalization, Σ is the singular locus, $\tilde{\Sigma}=\pi^{-1}(\Sigma)$, and $\delta_*=i_*\oplus(-\pi_*)$. Define groups

$$\mathscr{V}_Y=\text{kernel}\{\pi_*:H_0(\tilde{\Sigma})\longrightarrow H_0(\Sigma)\}$$
$$\mathscr{Z}_Y=\text{kernel}\{i_*:\mathscr{V}_Y\longrightarrow H_0(\tilde{Y})\}$$

and filter $H_1(Y)$ by

$$W_0=H_1(Y,\mathbf{Z})$$
$$W_{-1}=\pi_*H_1(\tilde{Y},\mathbf{Z}).$$

Then the boundary operator gives an isomorphism

$$L^0H_1(Y)\longrightarrow\mathscr{Z}_Y.$$

Note that $\mathscr{Z}_Y$ consists of zero-cycles on $\tilde{\Sigma}$ which have degree zero on each component of Y and on each fiber of π.

Next, define a bilinear form Λ_Y on $H_0(\tilde{\Sigma})$ such that

$$\Lambda_Y(x,x)=\tfrac{1}{2}$$
$$\Lambda_Y(x,x')=0$$

for arbitrary points $x,x'\in\tilde{\Sigma}$, provided that $x\neq x'$. Set

$$\hat{x}=x'-x''$$

and observe that the $\hat{x}$ give an orthonormal basis for $\mathscr{V}_Y$. The restriction of Λ_Y to $\mathscr{Z}_Y$ gives the *distinguished polarization* on the weight zero component of H_1.

Define also the rank-one forms Λ_x by

$$\Lambda_x(\xi, \xi') = \Lambda_Y(\hat{x}, \xi)\Lambda_Y(\hat{x}, \xi'),$$

and let $\Lambda_i = \Lambda_{x_i}$.

(2.1) PROPOSITION: *The versal cone $\sigma(Y)$ is naturally isomorphic to the cone of positive forms on $\mathscr{X}_Y$ generated by the Λ_i, and the distinguished polarization is given by*

$$\Lambda_Y = \sum \Lambda_i.$$

REMARK: Define the subordinates of Λ to be the integral forms satisfying

$$0 \leqq \Lambda' \leqq \Lambda.$$

Let $\sigma(\Lambda)$ be the cone generated by the subordinates of Λ. It is evident from the proposition that Λ_x is subordinate to Λ_Y, so that

$$\sigma(Y) \subseteq \sigma(\Lambda_Y).$$

In many, but not all cases, equality holds, giving an arithmetic characterization of $\sigma(Y)$ in terms of the polarization.

(2.2) EXAMPLE 1: Suppose Y is irreducible. Then $\mathscr{V}_Y = \mathscr{X}_Y$, so that in the basis $\hat{x}_i$, Λ_Y is given by the identity matrix. The rank-one subordinates are exactly the Λ_i, which are represented by diagonal matrices with a one in the i-th position.

(2.3) EXAMPLE 2: Let Y be the stable model of an irreducible curve with an ordinary triple point:

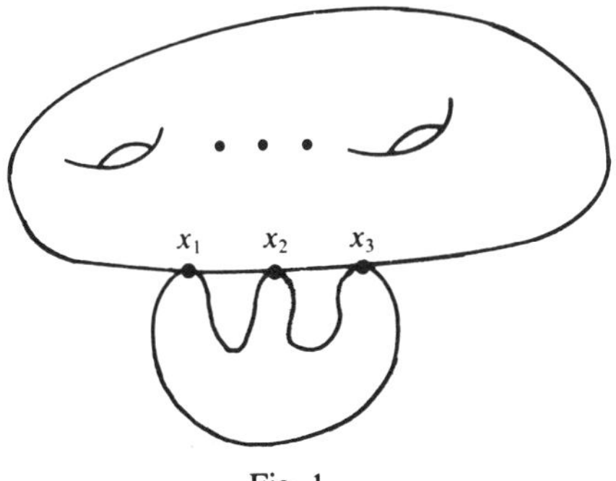

Fig. 1

Then $\hat{x}_2 - \hat{x}_1$ and $\hat{x}_3 - \hat{x}_2$ generate $\mathscr{X}_Y$, and in that basis

$$\Lambda_Y = \begin{bmatrix} 2 & -1 \\ -1 & 2 \end{bmatrix},$$

the matrix of the Dynkin diagram A_2. The forms associated to the double points are

$$\Lambda_1 = \begin{bmatrix} 1 & 0 \\ 0 & 0 \end{bmatrix} \quad \Lambda_2 = \begin{bmatrix} 1 & -1 \\ -1 & 1 \end{bmatrix} \quad \Lambda_3 = \begin{bmatrix} 0 & 0 \\ 0 & 1 \end{bmatrix},$$

and these are precisely the subordinates of Λ_Y.

The class of stable curves for which $\sigma(Y) = \sigma(\Lambda_Y)$ is large but not exhaustive: it includes the stable models of all irreducible curves with ordinary multiple points, but it excludes curves such as the one below:

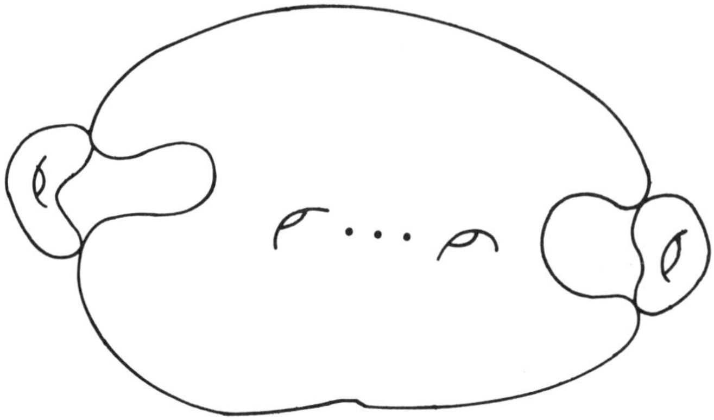

Fig. 2

In this case the distinguished polarization is twice the identity, and its four subordinates are the three matrices listed above, together with the matrix

$$\begin{bmatrix} 1 & 1 \\ 1 & 1 \end{bmatrix}.$$

On the other hand, $\sigma(Y)$ is given by the diagonal matrices with non-negative integer entries.

PROOF OF THE PROPOSITION:

(1) *Generic degenerations.* Let Y be stable with double points $x_1, \ldots, x_r$. We shall say that $f: \mathscr{X} \to \Delta^r$ is *generic* if (i) $\mathscr{X}$ is smooth (ii) there are functions u_j, v_j on $\mathscr{X}$ such that Y is given near x_j by $u_j v_j = t_j$, where t_j is a coordinate for the j-th factor of Δ^r. The cone of a generic deformation is isomorphic to the versal cone; indeed the versal fiber space is homotopy equivalent to that of a generic deformation. We may therefore confine our calculations to generic deformations.

(2) *Transport of $\sigma(Y)$ to $\mathscr{Z}_Y$.* The cone $\sigma(Y)$ is defined on the lattice of integral (1, 1) cohomology,

$$L^1H^1(X_t) = H^1(X_t, \mathbf{Z})/W_1.$$

Cup-product gives a perfect pairing of L^1 with the type (0, 0) lattice,

$$L^1H^1 \times L^0H^1 \longrightarrow \mathbf{Z},$$

hence an isomorphism of L^1H^1 with the dual of L^0H^1, that is, with $L^0H_1(X_t)$. Because $\mathscr{X}$ is generic, there is a deformation retraction

$$k : \mathscr{X} \longrightarrow Y$$

which induces a morphism

$$k_* : H_1(X_t) \longrightarrow H_1(Y).$$

Its kernel is W_{-2}, and it induces an isomorphism on the level of L^0. The composition

$$L^1H^1(X_t) \longrightarrow L^0H_1(X_t) \xrightarrow{k_*} L^0H_1(Y) \xrightarrow{\partial} \mathscr{Z}_Y$$

yields an identification by means of which we may view $\sigma(Y)$ as naturally defined on $\mathscr{Z}_Y$.

(3) *The Lefschetz formula.* The collapsing map k can be chosen so that

$$\delta_j = k^{-1}(x_j) \cap X_t$$

is an imbedded circle, the *vanishing cycle* attached to x_j. The monodromy transformation associated with the loop δ_j is then

$$T_j(\gamma) = \gamma + \phi(\delta_j, \gamma)\delta_j,$$

and its logarithm is

$$N_j(\gamma) = \phi(\delta_j, \gamma)\phi_j.$$

Consequently W_{-2} is generated by the vanishing cycles. Since Y is obtained, up to homotopy, by attaching disks to X_t along the δ_j, this shows that W_{-2} is indeed the kernel of k_*.

(4) *Completion of the proof.* The positive form associated to N_j is

$$\phi_j(\gamma, \gamma') = \phi(\delta_j, \gamma)\phi(\delta_j, \gamma').$$

Thus it suffices to show that

$$\phi_j(\gamma, \gamma') = \Lambda_j(\partial k_*\gamma, \partial k_*\gamma').$$

Equivalently, we must show that

$$\phi(\delta_j, \gamma) = \pm\Lambda_Y(\hat{x}_j, \partial k_*\gamma). \tag{$*$}$$

But this is clear: First, suppose that γ is a loop which meets δ_j transversely in a single point. Then $k_*\gamma$ is a cycle whose boundary is $\pm\hat{x}_j$, so that both sides of $(*)$ equal ± 1. Second, suppose that γ does not meet δ_j. Then $k_*\gamma$ does not meet x_j, so both sides of $(*)$ vanish. Since $H_1(Y, \mathbf{Z})$ is generated by cycles satisfying one of the above cases, the proof is complete.

(2.4) REMARKS: (1) For stable curves, the group $\mathcal{Z}_Y$ is naturally isomorphic to $H_1(\Gamma)$, where Γ is the dual graph of Y (a vertex for each component, an edge for each double point). Indeed, we can construct a natural exact commutative diagram in which all vertical maps are isomorphisms:

$$\begin{array}{ccccccccc} 0 \longrightarrow & \mathcal{Z}_Y & \longrightarrow & \mathcal{V}_Y & \xrightarrow{i_*} & H_0(\tilde{Y}) & \longrightarrow & H_0(Y) & \longrightarrow 0 \\ & \downarrow & & \downarrow & & \downarrow & & \downarrow & \\ 0 \longrightarrow & H_1(\Gamma) & \longrightarrow & C_1(\Gamma) & \longrightarrow & C_0(\Gamma) & \longrightarrow & H_0(\Gamma) & \longrightarrow 0. \end{array}$$

It is therefore easy to compute Λ_Y and the Λ_i from an orientation of Γ: If ξ, ξ' define simple closed curves in Γ, then

$$\Lambda_Y(\xi, \xi) = \text{number of edges in } \xi$$

$$\Lambda_Y(\xi, \xi') = \sum \epsilon_i(\xi, \xi').$$

The symbol ϵ_i is ± 1 if the i-th edge is common to both cycles and is zero otherwise. The plus sign is taken if ξ and ξ' assign the same orientation to the i-th edge, the minus sign is taken in the contrary case. Finally, each edge defines a generator of $\sigma(Y)$ by $\Lambda_i = \epsilon_i$.

(2) We say that $\sigma(Y)$ *separates* ξ and ξ' in $\mathcal{Z}_Y$ if $\Lambda(\xi, \xi') = 0$ for all $\Lambda \in \sigma(Y)$. Zero-cycles are separated by $\sigma(Y)$ if and only if their supports are disjoint. This fundamental property of the versal cone is essential to our applications.

2*b*. *The one-motifs*

We shall now give a direct interpretation of the nilpotent orbit associated to a curve degeneration. To do this, observe first that an element in $F^1H^1(X_t)$, for any mixed Hodge structure in the orbit, is represented by a logarithmic differential on $\tilde{Y}$ whose residue cycle lies in $\mathcal{Z}_Y \otimes \mathbf{C}$. Consequently there is a natural isomorphism

$$\text{res}: L^1(f) \longrightarrow \mathcal{Z}_Y.$$

Next, observe that the collapsing map gives a canonical isomorphism

$$k^*: J^1(f) \longrightarrow J^1(Y),$$

where

$$J^1(Y) = \frac{H^1(Y, \mathbf{C})}{F^1H^1(Y) + H^1(Y, \mathbf{Z})}$$

is the Hodge-theoretic Jacobian. Consequently we may transpose the original nilpotent orbit of one-motifs to the $\sigma(Y)$-orbit

$$\{u_Y\}: \mathcal{Z}_Y \longrightarrow J^1(Y)$$

using the correspondence

$$u_Y = k^{*-1} \circ u_f \circ \mathrm{res}^{-1}.$$

To make sense out of the values of u_Y, we must introduce the following groups:

$$M(\xi) = \{\gamma \in H_1(Y, \mathbf{Z}) \mid \Lambda(\partial\gamma, \xi) = 0 \text{ for all } \Lambda \in \sigma_{\mathbf{C}}(Y)\}$$

$$\sigma_{\mathbf{C}}(\xi) = \{\text{the linear functionals } \gamma \to \Lambda(\partial\gamma, \xi), \Lambda \in \sigma_{\mathbf{C}}(Y)\}.$$

Because $\sigma_{\mathbf{C}}(\xi)$ annihilates the kernel of ∂, it lies in $W_0H^1(Y)$. From this one derives the exact sequence of mixed Hodge structures

$$0 \longrightarrow \sigma_{\mathbf{C}}(\xi) \longrightarrow H^1(Y, \mathbf{C}) \longrightarrow M^*(\xi) \otimes \mathbf{C} \longrightarrow 0.$$

Passing to Jacobians, we obtain

$$0 \longrightarrow J^1\sigma_{\mathbf{C}}(\xi) \longrightarrow J^1(Y) \longrightarrow J^1M^*(\xi) \longrightarrow 0,$$

where

$$J^1\sigma_{\mathbf{C}}(\xi) = \sigma_{\mathbf{C}}(\xi)/\sigma_{\mathbf{Z}}(\xi)$$

is a subgroup of the multiplicative part of the canonical extension of $J^1(Y)$. *The point of all this is that*

$$u_Y(\xi) \in J^1M^*(\xi)$$

is well-defined, independent of the motivic homomorphism chosen from the orbit.

Let us therefore calculate $u_Y(\xi)$: Choose a logarithmic differential ω_ξ whose residue is ξ and define a linear functional by integration:

$$u_\xi : \gamma \longrightarrow \int_\gamma \omega_\xi.$$

By the final remark of the previous section, cycles in $M(\xi)$ can be moved off the polar locus of ω_ξ, so that the integral converges. However, because the value of the integral depends on the homology class of γ in Y minus the support of ξ, it is well-defined only up to a sum of residues, hence up to an integer. It follows that u_ξ is well-defined in the multiplicative group $M^*_{\mathbf{C}}(\xi)/M^*_{\mathbf{Z}}(\xi)$. Furthermore, the differential in ω_ξ is defined by its residue only up to abelian differentials on Y-elements of $F^1H^1(Y)$. Therefore u_ξ determines an element of the generalized torus $J^1M^*(\xi)$ which is independent of all choices. To summarize, we have obtained the following result:

(2.5) PROPOSITION: *Let Y be a stable curve, $\{u_Y\}$ the canonical nilpotent orbit of a generic degeneration, transposed to Y. Then $u_Y(\xi)$ projects to u_ξ, the linear functional given by integration, in the canonical quotient torus associated to ξ.*

(2.6) REMARK: In certain cases we can interpret u_ξ as giving a set of cross-ratios on Y. To explain this, note first that the canonical extension associated to $J^1(Y)$ is isomorphic to

$$0 \longrightarrow W_0 J^1(Y) \longrightarrow J^1(Y) \xrightarrow{\pi^*} J^1(\tilde{Y}) \longrightarrow 0$$

where $J^1(\tilde{Y})$ is the usual Jacobian. Therefore ξ is linearly equivalent to zero precisely when u_ξ lies in the multiplicative group $W_0 J^1$. In this case we may write

$$u_\xi = \mathrm{d}\log f_\xi / 2\pi\,\mathrm{i}$$

for a suitable meromorphic function on $\tilde{Y}$. Next, let $\langle\ ,\ \rangle$ denote the canonical pairing which integration gives between $W_0 H^1(Y)$ and $\mathscr{Z}_Y \cong \mathrm{Gr}_0 H_1(Y, \mathbf{Z})$, and let η be a primitive element of $\mathscr{Z}_Y$. Then the map defined on $W_0 H^1(Y, \mathbf{C})$ by

$$\alpha \longmapsto \exp(2\pi\,\mathrm{i}\langle \alpha, \eta\rangle)$$

produces a homomorphism

$$\eta : J^1 W_0 \longrightarrow \mathbf{C}^*.$$

If η is separated from ξ, we may therefore write

$$\langle u_\xi, \eta\rangle = \exp \int_{\partial^{-1}\eta} \mathrm{d}\log f_\xi,$$

where $\partial^{-1}\eta$ is any one-cycle on Y whose boundary is η. The fundamental theorem of calculus then yields

$$\langle u_\xi, \eta\rangle = \prod_p f_\xi(p)^{\nu_p(\eta)}$$

where $\nu_p(\eta)$ is the multiplicity of p in η. Moreover, the change-of-variables theorem in the calculus, coupled with the essential uniqueness of f_ξ, implies that $\langle u_\xi, \eta\rangle$ is a projective invariant of Y. Finally, if $\tilde{Y} \cong \mathbf{P}_1$, $\xi = b - a$, and $\eta = d - c$, then $\langle U_\xi, \eta\rangle$ is the cross-ratio of the four points a, b, c, d. In fact, let t be the unique coordinate such that $t(a) = \infty$, $t(b) = 0$, and $t(c) = 1$. Then $f_\xi = t$ and $\langle u_\xi, \eta\rangle = d$, which is precisely the asserted cross-ratio.

Let us end this discussion with an application to a Torelli-type problem.

(2.7) PROPOSITION: *Let Y be an irreducible rational curve with r nodes as its only singularities, and let $\mathscr{N}(Y)$ be the nilpotent orbit of a generic degeneration with Y as central fiber.*

(1) *If $r = 2$, then $\mathscr{N}(Y)$ determines Y up to isomorphism.*
(2) *If $r = 3$, then the map $Y \mapsto \mathscr{N}(Y)$ is generically two-to-one.*
(3) *If $r \geq 4$, then the map $Y \mapsto \mathscr{N}(Y)$ is generically one-to-one.*

PROOF: (1) *Let* $x_1, \ldots, x_r$ be the nodes of Y, and let $\pi^{-1}(x_i) = (a_i, b_i)$ in $\tilde{Y}$. The pairs (a_i, b_i), which depend on $2r-3$ projective moduli, determine Y. We claim first that $\mathcal{N}(Y)$ determines the cross-ratios

$$\mu_{ij} = \frac{b_j - b_i}{b_j - a_i} \cdot \frac{a_j - a_i}{a_j - b_i}$$

in a canonical way, and second that the degree of

$$\kappa : \frac{\{(a_1, b_1); \ldots ; (a_r, b_r)\}}{\text{Action of PGL}(1)} \longrightarrow \{(\mu_{ij}) \mid i > j\}$$

is at most two:

(2.8) LEMMA: *The map κ is an isomorphism if $r = 2$, is generically two-to-one onto its image if $r = 3$ and is generically one-to-one onto its image if $r > 3$.*

The proof follows jointly from the two claims.

(2) *Proof of first claim.* Recall that $\mathcal{N}(Y)$ gives a cone $\sigma(Y)$ of polarizing forms, and that its central element $\Sigma\, \Lambda_i$ is the distinguished polarization, Λ_Y. Furthermore, a Λ_Y-orthonormal basis $\{\xi_i\}$ for the weight-zero lattice may be identified with a basis $\{b_i - a_i\}$ for $\mathscr{X}_Y$, given a suitable ordering of the nodes and the corresponding fibers of π. By the remarks preceding the proposition, we obtain

$$\mu_{ij} = \langle u_{\xi_i}, \xi_j \rangle,$$

as desired.

(3) *Proof of second claim.* If $k = 2$, there is only one cross-ratio, and this is certainly an invariant of the quadruple (a_1, b_1, a_2, b_2). If $k > 2$, we attempt to construct an inverse for κ as follows: First, move (a_1, b_1, a_2) to $(\infty, 0, 1)$ by the unique projective transformation which does this. Determine b_2 by the equation

$$b_2 = \mu_{21}.$$

Second, adjoin the $2r-4$ equations

$$\frac{b_k}{a_k} = \mu_{k1}$$

$$\frac{b_k - \mu_{21}}{b_k - 1} \cdot \frac{a_k - 1}{a_k - \mu_{21}} = \mu_{k2},$$

where $k \geqq 3$.

Each pair, consisting of a linear and a quadratic equation determines a pair of conjugate solutions (a_j, b_j) and $(\bar{a}_j, \bar{b}_j)$ which have the indicated cross-ratios. If $r = 3$, there is nothing left to do. If $r > 3$, we must consider the supplementary equations imposed by the cross-ratios μ_{ji}, where $j > 3$ and $i > 2$. As the reader may verify with a modest amount of calculation, these are not satisfied generically by any choice of solutions other than $\{(a_i, b_i)\}$. This completes the proof.

3. Polyhedral cones and boundary components

In this section we present a Hodge-theoretic construction of the rational boundary components of $\mathcal{H}_g$. As mentioned in the introduction, these will be classifying spaces for nilpotent orbits of mixed Hodge structures, or equivalently, of one-motifs.

As before, we fix a rank $2g$ lattice $H_{\mathbf{Z}}$ and an integral, nonsingular skew form ϕ. Then the Siegel space classifies ϕ-polarized Hodge structures:

$$\mathcal{H}_g = \{F^1 \in \operatorname{Grass}(g, H_{\mathbf{C}}) \mid \phi(F^1, F^1) = 0, i\phi(F^1, \bar{F}^1) > 0\}. \tag{3.1}$$

Its compact dual is

$$\check{\mathcal{H}}_g = \{F^1 \in \operatorname{Grass}(g, H_{\mathbf{C}}) \mid \phi(F^1, F^1) = 0\}, \tag{3.2}$$

a nonsingular subvariety of the Grassmannian, homogeneous under the complex symplectic group

$$G_{\mathbf{C}} = \operatorname{Sp}(H_{\mathbf{C}}, \phi).$$

Moreover $\mathcal{H}_g$ sits inside $\check{\mathcal{H}}_g$ as an open set, homogeneous under the action of the real form

$$G = \{g \in G_{\mathbf{C}} \mid g(H_{\mathbf{R}}) = H_{\mathbf{R}}\}.$$

Now let

$$\mathfrak{g}_0 = \{X \in g\ell(H_{\mathbf{R}}) \mid \phi(Xu, v) + \phi(u, Xv) = 0\}, \tag{3.3}$$

let

$$\{0\} = S_0 \subset S_1 \subset \cdots \subset S_g$$

be a maximal flag of rational ϕ-isotropic subspaces of $H_{\mathbf{R}}$, and define abelian subalgebras of $\mathfrak{g}_0$ by

$$\eta_i = \{N \in \mathfrak{g}_0 \mid \operatorname{Im} N \subseteq S_i\}. \tag{3.4}$$

For each $N \in \eta_i$, the form ϕ_N is symmetric and its null space contains the ϕ-orthogonal complement $S_i^\perp$ of S_i. We can therefore consider the open real cone

$$\eta_i^+ = \{N \in \eta_i \mid \phi_N > 0 \text{ on } H_{\mathbf{R}}/S_i^\perp\} \tag{3.5}$$

and its closure $\operatorname{Cl}(\eta_i^+)$.

Given linearly independent elements $N_1, \ldots, N_r$ in $\operatorname{Cl}(\eta_i^+)$, let

$$\sigma = \sigma(N_1, \ldots, N_r)$$

be the closed polyhedral cone they generate, let $\operatorname{Int}(\sigma)$ be the interior, and for a subring $\Lambda \subset \mathbf{C}$ let

$$\sigma_\Lambda = \left\{ \sum \lambda_i N_i \mid \lambda_i \in \Lambda \right\}. \tag{3.6}$$

Because the proof of (1.1) depends only on the positivity properties of ϕ_N, we have the following:

(3.7) Lemma: *For any* $N \in \mathrm{Int}(\sigma)$, *we have*

(i) $W_1(N) = \cap \ker N_i$

(ii) $W_0(N) = \sum \text{image } N_i$.

Consequently the cone itself defines a weight filtration,

$$W_*(\sigma) = W_*(N),$$

where N is an interior element of σ. Moreover, there is a correspondence between nilpotent orbits and mixed Hodge structures:

(3.8) Proposition: *Let* $F^1 \in \check{\mathscr{H}}_g$ *and let* $\sigma \subset \mathrm{Cl}(\eta_i^+)$ *be a polyhedral cone. Then the following are equivalent*:

(i) *The two filtrations* $F^* = \{H_{\mathbf{C}} \supset F^1 \supset \{0\}\}$ *and* $W_*(\sigma)$ *define a* ϕ_N-*polarized mixed Hodge structure for all* $N \in \mathrm{Int}(\sigma)$.

(ii) *For any* $N \in \mathrm{Int}(\sigma)$, *the map*

$$z \longmapsto (\exp zN)F^1$$

defines a nilpotent orbit, i.e., there exist $\alpha \in \mathbf{R}$ *such that*

$$\exp(zN) \cdot F^1 \in \mathscr{H}_g \text{ for } \mathrm{Im}(z) > \alpha.$$

Proof: Assuming (i) we can consider the generalized Hodge decomposition of $H_{\mathbf{C}}$

$$H_{\mathbf{C}} = I^{1,1} \oplus I^{1,0} \oplus I^{0,1} \oplus I^{0,0}$$

where

$$I^{1,1} = F^1 \cap (\bar{F}^1 + W_0)$$

$$I^{1,0} = F^1 \cap W_1$$

$$I^{0,1} = \bar{F}^1 \cap W_1$$

$$I^{0,0} = W_0.$$

Thus $I^{1,0}$ is complex conjugate to $I^{0,1}$, $I^{0,0}$ is self-conjugate, but $I^{1,1}$ is not in general self-conjugate. Define hermitian forms

$$\psi(x, y) = i\phi(x, \bar{y})$$

$$\psi_N(x, y) = \phi(x, N\bar{y})$$

and observe that for interior elements of σ, ψ_N is positive definite on $I^{1,0}$, whereas ψ is positive definite on $I^{1,1}$ and identically zero on $I^{1,0}$.

Now given $f \in F^1$ we want to show

$$\mathrm{i}\phi(\exp(\mathrm{i}yN)\cdot f, \exp(-\mathrm{i}yN)\cdot \bar{f}) > 0$$

for y sufficiently large. A standard compactness argument will then imply (ii). But

$$\begin{aligned}\mathrm{i}\phi(\exp(\mathrm{i}yN)\cdot f, \exp(-\mathrm{i}yN)\cdot \bar{f}) &= \mathrm{i}\phi(f, \bar{f} - 2\mathrm{i}yN\bar{f})\\ &= \mathrm{i}\phi(f, \bar{f}) + 2y\phi(f, N\bar{f})\end{aligned}$$

and the last expression is clearly greater than zero for $y \gg 0$.

Conversely, let $\exp(zN)\cdot F^1$ be a nilpotent orbit. We first notice that $F^1 \cap W_0(\sigma) = \{0\}$; in fact, let $f \in F^1 \cap W_0(\sigma)$, in particular $f \in \mathrm{Im}(N)$ and therefore

$$\mathrm{i}\phi(\exp(zN)\cdot f, \exp(\bar{z}N)\cdot \bar{f}) \equiv 0.$$

Hence F^* induces a Hodge structure of weight zero on $\mathrm{Gr}_0(W_*(\sigma))$. In order to prove the corresponding statement for $\mathrm{Gr}_1(W_*(\sigma))$ it is enough to show that

$$W_1(\sigma) = W_0(\sigma) \oplus (W_1(\sigma) \cap F^1) \oplus (W_1(\sigma) \cap \bar{F}^1). \tag{3.9}$$

If $f \in F^1 \cap W_1(\sigma)$ is such that $f + \bar{f} \in W_0(\sigma)$ then

$$\phi(f, \bar{f}) = \phi(f, f + \bar{f}) = 0$$

but, clearly $\phi_N(f, \bar{f})$ also vanishes and thus $\mathrm{i}\phi(\exp(zN)\cdot f, \exp(\bar{z}N)\cdot \bar{f}) \equiv 0$. This implies $f = 0$ and consequently the sum (3.9) is direct.

We now have

$$\begin{aligned}g = \dim F^1 &= \dim(F^1 \cap \mathrm{Ker}\, N) + \dim(N(F^1))\\ &\leq \dim(F^1 \cap \mathrm{Ker}\, N) + \dim W_0(\sigma)\end{aligned}$$

from which it follows that the dimension of the right-hand side in (3.9) is greater than or equal to $2g - \dim W_0(\sigma) = \dim W_1(\sigma)$. This proves (3.9). Moreover, we also have $N(F^1) = \mathrm{Im}(N)$, which implies

$$H_{\mathbb{C}} = W_1(\sigma) + F^1 = W_1(\sigma) + \overline{F^1}$$

which is equivalent to the statement that F^* defines a Hodge structure of weight two, and pure type (1, 1), on $\mathrm{Gr}_2(W_*(\sigma))$. Since the polarization statements are clear, this completes the proof of (3.8).

We are now in the position to define the notion of rational boundary components.

(3.10) Definition: *Let $N_1, \ldots, N_r \in \mathrm{Cl}(\eta_i^+)$ be rational elements, $\sigma = \sigma(N_1, \ldots, N_r)$ the polyhedral cone they generate. The subset $B(\sigma) \subset \check{\mathcal{H}}_g$ of all $F^* \in \check{\mathcal{H}}_g$ which, together with $W_*(\sigma)$, define a polarized mixed Hodge* *structure will be called the pre-boundary component associated to σ. We also define the boundary component associated to σ as the quotient*

$$\mathbf{B}(\sigma) = B(\sigma)/\exp(\sigma_{\mathbf{C}}).$$

We note that (3.8) means that $B(\sigma) = \exp(\sigma_{\mathbf{C}}) \cdot \mathcal{H}_g$ and is therefore an open subset of $\check{\mathcal{H}}_g$ which contains $\mathcal{H}_g$. We shall sometimes identify $\mathbf{B}(\{0\})$ with $\mathcal{H}_g$.

(3.11) PROPOSITION: *Let σ_1 and σ_2 be rational polyhedral cones and assume that $\sigma_1 < \sigma_2$ (i.e. σ_1 is a face of σ_2). Then*

(i) *$B(\sigma_1) \subset B(\sigma_2)$ and*

(ii) *There exists a projection $p : \mathbf{B}(\sigma_1) \to \mathbf{B}(\sigma_2)$ such that the diagram*

$$\begin{array}{ccc} B(\sigma_1) & \hookrightarrow & B(\sigma_2) \\ \downarrow{\scriptstyle \pi_1} & & \downarrow{\scriptstyle \pi_2} \\ \mathbf{B}(\sigma_1) & \xrightarrow{p} & \mathbf{B}(\sigma_2) \end{array}$$

commutes, where π_i, $i = 1, 2$, are the natural projections.

PROOF: The only statement that needs to be checked is the surjectivity of p. Let $F^1 \in B(\sigma_2)$ and $N \in \mathrm{Int}(\sigma_2)$ then for some $y \in \mathbf{R}$, $F^1_y = \exp(iyN)$. $F^1 \in B(\sigma_1)$ and $\pi_2(F^1) = \pi_2(F^1_y)$.

Recall [12] that the Satake rational boundary components of $\mathcal{H}_g$ are in one to one correspondence with the rational totally isotropic subspaces of $H_{\mathbf{C}}$. In fact, given such a subspace U, the corresponding Satake boundary component is given by

$$\mathbf{B}_S(U) \cong \mathcal{H}(U^{\perp}/U, \tilde{\phi}) \tag{3.12}$$

where as before $U^{\perp}$ denotes the ϕ-annihilator of U and $\tilde{\phi}$ the non-degenerate skew-symmetric form induced by ϕ on $U^{\perp}/U$.

Given a polyhedral cone $\sigma \subset \mathrm{Cl}(\eta_i^+)$, defined over $\mathbf{Q}$, then $W_0(\sigma)$ is a totally isotropic subspace and $W_1(\sigma) = W_0(\sigma)^{\perp}$. Moreover, the map $\hat{\zeta} = \hat{\zeta}_\sigma$

$$\hat{\zeta} : B(\sigma) \longrightarrow \mathbf{B}_S(W_0(\sigma)) \tag{3.13}$$

which assigns to each $F^* \in B(\sigma)$, the polarized Hodge structure of weight one defined by F^* in $\mathrm{Gr}_1(W_*(\sigma))$, is onto and smooth relative to the natural differentiable structures. Note also that since $\exp(\sigma_{\mathbf{C}})$ acts trivially on the first graded quotient $\mathrm{Gr}_1(W_*(\sigma))$, the map $\hat{\zeta}$ factors through a map

$$\zeta : \mathbf{B}(\sigma) \longrightarrow \mathbf{B}_S(W_0(\sigma)). \tag{3.14}$$

If σ_1 and σ_2 are rational polyhedral cones and σ_1 is a face of σ_2 (denoted by $\sigma_1 < \sigma_2$), then $W_0(\sigma_1) \subset W_0(\sigma_2)$ and $W_1(\sigma_1) \supset W_1(\sigma_2)$. Hence, $W_0(\sigma_2)$ defines a $\tilde{\phi}$-totally isotropic rational subspace $\tilde{W}_0(\sigma_2)$, in

$\mathrm{Gr}_1(W_*(\sigma_1))$. Moreover $\mathrm{Gr}_1(W_*(\sigma_2))$ may be identified with the quotient space $\tilde{W}_0^{\perp}(\sigma_2)/\tilde{W}_0(\sigma_2)$. Hence the Siegel upper half space $\mathcal{H}(\mathrm{Gr}_1(W_*(\sigma_2)), \tilde{\phi})$ may be thought of as a Satake boundary component in $\mathcal{H}(\mathrm{Gr}_1(W_*(\sigma_1)), \tilde{\phi})$. Also, the following diagram commutes

$$\begin{array}{ccc} \mathbf{B}(\sigma_1) & \xrightarrow{p} & \mathbf{B}(\sigma_2) \\ \downarrow{\scriptstyle \zeta_1} & & \downarrow{\scriptstyle \zeta_2} \\ \mathbf{B}_S(W_0(\sigma_1)) & \xrightarrow{w} & \mathbf{B}_S(W_0(\sigma_2)) \end{array} \tag{3.15}$$

where w is the projection from a Siegel upper half space to one of its Satake boundary components [12].

(3.16) PROPOSITION: *The fibration $\hat{\zeta}: B(\sigma) \to \mathbf{B}_S(W_0(\sigma))$ is C^∞-trivial. The fiber is isomorphic to the subset of* $M = \mathrm{Grass}(w_1 - w_0, H_{\mathbf{R}}) \times \mathrm{Grass}(2g - w_1, H_{\mathbf{C}})$, $w_i = \dim_{\mathbf{C}} W_i(\sigma)$, *given by*

$$A = \left\{ \begin{array}{c} (U_1, U_2) \in M : W_1(\sigma)_{\mathbf{R}} = W_0(\sigma)_{\mathbf{R}} \oplus U_1, H_{\mathbf{C}} = W_1(\sigma) \oplus U_2 \\ \phi(U_1, U_2) = \phi(U_1, \bar{U}_2) = \phi(U_2, U_2) = 0 \end{array} \right\}.$$

PROOF: Given $F^* \in B(\sigma)$, let $H_{\mathbf{C}} = \bigoplus_{p,q \geq 0} I^{p,q}$ be the generalized Hodge decomposition. Set

$$U_1 = (I^{1,0} \oplus I^{0,1})_{\mathbf{R}}, \quad U_2 = I^{1,1}.$$

We need to check that $\phi(I^{1,0}, \overline{I^{1,1}}) = 0$, but $I^{1,0} = F^1 \cap W_1(\sigma)$ and $\overline{I^{1,1}} = \overline{F^1} \cap (F^1 + W_0(\sigma))$ and $\phi(W_0(\sigma), W_1(\sigma)) = 0$.

Conversely, given $(U_1, U_2) \in A$ and $\Omega \in \mathbf{B}_S(W_0(\sigma)) = \mathcal{H}(\mathrm{Gr}_1(W_*(\sigma)), \tilde{\phi})$, let $I^{1,0} \subset W_1(\sigma)$ be the unique subspace such that

$$U_1 = (I^{1,0} \oplus \overline{I^{1,0}})_{\mathbf{R}} \quad \text{and} \quad (I^{1,0} + W_0(\sigma))/W_0(\sigma) = \Omega.$$

We remark that if $(\Omega, U_1, U_2) \in B(\sigma)$ and $g \in \exp(\sigma_{\mathbf{C}})$ then $g(\Omega, U_1, U_2) = (\Omega, U_1, g(U_2))$, hence

$$\mathbf{B}(\sigma) \xrightarrow{\zeta} \mathbf{B}_S(W_0(\sigma)) \tag{3.17}$$

is also a trivial fibering, with fiber $A/\exp(\sigma_{\mathbf{C}})$.

We end this section with an example which shall illustrate some of the objects introduced above. Geometrically, the situation described below corresponds to that discussed in (2.7). Let $H_{\mathbf{Z}}$ be a lattice of rank $2g$, $\xi = \{e_1, \ldots, e_g, f_1, \ldots, f_g\}$ a $\mathbf{Z}$-basis of $H_{\mathbf{Z}}$ and ϕ the skew-symmetric form in $H_{\mathbf{Z}}$ defined by

$$\begin{aligned} \phi(e_i, e_j) &= \phi(f_i, f_j) = 0 \\ \phi(e_i, f_j) &= -\phi(f_i, e_j) = -\delta_{ij}. \end{aligned} \tag{3.18}$$

We set $S_i = \text{span}_{\mathbf{R}}\{e_1, \dots, e_i\}$ and let $N_j \in \text{Cl}(\eta_g^+)$ be given, relative to ξ, by

$$N_j = \left[\begin{array}{c|c} 0 & \Lambda_j \\ \hline 0 & 0 \end{array}\right]$$

where Λ_j is the diagonal matrix with 1 in the (j, j) position and zeroes elsewhere. The set

$$\sigma_r = \left\{N \in \text{Cl}(\eta_g^+) : N = \sum_{j=1}^{r} \lambda_j N_j, \lambda_j \geq 0\right\}.$$

is then a rational polyhedral cone in $\text{Cl}(\eta_g^+)$, with monodromy weight filtration:

$$W_0(\sigma_r) = S_r, \quad W_1(\sigma_r) = S_r^\perp = \text{span}_{\mathbf{R}}\{e_1, \dots, e_g, f_{r+1}, \dots, f_g\}.$$

If $F^1 \in \check{\mathscr{H}}$, then it may be represented as the row space of a matrix

$$P = [P_1, P_2]$$

where P_1 and P_2 are $g \times g$ complex matrices, and P is well defined up to left-action by $\mathbf{GL}(g, \mathbf{C})$. If in addition $F^1 \in B(\sigma_r)$, then for any $N \in \text{Int}(\sigma_r)$, $\exp(zN) \cdot F^1 \in \mathscr{H}_g$, for $\text{Im}(z)$ sufficiently large, and this is easily seen to imply that P_1 must be non-singular. Hence F^1 can be (uniquely) represented as

$$F^1 = [I, Z] \tag{3.19}$$

and since $F^1 \in \check{\mathscr{H}}_g$ we must have that Z is a symmetric matrix whose imaginary part is positive semi-definite. Writing

$$F^1 = \begin{bmatrix} I_r & 0 & Z_{11} & {}^tZ_{21} \\ 0 & I_{g-r} & Z_{21} & Z_{22} \end{bmatrix}.$$

We see that the Hodge structure induced by F^1 in $\text{Gr}_1(W_*(\sigma_r))$ has a period matrix

$$[I_{g-r}, Z_{22}].$$

Hence $\text{Im}(Z_{22}) > 0$, and the projection $\hat{\zeta}$ defined in (3.13) is given by $\hat{\zeta}(Z) = Z_{22}$.

The terms in the generalized Hodge decomposition are defined by

$$I^{0,0} = W_0(\sigma) = S_r$$

$$I^{1,0} = \text{row space of } [0, I_{g-r}, Z_{21}, Z_{22}], \quad I^{0,1} = \overline{I^{1,0}}$$

$$I^{1,1} = \text{row space of } [Z_{11}, \text{Re}({}^tZ_{21}), I_r, -\text{Im}({}^tZ_{21})]$$

and these in turn determine the pair (U_1, U_2) as in the proof of (3.16). Finally, note that $\exp(\sigma_{\mathbf{C}})$ acts on $B(\sigma)$ by

$$\exp\left(\sum_{i=1}^{r} z_i N_i\right) \cdot Z = Z + D(z_1, \dots, z_r)$$

where

$$D(z_1, \ldots, z_r) = \left[\begin{array}{c|c} \begin{matrix} z_1 & & \\ & \ddots & \\ & & z_r \end{matrix} & 0 \\ \hline 0 & 0 \end{array}\right]$$

and thus, it is the information contained in the terms that are off the diagonal of Z_{11} that passes on to the boundary component $\mathbf{B}(\sigma_r)$, while the Satake component contains only the element Z_{22}.

4. Group actions

In this section we exhibit the boundary components in the form defined by Mumford in [1] and [10].

Let $\sigma = \sigma(N_1, \ldots, N_r) \subset \mathrm{Cl}(\eta_i^+)$ be a rational polyhedral cone and consider the subgroup of G_Λ

$$\mathrm{Norm}_\Lambda(\sigma) = \{g \in G_\Lambda : \mathrm{Ad}(g)N_j = N_j, 1 \leq j \leq r\} \tag{4.1}$$

and the subalgebra of $\mathfrak{g}_\mathrm{C}$

$$L = L(\mathrm{Norm}_\mathrm{C}(\sigma)) = \{X \in \mathfrak{g}_\mathrm{C} : [X, N_j] = 0, 1 \leq j \leq r\}. \tag{4.2}$$

In particular, if $X \in L$ it must preserve the monodromy weight filtration $W_*(\sigma)$. We then set for $\ell \geq 0$

$$W_{-\ell} = W_{-\ell}(L) = \{X \in L : X(W_j(\sigma)) \subset W_{j-\ell}(\sigma)\}.$$

This defines a "weight" filtration

$$\{0\} = W_{-3} \subset W_{-2} \subset W_{-1} \subset W_0 = L \tag{4.3}$$

on L which is compatible with the Lie algebra structure. More precisely

$$[W_{-r}, W_{-s}] \subset W_{-(r+s)}$$

and hence it follows that $W_{-\ell}$ consists of nilpotent elements for $\ell \geq 1$. By exponentiation we can then define a corresponding increasing filtration $W_{-\ell}(\mathrm{Norm}_\mathrm{C}(\sigma))$ in $\mathrm{Norm}_\mathrm{C}(\sigma)$. Finally, we set

$$\begin{aligned} V(\sigma) &= \mathrm{Norm}_\mathbf{R}(\sigma) \cdot W_{-2}(\mathrm{Norm}_\mathrm{C}(\sigma)), \\ U(\sigma) &= V(\sigma) \cap W_{-1}(\mathrm{Norm}_\mathrm{C}(\sigma)). \end{aligned} \tag{4.4}$$

It is easy to check that $U(\sigma)$ is the unipotent radical of $V(\sigma)$ and, thus, a normal subgroup of $V(\sigma)$. Moreover, since $N_j(W_\ell(\sigma)) \subset W_{\ell-2}(\sigma)$, we have

$$\exp(\sigma_\mathrm{C}) \subset W_{-2}(\mathrm{Norm}_\mathrm{C}(\sigma)) \subset U(\sigma).$$

In particular we have (cf. section 3) that

$$B(\sigma) = W_{-2}(\mathrm{Norm}_\mathrm{C}(\sigma)) \cdot \mathscr{H}_g, \tag{4.5}$$

and a linear algebra argument shows

(4.6) THEOREM: *The group $U(\sigma)$ acts simply transitively on the fibers of the fibration* (3.16).

If $S(\sigma)$ is the subgroup of $V(\sigma)$ leaving the subspaces U_i of (3.16) stable, then $S(\sigma)$ is semisimple and $V(\sigma) = S(\sigma) \cdot U(\sigma)$. We then obtain,

(4.7) COROLLARY: (i) *The group $V(\sigma)$ acts transitively on the boundary precomponent $B(\sigma)$.*

(ii) *There are smooth identifications $B(\sigma) \cong \mathbf{B}_S(W_0(\sigma)) \times U(\sigma)$ and $\mathbf{B}(\sigma) = \mathbf{B}_S(W_0(\sigma)) \times (U(\sigma)/\exp(\sigma_\mathbf{C}))$, where $\mathbf{B}_S(W_0(\sigma)) = \mathcal{H}(\mathrm{Gr}_1(W_*(\sigma)), \tilde{\phi})$ is the Satake boundary component associated to the rational, ϕ-isotropic subspace $W_0(\sigma)$.*

(4.8) REMARK: We have already noted that

$$B(\sigma) = \exp W_{-2}(\mathrm{Norm}_\mathbf{C}(\sigma)) \cdot \mathcal{H}_g;$$

in particular, we see that $B(\sigma)$ agrees with the open set $D(F)$ defined by Mumford in [1], and the second identity in (ii) of (4.7) is just the decomposition in ([1], p. 235). The group $\mathrm{Norm}_\mathbf{Z}(\sigma)$ acts properly discontinuously on $B(\sigma)$ and $\mathbf{B}(\sigma)$ and we can consider the fibration

$$\exp(\sigma_\mathbf{C})/\exp(\sigma_\mathbf{Z}) \longrightarrow B(\sigma)/\mathrm{Norm}_\mathbf{Z}(\sigma) \longrightarrow \mathbf{B}(\sigma)/\mathrm{Norm}_\mathbf{Z}(\sigma).$$

Since $\sigma_\mathbf{C}$ is abelian, this is a toroidal fibration and it is the basic ingredient in Mumford's construction.

Finally, we shall again use the example discussed at the end of Section 3 to clarify some of the groups introduced above. In order to simplify the notation we shall make a slight change in the basis ξ by exchanging the subsets $\{f_1, \ldots, f_r\}$ and $\{f_{r+1}, \ldots, f_g\}$. In the new basis then

$$\phi = \left[\begin{array}{c|c|c} & & -I_r \\ \hline & \tilde{\phi} & \\ \hline I_r & & \end{array}\right] \qquad \tilde{\phi} = \left[\begin{array}{c|c} & -I_{g-r} \\ \hline I_{g-r} & \end{array}\right].$$

We then have that $\mathrm{Norm}_\mathbf{C}(\sigma)$ consists of complex matrices of the form:

$$\begin{bmatrix} A_{11} & A_{12} & A_{13} \\ & A_{22} & A_{23} \\ & & {}^tA_{11}^{-1} \end{bmatrix} \tag{4.9}$$

satisfying the following conditions

$$\begin{aligned} &\text{(i) } A_{11}^{-1}\Lambda_j = \Lambda_j {}^tA_{11} \quad j = 1, \ldots, r, \\ &\text{(ii) } A_{22} \in \mathrm{Sp}(g-r, \mathbf{R}), \\ &\text{(iii) } A_{11}^{-1}A_{12} + {}^tA_{23}\tilde{\phi}A_{22} = 0, \\ &\text{(iv) } A_{11}^{-1}A_{13} + {}^tA_{23}\tilde{\phi}A_{23} - {}^tA_{13}{}^tA_{11} = 0. \end{aligned} \tag{4.10}$$

Notice that the subgroup of $\mathbf{GL}(r, \mathbf{C})$ of matrices A_{11} satisfying (i) above, is finite. Now, $V(\sigma)$ consists of those matrices in $\mathrm{Norm}_{\mathbf{C}}(\sigma)$, for which A_{11}, A_{12} and hence A_{23} are real, while $U(\sigma)$ is the subgroup of $V(\sigma)$ defined by $A_{11} = I_r$, $A_{22} = I_{2(g-r)}$. Finally, the isomorphism in (4.7.ii) is defined in the following way: let $F^1 \in B(\sigma)$ and write as in (3.19), $F^1 = [I, Z]$ with

$$Z = \begin{bmatrix} Z_{11} & {}^tZ_{21} \\ Z_{21} & Z_{22} \end{bmatrix}.$$

Then $F^1 \leftrightarrow (Z_{22}, g)$, where $Z_{22} \in \mathbf{B}_S(W_0(\sigma_r))$ and

$$g = \left[\begin{array}{c|c|c} I & \mathrm{Im}(Z_{21}) \quad \mathrm{Re}(Z_{21}) & Z_{11} \\ \hline & I & \begin{matrix} \mathrm{Re}({}^tZ_{21}) \\ -\mathrm{Im}({}^tZ_{21}) \end{matrix} \\ \hline & & I \end{array}\right] \in U(\sigma).$$

5. Construction of the compactification

Here we shall present, in its bare outlines, the construction of a compactification of $\mathscr{H}_g/\Gamma$, $\Gamma = G_{\mathbf{Z}}$, whose boundary components are discrete quotients of the spaces described in sections 3 and 4. This is done in the spirit of Satake's original compactification [12, 13], the end result is Mumford's compactification but with the "toroidal embeddings", by which the boundary components are "glued" to $\mathscr{H}_g/\Gamma$, made somewhat more explicit. We refer the reader to [1] and [10] for the details of Mumford's more general construction. Let us also remark that Namikawa [11] has considered the compactification resulting from the Delony–Voronoi decomposition and its applications to the study of degenerations of abelian varieties.

For simplicity we shall assume that the polarizing form ϕ admits an integral symplectic basis; this is of course the case in the geometric situation.

Let $\xi = (e_1, \ldots, e_g, f_1, \ldots, f_g)$ be a basis of $H_{\mathbf{Z}}$, such that ϕ satisfies conditions (3.18), and

$$\{0\} = S_0 \subset S_1 \subset \cdots \subset S_g \tag{5.1}$$

the maximal flag of rational totally isotropic subspaces defined by

$$S_r = \mathrm{span}_{\mathbf{R}}\{e_1, \ldots, e_r\}.$$

If $\mathrm{Norm}_{\mathbf{Z}}(S_r) = \{g \in G_{\mathbf{Z}} : g(S_r) = S_r\}$, then its adjoint action on $\mathfrak{g}_0$ restricts to an action on η_r which preserves the cone η_r^+: indeed if $N \in \eta_r$,

$$\mathrm{Im}(\mathrm{Ad}(g)N) = \mathrm{Im}(g^{-1}N) = g^{-1}(\mathrm{Im}(N)) \subset g^{-1}(S_r) = S_r$$

and thus $\mathrm{Ad}(g)N \in \eta_r$. On the other hand, if $N \in \eta_r^+$, then $\phi_N(x, y)$ is positive definite in $H_{\mathbf{R}}/S_r^{\perp}$ but then so is the form $\phi_N(gx, gy) = \phi_{\mathrm{Ad}(g)N}(x, y)$ which implies that $\mathrm{Ad}(g)N \in \eta_r^+$.

Let $\mathrm{Cent}_{\mathbf{Z}}(S_r) = \{g \in \mathrm{Norm}_{\mathbf{Z}}(S_r) : g|_{S_r} = I\}$ and

$$G_{\mathbf{Z}}(S_r) = \mathrm{Norm}_{\mathbf{Z}}(S_r)/\mathrm{Cent}_{\mathbf{Z}}(S_r). \tag{5.2}$$

If we identify η_r^+ (via the basis ξ) with the set $\mathcal{M}$ of $r \times r$ positive definite real symmetric matrices then the action of $G_{\mathbf{Z}}(S_r)$ on η_r^+ corresponds to the natural action of $\mathbf{SL}(r, \mathbf{Z})$ on $\mathcal{M}$. Recall also the construction of Siegel sets in $\mathcal{M}$. Given $X \in \mathcal{M}$ we can write (Babylonian reduction) $X = {}^tWDW$, where W is an upper triangular unipotent matrix and D is diagonal. The set

$$\mathcal{W}_u = \{X = {}^tWDW \in \mathcal{M} : |w_{ij}| < u, 0 < d_i < ud_{i+1}\} \tag{5.3}$$

is called a *Siegel set*, and for u sufficiently large it is a fundamental set for the action of $\mathbf{SL}(r, \mathbf{Z})$ on $\mathcal{M}$. We shall carry over these notions to η_r^+ and the $G_{\mathbf{Z}}(S_r)$-action.

We can now define one of the basic objects for the construction of the compactification (cf. [1]).

(5.4) DEFINITION: *A collection $\{\sigma_\alpha^{(r)}\}$ of rational polyhedral cones in $\mathrm{Cl}(\eta_r^+)$ is said to be a $G_{\mathbf{Z}}(S_r)$-admissible rational polyhedral decomposition of η_r^+ if:*

(i) *If $\sigma \in \{\sigma_\alpha^{(r)}\}$ and $\sigma' < \sigma$, then $\sigma' \in \{\sigma_\alpha^{(r)}\}$.*

(ii) *If $\sigma, \sigma' \in \{\sigma_\alpha^{(r)}\}$ then $\sigma \cap \sigma' < \sigma$ and $\sigma \cap \sigma' < \sigma'$.*

(iii) *If $\sigma \in \{\sigma_\alpha^{(r)}\}$, $\gamma \in G_{\mathbf{Z}}(S_r)$ then $\gamma(\sigma) \in \{\sigma_\alpha^{(r)}\}$.*

(iv) *There exist finitely many cones $\sigma_1, \ldots, \sigma_n \in \{\sigma_\alpha^{(r)}\}$ such that for any $\sigma \in \{\sigma_\alpha^{(r)}\}$, there is $\gamma \in G_{\mathbf{Z}}(S_r)$ with $\gamma(\sigma) = \sigma_i$ for some $i = 1, \ldots, n$.*

(v) $\eta_r^+ = \bigcup_\alpha (\sigma_\alpha^{(r)} \cap \eta_r^+)$.

(vi) *There exists a Siegel set $\mathcal{W}_u \subset \eta_r^+$, such that for any $\sigma \in \{\sigma_\alpha^{(r)}\}$ there exists $\gamma \in G_{\mathbf{Z}}(S_r)$ with $\sigma_\alpha \cap \eta_r^+ \subset \gamma(\mathcal{W}_u)$.*

If $r < t$, then $S_r \subset S_t$, $\eta_r \subset \eta_t$ and $\mathrm{Cl}(\eta_r^+) \subset \mathrm{Cl}(\eta_t^+)$. Hence if $\sigma \subset \mathrm{Cl}(\eta_t^+)$ is a rational polyhedral cone, then so is $\sigma \cap \mathrm{Cl}(\eta_r^+)$. Notice moreover that if $\sigma \subset \mathrm{Cl}(\eta_t^+)$, $\sigma' < \sigma$ and $\sigma' \cap \mathrm{Cl}(\eta_r^+) \neq \emptyset$, then $\sigma' \subset \mathrm{Cl}(\eta_r^+)$.

(5.5) DEFINITION: *A collection $\{\sigma_\alpha\}$ of rational polyhedral cones in $\mathrm{Cl}(\eta_g^+)$ is said to be Γ-admissible if:*

(i) *For any $r = 1, \ldots, g$, the collection $\{\sigma_\alpha^{(r)}\} = \{\sigma_\alpha \cap \mathrm{Cl}(\eta_r^+)\}$ is a $G_{\mathbf{Z}}(S_r)$-admissible rational polyhedral decomposition of η_r^+.*

(ii) *Given r, t $(1 \le r < t \le g)$, then*

$$\{\sigma_\alpha^{(r)}\} = \{\sigma_\alpha^{(t)} \cap \mathrm{Cl}(\eta_r^+)\}.$$

The existence of Γ-admissible polyhedral decompositions has been proved by A. Ash in [1].

It is well known that any two totally isotropic rational subspaces of $H_{\mathbf{R}}$, of the same dimension, are conjugate under the action of $\Gamma = \mathrm{Sp}(n, \mathbf{Z})$.

Moreover, the same is true of maximal flags of such subspaces. In particular, if

$$\{0\} = S'_0 \subset S'_1 \subset \cdots \subset S'_g$$

is a maximal flag, then there exists $\gamma \in \Gamma$, such that

$$\{\gamma\sigma_\alpha\} \subset \mathrm{Cl}((\eta')^+_g)$$

is a Γ-admissible polyhedral decomposition.

(5.6) DEFINITION: *Given a Γ-admissible polyhedral decomposition $\{\sigma_\alpha, \alpha \in A\}$, we will denote by $\mathcal{H}^*_g$ the union of the boundary components $\gamma\sigma_\alpha$ for $\gamma \in \Gamma$, $\alpha \in A$ (we identify $\mathcal{H}_g$ with the boundary component $\mathbf{B}(\{0\})$).*

The next step is to construct a fundamental set for the action of Γ on $\mathcal{H}^*_g$. Given $\sigma \in \{\gamma\sigma_\alpha; \gamma \in \Gamma, \alpha \in A\}$ we shall denote by

$$\begin{aligned} &\Gamma_V(\sigma) = \Gamma \cap V(\sigma) = \mathrm{Norm}_{\mathbf{Z}}(\sigma), \quad \Gamma_U(\sigma) = \Gamma \cap U(\sigma) \\ &\Gamma_S(\sigma) = \Gamma \cap S(\sigma) \end{aligned} \tag{5.7}$$

where $V(\sigma)$, $U(\sigma)$ and $S(\sigma)$ are the subgroups of $\mathrm{Norm}_{\mathbf{C}}(\sigma)$ defined in section 4. Note that $\Gamma_V(\sigma)$ acts almost effectively and properly discontinuously on the pre-boundary component $B(\sigma)$, with kernel $Z(\sigma) \subset S(\sigma)$. We can identify $S(\sigma)/Z(\sigma)$ with a subgroup $\tilde{S}(\sigma) \subset S(\sigma)$ and $\tilde{S}(\sigma) \cong \mathrm{Sp}(g-r, \mathbf{R})$, $r = \dim W_0(\sigma)$. We know that $\Gamma_V(\sigma)$ can be expressed as a semi-direct product

$$\Gamma_V(\sigma) = \Gamma_S(\sigma) \cdot \Gamma_U(\sigma)$$

and, if we denote by $\Gamma_{-2}(\sigma) = \Gamma_V(\sigma) \cap W_{-2}(\mathrm{Norm}_{\mathbf{C}}(\sigma))$ then $\Gamma_{-2}(\sigma)$ is a normal subgroup of $\Gamma_V(\sigma)$ contained in $\Gamma_U(\sigma)$, and the quotient $\Gamma_U(\sigma)/\Gamma_{-2}(\sigma)$ is abelian. We also note that $\Gamma_{-2}(\sigma)$ contains the abelian group $\exp(\sigma_{\mathbf{Z}})$ and that

$$\Gamma(F(\sigma)) = \Gamma_V(\sigma)/Z(\sigma) \cdot \exp(\sigma_{\mathbf{Z}})$$

acts effectively and properly discontinuously on $\mathbf{B}(\sigma)$. Moreover, if $\Omega_S(\sigma) \subset \mathbf{B}_S(W_0(\sigma))$, $\Omega_U(\sigma) \subset U(\sigma)$ are fundamental sets for the action of $\Gamma_S(\sigma)$ and $\Gamma_U(\sigma)$, respectively, then

$$\Omega(\sigma) = \Omega_U(\sigma) \times \Omega_S(\sigma) \tag{5.8}$$

is a fundamental set for the action of $\Gamma_V(\sigma)$ on $B(\sigma)$. Furthermore $\Omega_U(\sigma)$ may be constructed as a product

$$\Omega_U(\sigma) = \Omega_{-2}(\sigma) \times j(\Omega_{-1}(\sigma))$$

where $\Omega_{-2}(\sigma) \subset W_{-2}(\sigma)$, $\Omega_{-1}(\sigma) \subset U(\sigma)/W_{-2}(\sigma)$ are fundamental sets for the action of $\Gamma_{-2}(\sigma)$ and $\Gamma_U(\sigma)/\Gamma_{-2}(\sigma)$, respectively,and $j: U(\sigma)/W_{-2}(\sigma) \to U(\sigma)$ is a global section. The projection

$$\tilde{\Omega}(\sigma) = (\Omega_U(\sigma)/\exp(\sigma_{\mathbf{C}})) \times \Omega_S(\sigma) \subset \mathbf{B}(\sigma)$$

is, moreover, a fundamental set for the action of $\Gamma(\mathbf{B}(\sigma))$.

We fix now a Γ-admissible polyhedral decomposition $\{\sigma_\alpha\}$ as in (5.5), and let $\sigma_1, \ldots, \sigma_n \in \{\sigma_\alpha\}$ be a set of representatives of $G_{\mathbf{Z}}(S_g)$-equivalence classes. We can then write

$$\overline{(\mathcal{H}_g/\Gamma)} = \mathcal{H}_g^*/\Gamma = \bigcup_{i=1}^{n} (\mathbf{B}(\sigma_i)/\Gamma(\mathbf{B}(\sigma_i))) \tag{5.9}$$

Associated to the choice of a maximal flag of rational totally isotropic subspaces, there is a choice of Siegel sets in $\mathcal{H}_g$; namely, if as in (3.19) we view $\mathcal{H}_g$ as

$$\{Z \in \mathcal{M}_g(\mathbf{C}) : {}^tZ = Z, \operatorname{Im}(Z) > 0\}$$

and we write $X = \operatorname{Re}(Z)$, $Y = \operatorname{Im}(Z) = {}^tWDW$, where W is upper triangular unipotent and D is diagonal, then the subset $\Omega_t \subset \mathcal{H}_g$ defined by the conditions

$$|x_{ij}| \leq t, \quad |w_{ij}| \leq t, \quad 1 \leq td_1, \quad d_i \leq td_{i+1} \tag{5.10}$$

is a Siegel set in $\mathcal{H}_g$ and for $t > 0$, sufficiently large, Ω_t is a Γ-fundamental set in $\mathcal{H}_g$. Moreover, if $p_i : \mathcal{H}_g \to \mathbf{B}(\sigma_i)$, $\zeta_i : \mathcal{H}_g \to \mathbf{B}_S(W_0(\sigma_i))$ are the projections defined in (3.11) and (3.14), then:

(5.11) *There exists a Siegel set $\Omega_t \subset \mathcal{H}_g$ such that for $i = 1, \ldots, n$, $\zeta_i(\Omega_t)$ is a Siegel set in $\mathbf{B}_S(W_0(\sigma_i))$ and $p_i(\Omega_t)$ is a $\Gamma(\mathbf{B}(\sigma_i))$-fundamental set in $\mathbf{B}(\sigma_i)$.*

In particular the set

$$\Omega^* = \Omega_t \cup \left(\bigcup_{i=1}^{n} p_i(\Omega_t)\right) \tag{5.12}$$

where $t > 0$ is large enough so that Ω_t is a Γ-fundamental set in $\mathcal{H}_g$ and (5.11) is satisfied, is a Γ-fundamental set in $\mathcal{H}_g^*$.

We can now define a topology in Ω^* in the following way: Let $x \in p_j(\Omega_t)$, U_j an open neighborhood of x in $p_j(\Omega_t)$ relative to the natural topology of $\mathbf{B}(\sigma_j)$. For each $j = 1, \ldots, n$, let $I(j) = \{i : \sigma_i < \sigma_j\}$ and for $i \in I(j)$ let

$$p_{ij} : \mathbf{B}(\sigma_i) \longrightarrow \mathbf{B}(\sigma_j)$$

be the projection in (3.11). Finally, for each $\lambda > 0$, let $\sigma_i(\lambda) = \{N = \Sigma\, \lambda_k N_k \in \sigma_i : \lambda_k > \lambda\}$ and set

$$U_i(\lambda) = p_{ij}^{-1}(U_j) \cap (\exp(\sigma_j(\lambda)) \cdot p_i(\Omega_t)) \subset \mathbf{B}(\sigma_i);\ i \in I(j)$$

$$U_0(\lambda) = p_j^{-1}(U_j) \cap (\exp(\sigma_j(\lambda)) \cdot \Omega_t) \subset \mathcal{H}_g.$$

Then the sets

$$\mathcal{V}(U_j, \lambda) = U_0(\lambda) \cup \left(\bigcup_{i \in I(j)} U_i(\lambda)\right)$$

where U_j varies on a basis of open sets in $\mathbf{B}(\sigma_j)$ and $\lambda \in \mathbf{R}$, $\lambda > 0$, define a basis of open sets for a topology $\mathcal{T}$ in Ω^*. Moreover, if we denote by Ω^*_S the fundamental set in Satake's extended space [12]:

$$\Omega_S{}^* = \Omega_t \cup \left(\bigcup_{i=1}^{n} \zeta_i(\Omega_t)\right)$$

then we have a continuous map

$$\zeta : \Omega^* \longrightarrow \Omega_S{}^*$$

where Ω^*_S is endowed with the Satake topology $\mathcal{T}_S$ [12]. It is then possible to carry the basic properties of $(\Omega^*_S, \mathcal{T}_S)$ to $(\Omega^*, \mathcal{T})$. In particular

(5.13)
(i) $(\Omega^*, \mathcal{T})$ *is compact and Hausdorff.*
(ii) $\mathcal{T}$ *induces the natural topology in* Ω_t *and* $p_i(\Omega_t)$, $i = 1, \ldots, n$.
(iii) *If* $\gamma \in \Gamma$, $x \in \Omega^*$ *and* $\gamma x \in \Omega^*$, *then for any* $\mathcal{T}$*-neighborhood* $\mathcal{V}'$ *of* γx, *there exists a* $\mathcal{T}$*-neighborhood* $\mathcal{V}$ *of* x, *such that* $\gamma\mathcal{V} \cap \Omega^* \subset \mathcal{V}'$.
(iv) *If* $\gamma \in \Gamma$, $x \in \Omega^*$ *and* $\gamma x \notin \Omega^*$, *then there exists a* $\mathcal{T}$*-neighborhood* $\mathcal{V}$ *of* x *such that* $\gamma\mathcal{V} \cap \Omega^* = \emptyset$.

We can now define a Satake topology in $\mathcal{H}^*_g$: a fundamental system of neighborhoods of $x \in \mathcal{H}^*_g$ consists of all sets $\mathcal{U} \subset \mathcal{H}^*_g$ which are invariant under the action of $\Gamma_x = \{\gamma \in \Gamma : \gamma x = x\}$ and satisfy:

(5.14) *If* $\gamma x \in \Omega^*$, $\gamma \in \Gamma$, *then* $\gamma\mathcal{U} \cap \Omega^*$ *is a* $\mathcal{T}$*-neighborhood of* γx *in* Ω^*.

It then follows from Theorem 1′ in [13] that the quotient topology in $(\overline{\mathcal{H}_g/\Gamma}) = \mathcal{H}^*_g/\Gamma$ is the unique topology satisfying:

(i) *It induces the natural topology on* Ω_t *and on* $p_i(\Omega_t)$, $i = 1, \ldots, n$.
(ii) *The operations of* $\gamma \in \Gamma$ *are continuous.*
(iii) $(\overline{\mathcal{H}_g/\Gamma})$ *is compact and Hausdorff.*
(iv) *For each* $x \in \mathcal{H}^*_g$, *there exists a fundamental system of neighborhoods* $\{\mathcal{U}\}$ *of* x *such that* $\gamma\mathcal{U} = \mathcal{U}$ *for* $\gamma \in \Gamma_x$, $\gamma\mathcal{U} \cap \mathcal{U} = \emptyset$ *for* $\gamma \notin \Gamma_x$.

REFERENCES

[1] A. Ash, D. Mumford, M. Rapoport and Y. Tai: Smooth compactifications of locally symmetric varieties. Math. Sci. Press (1975).

[2] W.L. Baily Jr. and A. Borel: Compactifications of arithmetic quotients of bounded symmetric domains. Ann. of Math. (2) *84* (1966) 442–528.

[3] J. Carlson: Extensions of mixed Hodge structures, this volume.

[4] E. Cattani and A. Kaplan: Extension of period mappings for Hodge structures of weight two, Duke Math. J., *44* (1977) 1–43.

[5] E. CATTANI and A. KAPLAN: The monodromy weight filtration for a several variables degeneration of Hodge structures of weight two, Inventiones Math. *52* (1979) 131–142.
[6] P. DELIGNE: Theorie de Hodge, II, III, Publ. de l'Inst. des Hautes Etudes Scient. *40* (1972) 5–57 and *44* (1975) 6–77.
[7] P.A. GRIFFITHS: Periods of integrals on algebraic manifolds, I, II, Amer. J. Math., *90* (1968) 585–626; 805–865.
[8] P.A. GRIFFITHS: Periods of integrals on algebraic manifolds: Summary of main results and discussion of open problems, Bull. Amer. Math. Soc. *76* (1970) 282–296.
[9] P.A. GRIFFITHS and W. SCHMID: Recent developments in Hodge Theory, Proc. International Colloquium on Discrete Subgroups of Lie Groups, Oxford U. Press, 31–127 (1975).
[10] D. MUMFORD: A new approach to compactifying locally symmetric varieties, Proc. International Colloquium on Discrete Subgroups of Lie Groups, Oxford U. Press, 211–224 (1975).
[11] Y. NAMIKAWA: A new compactification of the Siegel space and degeneration of abelian varieties, I, II, Math. Ann., *221* (1976) 97–141 and 201–241.
[12] I. SATAKE: Compactification des espaces quotients de Siegel, I, Séminaire Henri Cartan, 1957–58, Exposé 12.
[13] I. SATAKE: On compactification of the quotient spaces for arithmetically defined discontinuous subgroups, Ann. of Math. (2) *72* (1960) 555–580.
[14] I. SATAKE: On the arithmetic of tube domains, Bull. Amer. Math. Soc. *79* (1973) 1076–1094.
[15] W. SCHMID: Variation of Hodge structures: the singularities of the period mapping, Inventiones Math. *22* (1973) 211–319.

E. Cattani and A. Kaplan
Dept. of Math.
University of Massachusetts
Amherst, Mass. 01003 (USA)

J. Carlson
Dept. of Math.
University of Utah
Salt Lake City, Utah 84112 (USA)

EXTENSIONS OF MIXED HODGE STRUCTURES

James A. Carlson*

1. Introduction

According to Deligne, the cohomology groups of a complex algebraic variety carry a generalized Hodge structure, or, in precise terms, a mixed Hodge structure [2]. The purpose of this paper is to introduce an abstract theory of extensions of mixed Hodge structures which has proved useful in the study of low-dimensional varieties [1]. To give the theory meaning, we will give one simple, but illustrative application:

THEOREM A: *Let X be an irreducible projective algebraic curve whose singularities are ordinary and whose normalization is non-hyperelliptic (no two-sheeted cover of the Riemann sphere exists). Then X is determined by the polarized mixed Hodge structure on $H^1(X)$.*

In very rough outline, the proof is as follows: Let $\pi:\tilde{X}\to X$ be the normalization, and observe that $H^1(X)$ is an extension of a Hodge structure of pure weight one by a Hodge structure of pure type $(0,0)$:

$$0 \longrightarrow W_0 \longrightarrow H^1(X) \xrightarrow{\pi^*} H^1(\tilde{X}) \longrightarrow 0.$$

The classical Torelli theorem applied to the canonical quotient of weight one yields $\tilde{X}$. The geometric information contained in the extension then tells how to contract the fibers of π inside $\tilde{X}$ to obtain X. In technical language, this information is carried by the so-called one-motif of $H^1(X)$, a canonical homomorphism from a lattice to an abelian variety,

$$u: Gr_0H_1(X,\mathbf{Z}) \longrightarrow J^0W_{-1}H_1(X),$$

and by the polarizing form on the weight zero lattice.

As an illustration of the applications in higher dimensions, we mention the following result on singular K-3 surfaces:

THEOREM B: *Let X be the union of a plane and a cubic surface in* $\mathbf{P}_3$,

*Research supported in part by NSF Grant MCS 79-02753 and by an Alfred P. Sloan Fellowship.

simply tangent at no more than one point, and elsewhere in general position. Then X is determined by the polarized mixed Hodge structure on the primitive part of $H^2(X)$.

An account of these and other applications will appear in the future; for a preliminary report, see [1].

2. Extensions of mixed Hodge structures

a. Preliminary definitions

A *Hodge structure H of weight ℓ* is a pair consisting of a lattice $H_{\mathbf{Z}}$ and a decreasing filtration $F^{\cdot}$ of $H_{\mathbf{C}} = H_{\mathbf{Z}} \otimes \mathbf{C}$ such that

$$H_{\mathbf{C}} = F^p \oplus \bar{F}^q$$

when $p + q = \ell + 1$. Barring denotes conjugation with respect to $H_{\mathbf{R}} = H_{\mathbf{Z}} \otimes \mathbf{R}$. The relation above implies that the subspaces

$$H^{p,q} = F^p \cap \bar{F}^q$$

give a decomposition

$$H_{\mathbf{C}} = \bigoplus_{p+q=\ell} H^{p,q}$$

where

$$H^{p,q} = \overline{H^{q,p}}.$$

A *polarization* of H is a nonsingular, rational bilinear form S on $H_{\mathbf{Q}} = H_{\mathbf{Z}} \otimes \mathbf{Q}$ such that

(i) $S(x, y) = (-1)^{\ell} S(y, x)$
(ii) $S(x, y) = 0$ if $x \in F^p$, $y \in F^q$, and $p + q > \ell$
(iii) $i^{p-q} S(x, \bar{y})$ is hermitian positive definite on $H^{p,q}$.

A mixed Hodge structure H is a triple consisting of

(i) a lattice $H_{\mathbf{Z}}$
(ii) an increasing filtration W of $H_{\mathbf{Q}}$
(iii) a decreasing filtration $F^{\cdot}$ of $H_{\mathbf{C}}$.

These, the weight and Hodge filtrations, are compatible in the sense that $F^{\cdot}$ induces a Hodge structure of weight ℓ on each of the graded pieces

$$Gr_{\ell}^{W} = W_{\ell}/W_{\ell-1}.$$

For the sake of brevity, we write

$$H^{p,q} = (Gr_{p+q}^{W} \otimes \mathbf{C})^{p,q}$$

and we introduce the lattice

$$L^pH = H^{p,p} \cap Gr^W_{2p,\mathbf{Z}}.$$

A *polarization* of H is a set of bilinear forms $\{S_\ell\}$ which polarize the graded pieces individually.

Two measures of the complexity of a mixed Hodge structure, the *length* and *level*, are useful: To define the first, let $[a, b]$ be the smallest interval such that $W_\ell/W_{\ell-1} = 0$ for $\ell \notin [a, b]$. Then a and b are the lowest and highest weights, and $b - a$ is the length. A mixed Hodge structure of length zero is a Hodge structure of pure weight. To define the second, let $[c, d]$ be the smallest interval such that $F^p/F^{p+1} = 0$ for $p \notin [c, d]$. Then $d - c$ is the level. A level zero structure is necessarily of pure type (p, p). As such, it is rigid: there is a unique admissible Hodge filtration, namely $F^p = H_{\mathbf{C}}$. The unique structure of type (p, p) on the lattice of integers will be denoted by $T\langle p\rangle$.

A *morphism* of mixed Hodge structures,

$$\phi \colon A \longrightarrow B$$

is given by a homomorphism of lattices which preserves both filtrations: there is a fixed integer m such that

$$\phi(F^pA) \subset F^{p+m}B$$

$$\phi(W_\ell A) \subseteq W_{\ell+2m}B$$

for all p and ℓ. The *weight* of ϕ is, by definition, the integer $2m$. The category of mixed Hodge structures is abelian, and it admits both duals and tensor products. Finally, the lattice

$$\mathrm{Hom}(B, A)_{\mathbf{Z}} = \mathrm{Hom}(B_{\mathbf{Z}}, A_{\mathbf{Z}})$$

supports a canonical mixed Hodge structure with

$$W_\ell \,\mathrm{Hom}_{\mathbf{Q}} = \{\phi : \phi(W_rA) \subseteq W_{r+\ell}B \text{ for all } r\}$$

$$F^p \,\mathrm{Hom}_{\mathbf{C}} = \{\phi : \phi(F^rA) \subseteq F^{r+p}B \text{ for all } r\}.$$

b. The group of extension classes

We can now introduce the fundamental object of study: an *extension* is an exact sequence of mixed Hodge structures,

$$0 \longrightarrow A \xrightarrow{\;i\;} H \xrightarrow{\;\pi\;} B \longrightarrow 0.$$

By analogy with principal fiber bundles, we say that H is an extension of B

by A, that a *section* is a morphism

$$s: B \longrightarrow H$$

such that $\pi \circ s = 1_B$, and that an extension which admits a section is *split*. A *morphism* of extensions is a commutative diagram of morphisms, as below:

$$\begin{array}{ccccccccc} 0 & \longrightarrow & A & \longrightarrow & H & \longrightarrow & B & \longrightarrow & 0 \\ & & {\scriptstyle\alpha}\downarrow & & {\scriptstyle\eta}\downarrow & & {\scriptstyle\beta}\downarrow & & \\ 0 & \longrightarrow & A' & \longrightarrow & H' & \longrightarrow & B' & \longrightarrow & 0. \end{array}$$

An isomorphism for which α and β are each the identity is called a *congruence*; a split extension is congruent to the trivial one given by the direct sum

$$H = A \oplus B.$$

Finally, let us define

$$\mathrm{Ext}(B, A)$$

to be the set of congruence classes of extension of B by A. Its group structure is given by the result below.

PROPOSITION 1: *Baer summation imposes the structure of an abelian group on* Ext(B, A), *with zero given by the class of split extensions. Moreover,* Ext($*$, $*$) *is a functor, contravariant in the first variable, and covariant in the second.*
The proof depends only on the fact that the category of mixed Hodge structures is abelian: see the proof of the same proposition for R-modules in [4], p. 63.

Now the extensions which arise naturally from a fixed mixed Hodge structure,

$$0 \longrightarrow W_m \longrightarrow H \longrightarrow H/W_m \longrightarrow 0.$$

are *separated* in the sense that the highest weight of A is less than the lowest weight of B. In this case Ext is naturally isomorphic to a generalized complex torus (the quotient of a complex vector space by a discrete subgroup). To describe it, we define the p-th Jacobian of a mixed Hodge structure to be

$$J^pH = H_{\mathbf{C}}/(F^pH + H_{\mathbf{Z}}).$$

Whenever

$$p > \tfrac{1}{2}(\text{highest weight of } H),$$

the lattice $H_{\mathbf{Z}}$ projects without kernel to a discrete subgroup of $H_{\mathbf{C}}/F^p$, so J^pH is indeed a generalized torus. The fine structure is then given by the result below:

PROPOSITION 2: *Let A and B be separated mixed Hodge structures. Then there is a canonical and functorial isomorphism*

$$\operatorname{Ext}(B, A) \cong J^0 \operatorname{Hom}(B, A).$$

EXAMPLE: Consider extensions of $T\langle p\rangle$ by $T\langle q\rangle$, where $p > q$. There is a canonical isomorphism

$$\operatorname{Hom}(T\langle p\rangle, T\langle q\rangle)_{\mathbf{C}} \longrightarrow T\langle r\rangle_{\mathbf{C}},$$

where $r = q - p$. Since $r < 0$, $F^0T\langle r\rangle = 0$, and so

$$J^0 \operatorname{Hom} = \frac{T\langle r\rangle_{\mathbf{C}}}{T\langle r\rangle_{\mathbf{Z}}} \cong \mathbf{C}^*.$$

We conclude that there is a nontrivial one-parameter family of extensions, even though $T\langle p\rangle$ and $T\langle q\rangle$ are themselves rigid.

PROOF OF PROPOSITION 2: We shall construct a natural action of $J^0 \operatorname{Hom}$ on Ext which is both transitive and effective. The resulting correspondence between $J^0 \operatorname{Hom}$ and the orbit of the class of split extensions then gives the required identification. To this end, define a *normalized extension* of B by A to be one with underlying lattice.

$$L_{\mathbf{Z}} = A_{\mathbf{Z}} \oplus B_{\mathbf{Z}}.$$

Because of the separation hypothesis, the weight filtration on $L_{\mathbf{Q}}$ is determined by that on A and B:

$$W_mL_{\mathbf{Q}} = W_mA_{\mathbf{Q}} \oplus W_mB_{\mathbf{Q}}$$

is the only choice which makes

$$0 \longrightarrow A_{\mathbf{Q}} \xrightarrow{\;i\;} L_{\mathbf{Q}} \xrightarrow{\;\pi\;} B_{\mathbf{Q}} \longrightarrow 0$$

an exact sequence of filtered objects. Consequently, a normalized extension is determined by a decreasing filtration on $L_{\mathbf{C}}$ such that

$$F^pL \cap A_{\mathbf{C}} = F^pA$$

$$\pi(F^pL) = F^pB.$$

We shall denote the set of normalized extensions by

$$\widetilde{\mathrm{Ext}}(B, A).$$

Next, we claim that an arbitrary extension is congruent to a normalized one. To see this, define an *integral retraction* to be a homomorphism

$$r_{\mathbf{Z}}: H_{\mathbf{Z}} \longrightarrow A_{\mathbf{Z}}$$

such that $r_{\mathbf{Z}} \circ i = \mathrm{id}$. Transport F^pH to a filtration of $L_{\mathbf{C}}$ using the isomorphism

$$(r, \pi): H_{\mathbf{Z}} \longrightarrow L_{\mathbf{Z}}.$$

This defines a normalized extension congruent to the given one, consequently determines a natural isomorphism

$$\mathrm{Ext}(B, A) \cong \widetilde{\mathrm{Ext}}(B, A)/\text{congruence}.$$

Therefore it suffices to define an action on the right-hand side. To do this, consider the automorphisms of $L_{\mathbf{C}} = L_{\mathbf{Z}} \otimes \mathbf{C}$ given by

$$g(\psi) = \begin{bmatrix} 1_A & \psi \\ 0 & 1_B \end{bmatrix}$$

where

$$\psi \in \mathrm{Hom}(B, A)_{\mathbf{C}}.$$

Then $\mathrm{Hom}_{\mathbf{C}}$ acts on $\widetilde{\mathrm{Ext}}$ by

$$F^{\cdot} \longrightarrow g(\psi)F^{\cdot}.$$

One easily verifies that the action is transitive and that the isotropy group of the trivial extension,

$$F_0^{\cdot} = F^{\cdot}A \oplus F^{\cdot}B,$$

is F^0 Hom. Consequently there is a natural identification

$$\widetilde{\mathrm{Ext}}(B, A) \cong \mathrm{Hom}(B, A)_{\mathbf{C}}/F^0\,\mathrm{Hom}(B, A).$$

To complete the proof, observe that congruences of normalized extensions are given by matrices $g(\psi)$ which are integrally defined. Consequently

$$\frac{\widetilde{\operatorname{Ext}}(B, A)}{\text{congruence}} \cong \frac{\operatorname{Hom}(B, A)_{\mathbf{C}}}{F^0 \operatorname{Hom}(B, A) + \operatorname{Hom}(B, A)_{\mathbf{Z}}},$$

as desired

REMARK: The group law agrees with that given by Baer summation. To see this, recall that in any abelian category, pushforwards and pullbacks of extensions exist, as do direct sums, and the diagonal and codiagonal maps:

$$\Delta(x) = (x, x)$$
$$\nabla(x, y) = x + y.$$

The Baer sum of extensions H_1 and H_2 is gotten by pulling back the direct sum along the diagonal, and pushing it forward along the codiagonal:

$$H_1 + H_2 = \nabla_* \Delta^*(H_1 \oplus H_2).$$

But the direct sum is represented by the matrix $g(\psi_1) \oplus g(\psi_2)$, and application of $\nabla_* \Delta^*$ yields $g(\psi_1 + \psi_2)$, as desired.

c. Motifs

Let us define a *weak motif* to be a homomorphism

$$u : L \longrightarrow J$$

of a lattice to a generalized complex torus. Morphisms of such are commutative diagrams of homomorphisms,

$$\begin{array}{ccc} L_1 & \longrightarrow & J_1 \\ {\scriptstyle f_L}\downarrow & & \downarrow{\scriptstyle f_J} \\ L_2 & \longrightarrow & J_2 \end{array}$$

in which f_J is holomorphic with closed image. The category of all such is abelian. Moreover, if L is torsion-free, then the set of weak motifs with values in J is itself a generalized complex torus: choose a basis $\lambda_1, \ldots, \lambda_m$ for L and send u to $(u(\lambda_1), \ldots, u(\lambda_m))$ to get

$$\operatorname{Hom}(L, J) \xrightarrow{\cong} (J)^m.$$

If in addition J is an extension of an abelian variety by a complex multiplicative group (product of $\mathbf{C}^*$'s), then u is called a *one-motif.* By definition, a morphism of such objects preserves the multiplicative extensions.

Let us now describe the weak motif defined by an extension of mixed Hodge structures:

PROPOSITION 3: *Let H be a separated extension. Then there is a functorially defined weak motif*

$$u_H : L^p B \longrightarrow J^p A$$

which depends only on the congruence class of H.

PROOF: By proposition 2, there is a homomorphism

$$\psi_H : B_{\mathbf{C}} \longrightarrow A_{\mathbf{C}}$$

which represents H, unique up to addition of elements in

$$F^0 \operatorname{Hom}(B, A) \longrightarrow \operatorname{Hom}(B, A)_{\mathbf{Z}}.$$

Consequently the formula

$$u_H(\beta) = \psi_H(\beta) + F^p A + A_{\mathbf{Z}}$$

gives a homomorphism, independent of the choice of ψ, with the required properties.

In fact, the construction yields more: First, the correspondence $H \rightarrow u_H$ determines a holomorphic homomorphism

$$\operatorname{Ext}(B, A) \longrightarrow \operatorname{Hom}(L^p B, J^p A).$$

Second, pullback of extensions along the inclusion of mixed Hodge structures

$$k : L^p B \longrightarrow B$$

yields a factorization

$$\begin{array}{ccc} \operatorname{Ext}(B, A) & \xrightarrow{k^*} & \operatorname{Ext}(L^p B, A) \\ & \searrow \qquad \swarrow {\scriptstyle \cong} & \\ & \operatorname{Hom}(L^p B, J^p A) & \end{array}$$

Lifting homomorphisms of L^pB from J^pA to A_C, and making use of proposition 2, one easily verifies that the right-hand map is an isomorphism of generalized tori. In particular, we see that an extension is determined by its weak motif if and only if B is of pure type (p, p).

REMARKS: (1) The last statement must be interpreted with some care: it does not express an isomorphism of functors, because not all morphisms of weak motifs are induced by morphisms of mixed Hodge structures. More precisely, there are complex homomorphisms $J^pA \to J^pA'$ which are not induced by morphisms $A \to A'$.

(2) For one-motifs, Deligne proved the stronger statement. Indeed, he exhibited an equivalence of categories between mixed Hodge structures of level one and one-motifs. (*cf.* [2] section 10).

(3) The dual extension

$$0 \longrightarrow \hat{B} \longrightarrow \hat{H} \longrightarrow \hat{A} \longrightarrow 0$$

defines a weak motif

$$\hat{u}_H : L^{-p}\hat{A} \longrightarrow J^{-p}\hat{B}.$$

Its value on any element γ is represented by a linear functional u_γ on B_C. To describe it, let $\langle\ ,\ \rangle$ denote the pairing between a module and its dual, and let $\hat{\psi}_H$ be a homomorphism representing the dual extension. Then u_γ acts by

$$\begin{aligned} u_\gamma(\beta) &= \langle \beta, \hat{\psi}_H(\gamma)\rangle \\ &= \langle \psi_H(\beta), \gamma\rangle. \end{aligned} \tag{*}$$

This remark, although trivial, will be useful in the applications.

d. A formula for ψ

To calculate the extensions which arise in geometry, we need a formula for the representing homomorphism ψ. This will follow from the following direct characterization of ψ. To this end, we define a *section of the Hodge filtration* to be a homomorphism

$$s_F : B_C \longrightarrow H_C$$

such that $\pi \circ s_F = id$ and such that

$$s_F(F^pB) \subseteq F^pH$$

for all p.

LEMMA 4: *Fix a separated extension H of B by A. Then all homomorphisms which represent H are of the form*

$$\psi = r_Z \circ s_F,$$

where r_Z is an integral retraction, where s_F is a section of the Hodge filtration.

There is a useful alternative form of the lemma which characterizes $\psi(\omega)$ as a linear functional on $\hat{A}$: Let $s_Z = \hat{r}_Z$ be the adjoint section of the dual lattice structure. Then $\psi(\omega)$ acts by

$$\gamma \longrightarrow \langle s_F(\omega), s_Z(\gamma)\rangle.$$

In a typical geometric situation, H is the k-th cohomology of an algebraic variety, $s_F(\omega)$ is represented by a suitable deRham element, and the canonical pairing is represented by integration:

$$\psi(\omega)(\gamma) = \int_{S_Z(\gamma)} s_F(\omega). \qquad (*)$$

Thus the extension is calculated in terms of the period matrix of a singular variety.

PROOF OF LEMMA 4: We must show that $g(\psi)$ carries the reference filtration on L to the filtration of a normalized extension congruent to H. To do this efficiently, view elements of L as column vectors ${}^t(a, b)$. Then the column-vector of homomorphisms ${}^t(r, \pi)$ defines a congruence between H and a normalized extension with filtration F^pL. Moreover, the row vector of homomorphisms (i, s) sends $L_{\mathbf{C}}$ isomorphically to $H_{\mathbf{C}}$, carrying the reference filtration to the given one on H, since

$$i(F^pA) + s(F^pB) \subseteq F^pH.$$

But

$$\begin{pmatrix} r \\ \pi \end{pmatrix} (i, s) = \begin{bmatrix} ri & rs \\ \pi i & \pi s \end{bmatrix} = \begin{bmatrix} 1_A & \psi \\ 0 & 1_B \end{bmatrix} = g(\psi).$$

Consequently

$$\begin{aligned} g(\psi)(F^pA \oplus F^pB) &= \begin{pmatrix} r \\ \pi \end{pmatrix} (i, s)(F^pA \oplus F^pB) \\ &= \begin{pmatrix} r \\ \pi \end{pmatrix} F^pH \\ &= F^pL. \end{aligned}$$

This completes the proof.

With the lemma in hand, we can give the formula. First, choose a basis $\{\gamma^i\}$ of A_Z, let $\{\gamma_i\}$ be the dual basis in $\hat{A}_Z$, and let $\{\Gamma_i\}$ be a lifting to $\hat{H}_Z$. Then

$$r_Z(\Omega) = \sum \gamma^i \langle \Omega, \Gamma_i \rangle$$

defines an integral retraction, and in fact all such are obtained in this way. Second, let $\omega \in F^p B$, and let Ω be a lifting to $F^p H$. Since there is always a section of the Hodge filtration which sends ω to Ω, there is a representing homomorphism with

$$\psi(\omega) = \sum \gamma^i \langle \Omega, \Gamma_i \rangle.$$

This is the desired formula. In the geometric case it becomes

$$\psi(\omega) = \sum \gamma^i \int_{\Gamma_i} \Omega,$$

where $\Omega = s_F(\omega)$, $\Gamma_i = s_Z(\gamma_i)$.

e. Jacobians

We conclude this section with a few remarks on Jacobians, beginning with a proof that $J^p H$ is indeed a generalized torus.

LEMMA 5: *Let H be a mixed Hodge structure of highest weight m. If $p > m/2$ then the natural map*

$$\rho : H_{\mathbf{R}} \longrightarrow H_{\mathbf{C}}/F^p$$

is an injection of real vector spaces.

PROOF: An element of the kernel of ρ lies in $F^p \cap H_{\mathbf{R}}$, hence in $F^p \cap \bar{F}^p$. Let ℓ be the weight of x, and consider the projection of x in Gr_ℓ^W. An element of $F^p \cap \bar{F}^p Gr_\ell$ is a sum of components with type (a, b), where $a \geqq p$ and $b \geqq p$. Therefore $\ell \geqq 2p > m$, and consequently $Gr_\ell^W = 0$. It follows that $x = 0$, as desired.

LEMMA 6: *If $p > m/2$, then the p-th Jacobian of H is a generalized torus.*

PROOF: Let K be a complement of $\rho(H_{\mathbf{R}})$ in the real vector space of $H_{\mathbf{C}}/F^p$. Then the decomposition

$$H_{\mathbf{C}}/F^p \cong H_{\mathbf{R}} \oplus K$$

yields an isomorphism of real manifolds

$$J^pH \cong (H_{\mathbf{R}}/H_{\mathbf{Z}}) \oplus K$$
$$\cong (S^1)^a \times \mathbf{R}^b.$$

Since $H_{\mathbf{Z}}$ acts discretely in $H_{\mathbf{R}}$, $\rho(H_{\mathbf{Z}})$ acts discretely in $H_{\mathbf{C}}/F^p$, as desired.

REMARKS: (1) If $F^p = 0$, then J^pH is of multiplicative type:

$$J^pH = H_{\mathbf{C}}/H_{\mathbf{Z}} \cong (\mathbf{C}^*)^m.$$

This is always the case when H has level zero and weight less than $2p$.

(2) If $H_{\mathbf{C}} = F^p \oplus \bar{F}^p$, then a dimension count shows that $b = 0$ in the above lemma, so that J^pH is compact. In particular, J^pH is compact when H is a Hodge structure of weight $2p - 1$.

(3) Let U be a complex vector space, $L \subset U$ a lattice, $J = U/L$ a generalized torus. The dual torus is then the quotient of the dual vector space by the dual lattice:

$$\hat{J} = \hat{U}/\hat{L}.$$

An integral unimodular bilinear form on U gives an isomorphism $U \to \hat{U}$ which carries L to $\hat{L}$, hence an isomorphism between J and $\hat{J}$.

To apply this to Jacobians, recall that the annihilator of F^pH in $\hat{H}$ is $F^{1-p}\hat{H}$. Consequently the torus dual to J^pH is $J^{1-p}\hat{H}$. Moreover, if H is a Hodge structure of weight $2p - 1$ with a unimodular polarization, then there is a canonical isomorphism

$$J^pH \longrightarrow J^{1-p}\hat{H}.$$

Note that the polarization makes J^pH an abelian variety when H is a level one structure.

Next, we claim that the Jacobian is filtered by the subobjects

$$W_\ell J^pH = (W_\ell \otimes \mathbf{C} + F^p + H_{\mathbf{Z}})/(F^p + H_{\mathbf{Z}}).$$

The lemma below shows that they are also generalized tori because it implies that

$$W_\ell J^pH = J^p W_\ell H.$$

Consequently there are holomorphic fibrations

$$0 \longrightarrow W_{\ell-1}J^pH \longrightarrow W_\ell J^pH \longrightarrow Gr_\ell J^pH \longrightarrow 0.$$

Because of the strictness property of morphisms ([3] p. 39), we also have

$$Gr_\ell J^p H \cong J^p Gr_\ell H.$$

EXAMPLE: Let H be a polarized mixed Hodge structure of level one and highest weight $2p$. Then the canonical fibration gives

$$0 \longrightarrow J^p W_{2p-2} \longrightarrow J^p W_{2p-1} \longrightarrow J^p Gr^W_{2p-1} \longrightarrow 0,$$

where the quotient is an abelian variety and the subobject is a torus of multiplicative type. The weak motif

$$u : L^p H \longrightarrow J^p W_{2p-1}$$

then carries the structure of a one-motif.

LEMMA 7: *Let H be a mixed Hodge structure of highest weight $2p-1$. Then*

$$W_\ell \cap (F^p + H_{\mathbf{Z}}) = (W_\ell \cap F^p) + (W_\ell \cap H_{\mathbf{Z}}).$$

PROOF: Write $x \in W_\ell \cap (F^p + H_{\mathbf{Z}})$ as $y+z$, where $y \in F^p$ and $z \in H_{\mathbf{Z}}$. Suppose $z \in W_m$, and consider the congruence $x \equiv y+z$ modulo W_{m-1}. If $m > \ell$, then $x \equiv 0$, hence $y+z \equiv 0$. Consider now the components of x, y, and z in the I^{rs} decomposition (see [3] p. 37), where $r+s=m$. Because $y_{rs} \equiv 0$ for $r<p$, we have also $z_{rs} \equiv 0$ for $r<p$. Conjugating the last relation, we obtain $z_{rs} \equiv 0$ for $s<p$ as well. But since $r+s<2p$, one of these alternatives must hold for each z_{rs}, so that z itself is congruent to zero modulo W_{m-1}. But then $y+z \equiv 0$ implies $y \equiv 0$. But now both y and z lie in W_{m-1}, provided that $m > \ell$. We conclude that both elements lie in W_ℓ, as desired.

3. Applications to curves

a. Topology

To prove the Torelli theorem for curves mentioned in the introduction, we need a very explicit description of the mixed Hodge structure. We begin by discussing the topology, an understanding of which will yield both the weight filtration and the polarizations.

Thus, let X be a curve, Σ its singular locus,

$$\pi : \tilde{X} \longrightarrow X$$

its normalization, $\tilde{\Sigma} = \pi^{-1}(\Sigma)$, and

$$i : \tilde{\Sigma} \longrightarrow \tilde{X}$$

the natural inclusion.

LEMMA 8: *There is an exact sequence of singular homology,*

$$0 \longrightarrow H_1(\tilde{X}) \xrightarrow{\pi_*} H_1(X) \xrightarrow{\partial} H_0(\tilde{\Sigma}) \xrightarrow{\delta_*} H_0(\tilde{X}) \oplus H_0(\Sigma),$$

where $\delta_* = i_* \oplus (-\pi_*)$.

PROOF: Consider the diagram

(*)

$$\begin{array}{ccccc} & & \tilde{\Sigma} & & \\ & {}^{i}\swarrow & & \searrow^{\pi} & \\ \tilde{X} & & & & \Sigma \\ & {}_{\pi}\searrow & & \swarrow_{i} & \\ & & X & & \end{array}$$

From a formal point of view, this represents the nerve of a generalized cover, and the sequence above is its Mayer–Vietoris sequence. To make this remark precise, consider the space $|X|$ obtaining by glueing $\tilde{\Sigma} \times [0, 1]$ to $\tilde{X}$ and Σ along the maps

$$i : \tilde{\Sigma} \times \{0\} \longrightarrow \tilde{X}$$

$$\pi : \tilde{\Sigma} \times \{1\} \longrightarrow \Sigma.$$

There are natural projections

$$p : |X| \longrightarrow [0, 1]$$

$$q : |X| \longrightarrow X,$$

and the fiber of q over $x \in X - \Sigma$ is a single point, whereas the fiber of q over $x \in \Sigma$ is a cone over $\pi^{-1}(x)$. In either case the fiber is contractible, so q is a homotopy equivalence. Now cover $|X|$ by the open sets

$$U_0 = p^{-1}(0 \leqq t < 1)$$

$$U_1 = p^{-1}(0 < t \leqq 1).$$

The nerve of this cover is

$$\begin{array}{ccc} & U_0 \cap U_1 & \\ \swarrow & & \searrow \\ U_0 & & U_1 \\ \searrow & & \swarrow \\ & U_0 \cup U_1 & \\ & \| & \\ & |X| & \end{array}$$

which is homotopy equivalent to the diagram (*). Applying these equivalences to the usual Mayer–Vietoris sequence of the cover, we obtain the Mayer–Vietoris sequence of the generalized cover.

Now each term in the generalized Mayer–Vietoris sequence corresponding to a smooth variety carries natural Hodge structure. Moreover, since the category of Hodge structures is abelian, the group

$$\mathscr{X}_X = \mathrm{kernel}(\delta_*)$$

carries a natural structure of pure type $(0, 0)$. Thus, we have an exact sequence

$$0 \longrightarrow H_1(\tilde{X}, \mathbf{Z}) \xrightarrow{\pi_*} H_1(X, \mathbf{Z}) \xrightarrow{\partial} \mathscr{X}_X \longrightarrow 0$$

with Hodge structures of weights -1 and 0 on the end terms. Consequently it is natural to define a weight filtration (over $\mathbf{Z}$) by

$$W_{-1} = \pi_* H_1(\tilde{X}, \mathbf{Z})$$
$$W_0 = H_1(X, \mathbf{Z}).$$

Next, let us define natural polarizations. On the weight -1 term this is clear: transport the intersection product of one-cycles on $\tilde{X}$ to X by means of the isomorphism

$$\pi_* : H_1(\tilde{X}) \xrightarrow{\cong} W_{-1}.$$

On the weight zero part, we proceed as follows: Let S_0 be the unique symmetric bilinear form on $H_0(\tilde{\Sigma}, \mathbf{Z})$ for which the classes of points form an orthonormal basis. Then carry S_0 back by the isomorphism

$$\partial : W_0 / W_{-1} \xrightarrow{\cong} \mathscr{X}_X.$$

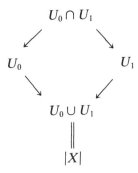

Implicit in this construction is the identification of $\mathscr{X}_X$ with a special group of zero cycles of degree zero on $\tilde{X}$. Such a cycle ξ is characterized by the following conditions:

(i) the support of ξ is in $\tilde{\Sigma}$
(ii) the degree of ξ on each component of $\tilde{X}$ is zero
(iii) the degree of ξ on each fiber of π is zero.

In general, the form S_0 is complicated. However, let us calculate it when X is an irreducible curve with ordinary singularities. To this end, let y be a singular point, and order the points of $\pi^{-1}(y)$ as $x_0, \ldots, x_{m-1}$, where m is the multiplicity. Define

$$\mathscr{X}(y) = \left\{ \sum n_i X_i \,\middle|\, \sum n_i = 0 \right\}.$$

An easy calculation shows that

$$\mathscr{X}_X = \bigoplus_{y \in \Sigma} \mathscr{X}(y).$$

Moreover, since distinct summands are orthogonal under S_0, it suffices to describe S_0 on one such. To do this, introduce the basis

$$e_i = x_i - x_{i-1},$$

where $i = 1, \ldots, m-1$. Then

$$\langle e_i, e_i \rangle = 2,$$
$$\langle e_i, e_{i-1} \rangle = -1, \text{ and}$$
$$\langle e_i, e_j \rangle = 0$$

if $|i-j| > 1$. Consequently the matrix of S_0 in this basis is

$$\begin{bmatrix} 2 & -1 & & & & \\ -1 & 2 & -1 & & & \\ & -1 & 2 & \cdot & & \\ & & & \cdot & & \\ & & & & \cdot & \\ & & & & & \cdot \end{bmatrix}$$

the matrix of the Dynkin diagram A_{m-1}. Note that the vectors e_i form a system of primitive, positive roots, and that any root vector is of the form $x_i - x_j$. These vectors are characterized by the fact that their length is minimal among all nonzero lattice vectors. Note also that any system of primitive, positive roots is conjugate to the given one by a general element of the orthogonal group of S_0 on $\mathscr{X}(y)$. Such elements are faithfully represented by permutations of the set $\pi^{-1}(y)$.

b. The mixed Hodge structure

Let us now describe the mixed Hodge structure on cohomology. The dual of our basic sequence gives

$$0 \longrightarrow \hat{\mathscr{X}}_X \xrightarrow{\hat{\partial}} H^1(X, \mathbf{Z}) \xrightarrow{\pi^*} H^1(\tilde{X}, \mathbf{Z}) \longrightarrow 0,$$

hence a weight filtration

$$W_0 = \text{kernel}(\pi^*)$$
$$W_1 = H^1(X).$$

It therefore remains to define an interpolating Hodge filtration on $H^1(X)$, compatible with the given filtrations on the end terms. To do this, we define an *abelian differential* on X to be a holomorphic form on $X - \Sigma$ which is square-integrable:

$$\int_{X-\Sigma} \omega \wedge \bar{\omega} < \infty.$$

The line integral of such a form over a closed path in X is defined and depends only on its homology class. Consequently the abelian differentials span a subspace of the complex cohomology, which we denote by F^1. This gives the desired filtration:

$$F^1 = \{\text{classes of abelian differentials}\}$$
$$F^0 = H^1(X, \mathbf{C}).$$

Let us now verify the necessary compatibility statements:

(a) $F^1 \cap W_0 = 0$

(b) $W_1/W_0 = F^1(W_1/W_0) \oplus \bar{F}^1(W_1/W_0)$.

These are jointly equivalent to the assertion that $F^{\cdot}$ induces Hodge structures of weight zero and one on subspace and quotient. Equivalently, we must verify

(a′) $F^1 \cap \ker \pi^* = 0$

(b′) $\pi^* F^1 H^1(X) = F^1 H^1(\tilde{X})$.

(Use (b′) and its conjugate statement to recover (b)). Consequently it suffices to prove that

$$\pi^*: F^1H^1(X) \longrightarrow F^1H^1(\tilde{X})$$

is an isomorphism. But this is easy in view of the relation of improper integrals

$$\int_{\tilde{X}-\tilde{\Sigma}} \pi^*\omega \wedge \overline{\pi^*\omega} = \int_{X-\Sigma} \omega \wedge \bar{\omega}.$$

Because square-integrable holomorphic forms on $\tilde{X} - \tilde{\Sigma}$ extend to abelian differentials on $\tilde{X}$, π^* has the asserted range. Because π is a homeomorphism on the complement of $\tilde{\Sigma}$, abelian differentials on $\tilde{X}$ define abelian differentials on X. This establishes surjectivity. Finally, suppose that $\pi^*\omega$ vanishes in cohomology. Then the L^2-norm of $\pi^*\omega$ vanishes, and so it vanishes as a differential form. Consequently ω also vanishes as a differential, and this establishes injectivity.

c. Comparison of motifs

The final tool which we need for the Torelli theorem is a comparison between the Hodge-theoretic motif and the Abel–Jacobi homomorphism

$$\mathcal{A}: \mathcal{Z}_X \longrightarrow \mathrm{Jac}(X).$$

PROPOSITION 9: *Let X be a projective algebraic curve. Then there is a natural isomorphism of one-motifs*

$$\hat{u}_H \xrightarrow{\cong} \mathcal{A}.$$

PROOF: The isomorphism is equivalent to the commutativity of the diagram

$$\begin{array}{ccc} L^0H_1(X) & \xrightarrow{\hat{u}_H} & J^0W_{-1}H_1(X) \\ \downarrow \partial & & \downarrow \cong \\ \mathcal{Z}_X & \xrightarrow{\mathcal{A}} & \mathrm{Jac}(\tilde{X}), \end{array}$$

where the right hand isomorphism is the composition of the map induced by projection,

$$\pi_*: J^0H_1(\tilde{X}) \xrightarrow{\cong} J^0W_{-1}H_1(X),$$

and the duality induced by integration,

$$J^0H_1(\tilde{X}) \cong H^{1,0}(\tilde{X})^*/H_1(\tilde{X}, \mathbf{Z}).$$

According to the prescriptions of section 2c the motivic functional is given by

$$u_\gamma : \omega \longmapsto \langle \psi(\omega), \gamma \rangle$$

where $\gamma \in L^0H_1(X)$, $\omega \in F^1Gr_1H^1(X)$. According to section 2*d*, this can be written

$$u_\gamma : \omega \longmapsto \int_{S_Z(\gamma)} s_F(\omega).$$

But the right-hand side is just the integral of ω, viewed as an abelian differential on $\tilde{X}$, over a chain whose boundary is the zero-cycle $\xi = \partial\gamma$. In other words, u_γ acts, under the appropriate identifications, as does the Abel–Jacobi functional,

$$\mathcal{A}_\xi : \omega \longrightarrow (\text{abelian sum of } \omega \text{ with respect to } \xi).$$

This completes the proof.

REMARKS: (1) Since $H^1(X)$ has level one, it is completely determined by its one-motif.

(2) This in turn implies that the group of extensions can be represented as a sum of Jacobians:

$$\begin{aligned} \operatorname{Ext}(H^1(\tilde{X}), W_0) &\cong \operatorname{Ext}(W_0/W_{-1}, H_1(\tilde{X}) \\ &\cong \operatorname{Hom}(\mathcal{Z}_X, H_1(\tilde{X})) \\ &\cong (\operatorname{Jac}(\tilde{X}))^m, \end{aligned}$$

where $m = \operatorname{rank} \mathcal{Z}_X$. All isomorphisms are canonical except the last, which depends on a choice of basis for $\mathcal{Z}_X$.

d. Proof of the Torelli theorem

We will now prove theorem A by using the one-motif to construct a diagram isomorphic to

$$\begin{array}{ccc} & \tilde{\Sigma} & \\ {}^{i}\swarrow & & \searrow^{\pi} \\ \tilde{X} & & \Sigma. \end{array}$$

This suffices, since X is the variety obtained by contracting $\tilde{\Sigma} \subset \tilde{X}$ along π to Σ:

$$X \cong \tilde{X} \bigcup_{\tilde{\Sigma}} \Sigma.$$

To get $\tilde{X}$, we apply the classical Torelli theorem to the polarized Hodge structure on W_1/W_0. To get Σ, we observe that ∂ places the indecomposable summands on $L^0H_1(X)$ in one-to-one correspondence with the points of Σ:

$$y \longleftrightarrow \mathscr{L}(y).$$

Finally, to construct $\tilde{\Sigma}$ and its associated maps, we employ the motivic homomorphism connecting the pieces of pure weight: Let Λ be an indecomposable summand of $L^0H_1(X)$, corresponding to a point $y \in \Sigma$, and let ξ be nonzero vector of minimal length in Λ. Then

$$\partial\xi = b - a,$$

with $a, b \in \pi^{-1}(y)$. Comparing the Hodge-theoretic motif with the Abel–Jacobi map, we see that $\hat{u}(\xi)$ determines the class of $b - a$ in the Jacobian of $\tilde{X}$. Now consider the Abel–Jacobi map on the Cartesian square which sends (u, v) to the class of $v - u$:

$$\mathscr{A}: \tilde{X} \times \tilde{X} \longrightarrow \mathrm{Jac}(\tilde{X}).$$

An easy argument using the fact that $\tilde{X}$ is nonhyperelliptic shows that $\mathscr{A}$ descends to an imbedding of the symmetric square of $\tilde{X}$, modulo the diagonal, which maps to zero. Consequently $\hat{u}(\xi)$ in fact determines the point-set $\{a, b\}$ in $\tilde{X}$. The union of these point-sets for all minimal vectors in Λ is just the fiber $\pi^{-1}(y)$, as a subset of $\tilde{X}$. Thus we have built the diagram

$$\begin{array}{ccc} & \pi^{-1}(y) & \\ \swarrow & & \searrow \\ \tilde{X} & & \{y\} \end{array}$$

from the one-motif

$$\hat{u}: \Lambda \longrightarrow J^0W_{-1}H_1(X),$$

together with the polarization of Λ. Amalgamating these diagrams in the obvious way, we obtain the desired presentation of X.

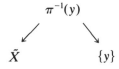

REMARKS: (1) Suppose that $\tilde{X}$ is hyperelliptic, but of genus at least two. In general, the points of $\pi^{-1}(y)$ are determined only up to the canonical involution, so that finitely many curves correspond to a given mixed Hodge structure.

(2) Suppose that $\tilde{X}$ is elliptic. Let $|\Sigma|$ and $|\tilde{\Sigma}|$ denote the number of points in the given set. Then X defines a marking of $\tilde{X}$ by $|\tilde{\Sigma}|$ points. Since $\tilde{X}$ has a one-parameter family of automorphisms, the number of moduli of X is $|\tilde{\Sigma}|$. On the other hand, the number of moduli of $H^1(X)$ is the number of moduli of its one-motif, which is

$$\text{genus}(\tilde{X}) + \text{rank}(\mathscr{L}_X) = 1 + |\tilde{\Sigma}| - |\Sigma|.$$

This equals the number of moduli of X precisely when there is one singular point. If there are more, then infinitely many curves correspond to the polarized mixed Hodge structure on $H^1(X)$.

(3) Suppose X is an elliptic curve with one ordinary singular point. Then Torelli holds: Pick a base point x_0 and identify $\tilde{X}$ with its Jacobian by $x \to$ class of $x - x_0$. Then choose a system of positive simple roots and define points $x_1, \ldots, x_n$ on $\tilde{X}$ inductively by

$$x_i = \hat{u}(\xi_i) + x_{i-1}.$$

Contraction of $\{x_0, \ldots, x_n\} \subset \tilde{X}$ to a point produces X.

(4) If $\tilde{X}$ is rational, then $H^1(X)$ is of level zero, hence deprived of moduli. On the other hand, X marks $|\tilde{\Sigma}|$ points on $\tilde{X}$, and so depends on $|\tilde{\Sigma}| - 3$ moduli. The Torelli theorem for $H^1(X)$ therefore fails completely.

REFERENCES

[1] JAMES A. CARLSON: The obstruction to splitting a mixed Hodge structure over the integers, I. University of Utah preprint, Spring 1977, revised edition, Summer 1979 (120 pp.).

[2] PIERRE DELIGNE: "Theorie de Hodge II, III," Publ. de l'Inst. des Hautes Études Scientifiques, *40* (1972) 5–57 and *44* (1975) 6–77.

[3] PHILLIP GRIFFITHS and WILFRIED SCHMID: "Recent developments in Hodge theory: A discussion of techniques and results," *Proc. International Colloquium on Discrete subgroups of Lie Groups* (Bombay).

[4] SAUNDERS MACLANE: "Homology", Springer-Verlag, Berlin–Heidelberg–New York (1975).

[5] WILFRIED SCHMID: "Variation of Hodge structure: The singularities of the period mapping," Inventiones Math., *22* (1973) 211–319.

Dept. of Math.
University of Utah
Salt Lake, Utah 84112 (USA)

THE HODGE CONJECTURE FOR FANO COMPLETE INTERSECTIONS OF DIMENSION FOUR

A. Conte and J. P. Murre

Let X be a smooth projective variety defined over the field $\mathbf{C}$ of complex numbers. Consider the Hodge decomposition:

$$H^i(X, \mathbf{C}) = \sum_{p+q=i} H^{p,q}(X).$$

Let $CH^p(X)$ denote the Chow group of algebraic cycles on X, modulo rational equivalence, of codimension p. Consider the standard map:

$$\lambda_X^p : CH^p(X) \otimes \mathbf{Q} \longrightarrow H^{2p}(X, \mathbf{Q}) \cap H^{p,p}(X).$$

The Hodge (p, p)-conjecture states that the map λ_X^p is surjective, i.e. that every rational cohomology class of type (p, p) comes from an algebraic cycle with rational coefficients. If $\dim X = n$, the Hodge (p, p)-conjecture is always true for $p = 0, 1, n-1, n$.

In [4] one of the authors showed that the Hodge $(2, 2)$-conjecture holds for all unirational fourfolds (fourfold = algebraic variety of dimension 4). In [2] we showed that the $(2, 2)$-conjecture holds for all fourfolds which can be covered by an algebraic family of rational curves or, what is the same, for all uniruled fourfolds (remember that an algebraic variety X is called *uniruled* if there exists a generically surjective rational map $\mathbf{P}^1 \times Y \to X$, where Y is an algebraic variety of dimension one less than the dimension of X).

Remember that an algebraic variety X is called a *Fano variety* if its anticanonical sheaf $-K_X$ is ample.

In this paper we show (theorem 1) that the complete intersections (of any dimension) which can be covered by a family of conics are exactly the ones which are Fano Varieties. It follows that the Hodge $(2, 2)$-conjecture holds for all Fano complete intersections of dimension four. This result was announced in (1).

The proof of theorem 1 is based on an idea which goes back to Predonzan [5] and which was also used by Tennison [6]. The method of proof is a generalization of the one by which we proved in [2] that all quintic fourfolds can be covered by conics.

Throughout the paper we will work over the field $\mathbf{C}$ of complex numbers.

1. Let Q be the variety (the Hilbert scheme) of all complete intersections $X \subseteq \mathbf{P}^r$, $r \geq 3$, of dimension $r-\nu$ and type $(n_1,\ldots,n_\nu)$, with $1 \leq \nu \leq r-2$ and $n_\alpha \geq 2$, $\alpha = 1,2,\ldots,\nu$. Note that Q is a smooth, irreducible variety; let d denote the dimension of Q. In case we have for the degrees n_α inequalities $n_1 < n_2 < \ldots. < n_\nu$ we have a sequence of morphisms

$$Q = Q_{n_1\ldots n_\nu} \longrightarrow \ldots\ldots\ldots\ldots \longrightarrow Q_{n_1 n_2} \longrightarrow Q_{n_1} \longrightarrow \mathrm{Spec}(\mathbf{C}),$$

where each morphism is a projective bundle. In the general case, i.e., when also equalities can occur, we have a similar kind of sequence for $Q = Q_{n_1\ldots n_\nu}$, but now the morphisms are Grassmann bundles.

Let V be the variety of conics in $\mathbf{P}^r$, defined as in [6]. V is a smooth, irreducible projective variety of dimension $3r-1$.

Let us denote by $\Omega \subseteq \mathbf{P}^r \times Q$ the incidence correspondence whose points are given by the couples $(p,f) \in \mathbf{P}^r \times Q$ such that $p \in f$. Ω is a closed subvariety of $\mathbf{P}^r \times Q$ and we have a diagram:

$$\begin{array}{ccc} \Omega & \xrightarrow{t} & Q \\ {\scriptstyle s}\downarrow & & \\ \mathbf{P}^r & & \end{array}$$

LEMMA 1: *Ω is irreducible and of dimension $r+d-\nu$ (recall that $d = \dim Q$).*

PROOF: For all $p \in \mathbf{P}^r$, the fibre $s^{-1}(p)$ is a variety of a nature similar as the nature of the varieties Q explained above. If $F_1 = \cdots = F_\nu = 0$ are the equations of the complete intersection X, then the conditions $F_1(p) = \ldots = F_\nu(p) = 0$ have as consequence that $\dim s^{-1}(p) = d-\nu$. It follows that Ω is irreducible and $\dim \Omega = r+d-\nu$.

Let now Z be the closed subvariety of $V \times \Omega$ whose points are the triples $(q,p,f) = (q,(p,f)) \in V \times \Omega$ such that $p \in q \subseteq f$, so that we will have a diagram of morphisms of varieties:

$$\begin{array}{ccc} Z & \xrightarrow{\varphi} & \Omega \\ {\scriptstyle \psi}\downarrow & & \\ V & & \end{array}$$

Putting $n = \sum_{\alpha=1}^{\nu} n_\alpha$, we have:

LEMMA 2: *Z is irreducible and of dimension $3r+d-2n-\nu$.*

PROOF: It is sufficient to show that, for all $q \in V$, the fibre $\psi^{-1}(q)$ is irreducible and of dimension $d-2n-\nu+1$.

Remember that every conic $q \subseteq \mathbf{P}^r$ lies in a unique 2-plane π and choose homogeneous coordinates $(x_0:\ldots:x_{r-3}:u:v:w)$ in $\mathbf{P}^r$ such that π has equations:

$$x_0 = \ldots = x_{r-3} = 0 \tag{1}$$

and $q \subseteq \pi$ has equations:

$$x_0 = \ldots = x_{r-3} = \chi(u,v,w) = 0 \tag{2}$$

where χ is a quadratic form in three variables.

If $f \in (t \circ \varphi)(\psi^{-1}(q))$ is a complete intersection of type $(n_1,\ldots,n_\nu)$ containing q, then f will have equations of the shape:

$$\chi(u,v,w)\Phi_\alpha(u,v,w) + \Psi_\alpha(x_0,\ldots,x_{r-3}, u,v,w) = 0,\ \alpha = 1,2,\ldots,\nu \tag{3}$$

where each Φ_α is homogeneous of degree $n_\alpha - 2$ and each Ψ_α is homogeneous of degree n_α and such that $\Psi_\alpha(0,\ldots,0, u,v,w) = 0$ for any u,v,w not all 0. On the other hand, if f^* is any complete intersection in Q, then f^* will have equations of the shape:

$$\Phi^*_\alpha(u,v,w) + \Psi^*_\alpha(x_0,\ldots,x_{r-3}, u,v,w) = 0,\ \alpha = 1,2,\ldots,\nu, \tag{4}$$

where Φ^*_α, Ψ^*_α are homogeneous of degree n_α and moreover $\Psi^*_\alpha(0,\ldots,0, u,v,w) = 0$ for any u,v,w not all 0. Since the number of coefficients in Φ_α is $\binom{n_\alpha}{2}$ whilst the number of coefficients in Φ^*_α is $\binom{n_\alpha+2}{2}$, it follows that q gives $2n_\alpha + 1$ conditions on the hypersurfaces of degree n_α of $\mathbf{P}^r$ containing it, so that $(t \circ \varphi)(\psi^{-1}(q))$ is an irreducible subspace of q of codimension $\sum_{\alpha=1}^{\nu} (2n_\alpha + 1) = 2n + \nu$. It follows that $\psi^{-1}(q) \simeq (t \circ \varphi)(\psi^{-1}(q)) \times q$ is irreducible and of dimension $d-2n-\nu+1$.

We can now prove our main result:

THEOREM 1: *On every $X \in Q$ there exists a family of conics covering it if and only if:*

$$n = \sum_{\alpha=1}^{\nu} n_\alpha \leq r.$$

PROOF: Note that showing that on every $X \in Q$ there exists a family of conics covering it is equivalent to show that φ is surjective since this means that for every $X \in Q$ and $p \in X$ there exists $q \in V$ such that $p \in q \subseteq X$. The condition $n \leq r$ is therefore necessary since, if φ is surjective, for $(p, X) \in \Omega$ generic, lemmas 1 and 2 imply that:

$$\dim \varphi^{-1}[(p, X)] = \dim Z - \dim \Omega = 2(r-n) \geq 0.$$

To show that $n \leq r$ is also sufficient, put $\varphi(Z) = \Omega_1 \subseteq \Omega$ and let $\dim \Omega_1 = \dim \Omega - \epsilon$, so that the proof will be complete if we can show that $\epsilon = 0$. Assume therefore, by contradiction, that:

$$\epsilon > 0, \tag{*}$$

so that, for all $(p, f) \in \Omega_1$:
$\dim \varphi^{-1}[(p, f)] \geq \dim Z - \dim \Omega_1 = \dim Z - \dim \Omega + \epsilon = 2(r-n) + \epsilon.$

Let now $q_1 \in V$ be a fixed non-degenerate conic, π_1 the unique 2-plane containing it and:

$$\Omega' = \{(p, f) \in \Omega_1 | p \in q_1 \subseteq f\}.$$

Put $Z' = \varphi^{-1}(\Omega')$ and let Z_0' be an irreducible component of Z' of maximal dimension and $V' = \psi(Z_0')$. We will have a diagram:

$$\begin{array}{ccccc} Z_0' & \hookrightarrow & Z' & \hookrightarrow & \Omega' \\ \Big\downarrow & & \downarrow & & \downarrow \\ \Big\downarrow{\scriptstyle \psi'} & & Z & \xrightarrow{\varphi} & \Omega \\ \Big\downarrow & & \Big\downarrow{\scriptstyle \psi} & & \\ V' & \hookrightarrow & V & & \end{array}$$

where $\psi' = \psi | Z_0'$.

Now, $\Omega' \simeq \psi^{-1}(q_1)$ under the map $(p, f) \mapsto (q_1, p, f)$, so that $\dim \Omega' = d - 2n - \nu + 1$. It follows that, for all $(p, f) \in \Omega'$, $\dim \varphi^{-1}[(p, f)] \geq \dim Z' - \dim \Omega'$ and, for almost all $(p, f) \in \Omega'$, $\dim \varphi^{-1}[(p, f)] = \dim Z' - \dim \Omega'$. On the other hand, $\dim \varphi^{-1}[(p, f)] \geq 2(r-n) + \epsilon$ for all $(p, f) \in \Omega_1$, so that:

$$\begin{aligned} \dim Z_0' = \dim Z' &\geq \dim \Omega' + 2(r-n) + \epsilon \qquad (**) \\ &= \dim \psi^{-1}(q_1) + 2(r-n) + \epsilon \\ &> \dim \psi^{-1}(q_1) + 2(r-n) \quad \text{(by (*))}. \end{aligned}$$

We will split now the proof in several cases (and subcases):

Case 1. $V' = \{q_1\}$.

This implies that $Z_0' = \psi^{-1}(q_1)$, which contradicts (**), since $r \geq n$.

Case 2. All conics $q \in V'$ lie in π_1 and there exists $q \in V'$ such that $q \neq q_1$.

We will have $\dim V' \leq 5$; moreover:

$$\psi'^{-1}(q) = \{(q,p,f) \in Z_0' | q \subseteq f,\ q_1 \subseteq f,\ p \in q \cap q_1\},$$

so that $\psi'^{-1}(q)$ has the same dimension as the subvariety of all complete intersections $f \in Q$ going through q and q_1. On the other hand, we saw in the proof of lemma 2 that $q_1 \subseteq f$ gives $2n + \nu$ linearly independent conditions on f, whilst $q \subseteq f$ gives $2n - 3\nu$ more clearly linearly independent conditions on f, owing to the fact that q and q_1 have 4 common points. It follows that $\dim \psi'^{-1}(q) = d - 4n + 2\nu$, so that:

$$\dim Z_0' \leq d - 4n + 2\nu + 5, \tag{5}$$

whilst, by (**), we should have:

$$\dim Z_0' > d - 4n + 2r - \nu + 1, \tag{6}$$

so that:

$$2r < 3\nu + 4. \tag{7}$$

To see that this is impossible, note that from the hypothesis $n_\alpha \geq 2$ for $\alpha = 1,2,\ldots,\nu$ and from $n \leq r$ it follows that:

$$2\nu \leq r. \tag{8}$$

Adding up to (7), one gets:

$$r < \nu + 4 \tag{9}$$

or:

$$r - \nu < 4. \tag{10}$$

Therefore, if $\dim X = r - \nu \geq 4$ the proof is complete. Otherwise, note that from (8) it follows immediately that:

$$r \leq 2(r - \nu) = 2 \dim X. \tag{11}$$

The possible cases when $\dim X = 2, 3$ are therefore the following:

a) $r=3, \quad \nu=1, \quad n_1=2, \qquad \dim X=2$
b) $r=3, \quad \nu=1, \quad n_1=3, \qquad \dim X=2$
c) $r=4, \quad \nu=2, \quad n_1=n_2=2, \qquad \dim X=2$
d) $r=4, \quad \nu=1, \quad n_1=2, \qquad \dim X=3$
e) $r=4, \quad \nu=1, \quad n_1=3, \qquad \dim X=3$
f) $r=4, \quad \nu=1, \quad n_1=4, \qquad \dim X=3$
g) $r=5, \quad \nu=2, \quad n_1=n_2=2, \qquad \dim X=3$
h) $r=5, \quad \nu=2, \quad n_1=2, n_2=3, \qquad \dim X=3$
i) $r=6, \quad \nu=3, \quad n_1=n_2=n_3=2, \qquad \dim X=3.$

Of these, only in cases a), b), c) and i) the inequality (7) does not give a contradiction. However, it is well known that every complete intersection of this kind too is covered by conics, so that $\epsilon=0$ also in these cases. The proof of case 2 is therefore complete.

Case 3. *The plane* π *of the generic conic* $q \in V'$ *meets* π_1 *in a line* l.

3a) $q_1 \cap l$ *consists of two points.*

First of all, note that the family of 2-planes π in $\mathbf{P}^r$ meeting the fixed 2-plane π_1, in a line l, being the Schubert variety $[1,2,r]$, has dimension r. There are the following two possibilities for q:

(i) q *is non-degenerate and meets* q_1 *in two points.*

In π the family of non-degenerate conics going through two points has dimension 3, so that:

$$\dim V' \leq r+3. \tag{12}$$

Again, $\psi'^{-1}(q)$ has the same dimension as the subvariety of all $f \in Q$ going through q and q_1. We know also that $q_1 \subseteq f$ gives $2n+\nu$ conditions on f, whilst $q \subseteq f$ gives $2n-\nu$ more conditions on f, owing to the fact that q_1 and q have two common points. It is therefore sufficient to show that these conditions are linearly independent, since it will follow that dim $\psi'^{-1}(q)=d-4n$ and:

$$\dim Z_0' \leq d-4n+r+3 \tag{13}$$

which, together with (**), implies the impossible inequality:

$$r-\nu<2. \tag{14}$$

To see that, we may assume that the planes π and π_1 have respectively equations:

$$\pi : x_0 = \ldots = x_{r-3} = 0, \tag{15}$$

$$\pi_1 : x_0 = \ldots = x_{r-2} = u = 0, \tag{16}$$

so that the equations of l will be:

$$l : x_0 = \ldots = x_{r-3} = u = 0. \tag{17}$$

Moreover, we can choose the coordinate system in $\mathbf{P}^r$ in such a way that the conics q and q_1 have respectively equations:

$$q : x_0 = \ldots = x_{r-3} = u^2 + v^2 + w^2 = 0, \tag{18}$$

$$q_1 : x_0 = \ldots = x_{r-2} = u = x_{r-3}^2 + v^2 + w^2 = 0. \tag{19}$$

If $f \in Q$ has to contain q_1, then f will have equations of the shape:

$$(x_{r-3}^2 + v^2 + w^2)\Phi_\alpha(x_{r-3}, v, w) + \Psi_\alpha(x_0, \ldots, x_{r-3}, u, v, w) = 0, \ \alpha = 1, 2, \ldots, \nu, \tag{20}$$

where, more precisely, we can write, for all α:

$$\Phi_\alpha(x_{r-3}, v, w) = \Phi'_\alpha(v, w) + \Phi''_\alpha(x_{r-3}, v, w) \tag{21}$$

$$\Psi_\alpha(x_0, \ldots, x_{r-3}, u, v, w) = \tag{22}$$

$$= u(g'_{\alpha, n_\alpha - 1}(v, w) + g''_{\alpha, n_\alpha - 1}(x_{r-3}, v, w)) +$$

$$\cdots\cdots\cdots\cdots\cdots\cdots$$

$$+ u^{n_\alpha - 1}(g'_{\alpha,1}(v, w) + g''_{\alpha,1}(x_{r-3}, v, w)) +$$

$$+ cu^{n_\alpha} + \Psi'_\alpha(x_0, \ldots, x_{r-3}, u, v, w)$$

with Φ'_α, Φ''_α homogeneous of degree $n_\alpha - 2$, $g'_{\alpha,i}$, $g''_{\alpha,i}$ homogeneous of degree i for $i = 1, \ldots, n_\alpha - 1$, c a constant, Ψ'_α homogeneous of degree n_α and:

$$\Phi''_\alpha(0, v, w) = 0, \ g''_{\alpha,i}(0, v, w) = 0, \ i = 1, \ldots, n_\alpha - 1 \text{ for all } v, w; \tag{23}$$

$$\Psi'_\alpha(0, \ldots, 0, x_{r-3}, u, v, w) = 0 \text{ for all } x_{r-3}, u, v, w \text{ not all } 0. \tag{24}$$

Here the complex of terms containing only the variables u, v, w is:

$$(v^2 + w^2)\Phi'_\alpha(v, w) + ug'_{\alpha, n_\alpha - 1}(v, w) + \ldots + u^{n_\alpha - 1}g'_{\alpha,1}(v, w) + cu^{n_\alpha} \tag{25}$$

and the number of coefficients is:

$$(1+2+\ldots+n_\alpha)+n_\alpha-1=\tfrac{1}{2}(n_\alpha^2+3n_\alpha-2) \tag{26}$$

On the other hand, if f has to contain also q, its equations must be of the shape:

$$(u^2+v^2+w^2)A_\alpha(u,v,w)+B_\alpha(x_0,\ldots,x_{r-3},\ u,v,w)=0,\ \alpha=1,\ldots,\nu, \tag{27}$$

with A_α homogeneous of degree $n_\alpha-2$, B_α homogeneous of degree n_α and $B_\alpha(0,\ldots,0,u,v,w)=0$ for any u,v,w not all 0. Since the number of coefficients in A_α is $\binom{n_\alpha}{2}=\frac{1}{2}(n_\alpha^2-n_\alpha)$ and:

$$\tfrac{1}{2}(n_\alpha^2+3n_\alpha-2)-\tfrac{1}{2}(n_\alpha^2-n_\alpha)=2n_\alpha-1, \tag{28}$$

it follows that $q\subseteq f$ gives $2n-\nu$ more linearly independent conditions on f, as had to be shown.

(ii) *q is non-degenerate and meets q_1 in one point (which coincides necessarily with p).*

In π the family of non-degenerate conics going through one point has dimension 4, so that:

$$\dim\ V'\leq r+4. \tag{29}$$

By the same kind of argument used in the preceding subcase, one sees easily that $q_1\subseteq f$ and $q\subseteq f$ give $(2n+\nu)+2n=4n+\nu$ linearly independent conditions on f, so that $\dim\psi'^{-1}(q)=d-4n-\nu$ and $\dim\ Z_0'\leq d-4n-\nu+r+4$ which, together with (**), gives $r<3$ contradicting our hypothesis that $r\geq 3$.

(iii) *$q=l_1+l_2$ is degenerate, with l_1, l_2 lines such that $l_1\neq l_2$, $l_1\neq l$, $l_2\neq l$, and meets q_1 in two points.*

In π the family of conics degenerating in two distinct lines and going through two distinct points has dimension 2, so that:

$$\dim\ V'\leq r+2. \tag{30}$$

Moreover, $q_1\subseteq f$ gives as always $2n+\nu$ conditions on f, whilst $l_1\subseteq f$ gives n conditions (since l_1 has one point in common with q_1) and $l_2\subseteq f$ gives $n-\nu$ more conditions (since l_2 has one point in common with q_1 and one *distinct* point in common with l_1) which can be shown, as in case 3a(i), to be linearly independent, so that $\dim\psi'^{-1}(q)=d-4n$ and $\dim\ Z_0'\leq d-4n+r+2$ which, together with (**), gives the impossible inequality:

$$r-\nu<1. \tag{31}$$

(iv) *$q=l_1+l_2$ is degenerate, with l_1, l_2 lines such that $l_1\neq l_2$, $l_1\neq l$, $l_2\neq l$, and meets q_1 in one point.*

In π the family of conics degenerating in two distinct lines and going through one point has dimension 3, so that:

$$\dim V' \leq r+3. \tag{32}$$

Moreover, $q_1 \subseteq f$ gives as always $2n+\nu$ conditions on f, whilst $l_1 \subseteq f$ and $l_2 \subseteq f$ give both n more conditions, clearly linearly independent, so that $\dim \psi'^{-1}(q) = d-4n-\nu$ and $\dim Z_0' \leq d-4n-\nu+r+3$ which, together with (**), gives $r<2$, contradicting our hypothesis that $r \geq 3$.

(v) *$q = l_1 + l$, with l_1 line such that $l_1 \neq l$.*

In π the family of such conics has dimension 2, so that:

$$\dim V' \leq r+2. \tag{33}$$

Moreover, $q_1 \subseteq f$ gives as always $2n+\nu$ conditions on f, whilst $l \subseteq f$ gives $n-\nu$ conditions (since l meets q_1 in two points) and $l_1 \subseteq f$ gives n more conditions, clearly linearly independent, so that $\dim \psi'^{-1}(q) = d-4n$ and $\dim Z_0' \leq d-4n+r+2$ which, together with (**), gives the impossible inequality:

$$r-\nu<1. \tag{34}$$

(vi) *$q = 2l_1$, with l_1 line such that $l_1 \neq l$.*

Note that, according to Tennison [6], we should consider now the "conic" $q = 2l_1$ as a couple $(\pi, 2l_1)$. The family of such conics has dimension 2, since l_1 can be any line in π going through one of the two points $l \cap q_1$, so that:

$$\dim V' \leq r+2. \tag{35}$$

Moreover, $q_1 \subseteq f$ gives as always $2n+\nu$ conditions on f. To see how many conditions gives $2l_1 \subseteq f$, which must be interpreted in the sense that the intersection of f with π contains the line l_1 with multiplicity at least 2, let us take for π_1 and q_1 the same equations (16) and (19) as in the preceding case 3a(i), so that f, in order to contain q_1, will have equations of the shape (20) satisfying conditions (21) – (24) and the number of coefficients in the complex of terms (25) containing only the variables u,v,w will be equal as before, for each $\alpha = 1,\ldots,\nu$, to $\frac{1}{2}(n_\alpha^2 + 3n_\alpha - 2)$.

We may also assume that π has equations (15) and that l_1 has equations:

$$x_0 = \ldots = x_{r-3} = u + av + bw = 0, \tag{36}$$

where a,b are constants. If f has to contain $2l_1$, its equations must be of the shape:

$$(u + av + bw)^2 A'_\alpha(u,v,w) + B'_\alpha(x_0,\dots,x_{r-3},u,v,w) = 0, \quad \alpha = 1,\dots,\nu \qquad (37)$$

with A'_α homogeneous of degree $n_\alpha - 2$, B'_α homogeneous of degree n_α and $B'_\alpha(0,\dots,0,u,v,w) = 0$ for any u,v,w. Moreover, assuming that the point $q_1 \cap l_1$ is $(0,\dots,0,0,1,i)$, each A'_α must satisfy the extra condition:

$$A'_\alpha(0,1,-i) = 0, \quad \alpha = 1,\dots,\nu \qquad (38)$$

since f, containing q_1, must also contain the second point $q_1 \cap l$. Since the number of coefficients in A'_α is $\binom{n_\alpha}{2}$ it follows, taking care of condition (38) and by the same argument as in case 3a(i), that $2l_1 \subseteq f$ gives $2n$ more linearly independent conditions on f, so that $\dim \psi'^{-1}(q) = d - 4n - \nu$ and $\dim Z'_0 \leq d - 4n - \nu + r + 2$ which, together with (**), gives $r < 1$, contradicting our hypothesis that $r \geq 3$.

(vii) $q = 2l$.
Here we have:

$$\dim V' \leq r \qquad (39)$$

and, proceeding as before, one sees that $2l \subseteq f$ gives $2n - \nu$ more linearly independent conditions besides the $2n + \nu$ given by $q_1 \subseteq f$, so that $\dim \psi'^{-1}(q) = d - 4n$ and $\dim Z'_0 \leq d - 4n + r$ which, together with (**), gives the impossible inequality:

$$r - \nu < -1. \qquad (40)$$

3b) *$q_1 \cap l$ consists of one point.*
This means that l is tangent to q_1. Since the family of lines tangent to q_1 has dimension 1, it follows that the family of planes π has dimension $r - 1$. Moreover, q must go necessarily through the unique point $q_1 \cap l$ (and this is the point $p \in q \cap q_1$), so that in π the family of such conics has dimension 4. It follows that:

$$\dim V' \leq r + 3. \qquad (41)$$

As usual, $q_1 \subseteq f$ gives $2n + \nu$ conditions on f and the fact that l is tangent to q_1 implies that l is tangent also to f. By the same kind of argument used before, it follows that $q \subseteq f$ gives only $2n - \nu$ linearly independent conditions on f, so that $\dim \psi'^{-1}(q) = d - 4n$ and $\dim Z'_0 \leq d - 4n + r + 3$ which, together with (**), gives the impossible inequality:

$$r - \nu < 2. \qquad (42)$$

Case 4. The plane π of the generic conic $q \in V'$ meets π_1 in one point U.

Note first of all that $\pi \cap \pi_1 = \emptyset$ is impossible, since $p \in q \cap q_1$, so that also $p = U$.

The family of 2-planes in $\mathbf{P}^r$ meeting the fixed 2-plane π_1 in one point, being the Schubert variety $[2,r-1,r]$, has dimension $2r-2$. Therefore, the family of 2-planes π in $\mathbf{P}^r$ meeting the 2-plane π_1 in one point $U \in q_1$ has dimension $2r-3$. Moreover, q must go necessarily through the point $p = U$, so that in π the family of such conics has dimension 4. It follows that:

$$\dim V' \leq 2r+1. \tag{43}$$

As usual, $q_1 \subseteq f$ gives $2n+\nu$ conditions on f and one checks easily, in a way similar as in the proof of case 3a(i), that $q \subseteq f$ gives $2n$ more linearly independent conditions on f. It follows that $\dim \psi'^{-1}(q) = d - 4n - \nu$ and $\dim Z_0' \leq d - 4n - \nu + 2r + 1$ which, together with (**), gives the impossible inequality:

$$2r - \nu + 1 < 2r - \nu + 1. \tag{44}$$

It follows that $\epsilon = 0$ and the proof of the theorem is complete.

2. Let us now see which complete intersections satisfy the condition of theorem 1 when $\dim X = 4$. Note that from (11) it follows that:

$$5 \leq r \leq 8. \tag{45}$$

The complete intersections of dimension 4 which are covered by conics and hence for which the Hodge (2, 2)-conjecture holds are therefore the following (where X_d denotes an hypersurface of degree d):

$r = 5$

$$X_2, X_3, X_4, X_5 \hookrightarrow \mathbf{P}^5$$

$r = 6$

$$X_2 \cdot X_2', X_2 \cdot X_3, X_2 \cdot X_4, X_3 \cdot X_3' \hookrightarrow \mathbf{P}^6$$

$r = 7$

$$X_2 \cdot X_2' \cdot X_2'', X_2 \cdot X_2' \cdot X_3 \hookrightarrow \mathbf{P}^7$$

$r = 8$

$$X_2 \cdot X_2' \cdot X_2'' \cdot X_2''' \hookrightarrow \mathbf{P}^8.$$

Here $X_2 \subseteq \mathbf{P}^5$ is rational, whilst $X_3 \subseteq \mathbf{P}^5$, $X_2 \cdot X_2'$, $X_2 \cdot X_3 \subseteq \mathbf{P}^6$ and $X_2 \cdot X_2' \cdot X_2'' \subseteq \mathbf{P}^7$ are unirational. Moreover, the (2, 2)-conjecture for X_4, $X_5 \subseteq \mathbf{P}^5$ had already been proved in [2]. The other cases given by theorem 1 are new.

Note that if, instead of considering conics, one looks more generally for the existence of rational normal curves of degree m on a complete intersection $X \subseteq \mathbf{P}^r$ of type $(n_1, \ldots, n_\nu)$, $r \geq 3$, $n_\alpha \geq 2$ for $\alpha = 1, \ldots, \nu$, it is not difficult to see, taking care of the fact that the family of rational normal curves of degree m in $\mathbf{P}^r$ has dimension $(r-m)(m+1)+(m+3)(m-1)$, that the condition in order that X be covered by such curves is:

$$m(r-n+1) \geq 2, \tag{46}$$

from which one gets the condition of theorem 1 by putting $m = 2$. In principle, one could think of giving in this more general case a proof completely analogous to the one we just gave for conics. However, the number of cases to take care of is so higher that it does not seem worth while to fulfill such a penible task. Moreover, from the point of view of the Hodge conjecture this would not give nothing new. In fact, condition (46) is satisfied if and only if the bracket is ≥ 1 for $m \geq 2$ and ≥ 2 for $m = 1$, so that the widest class of complete intersections admitting a covering by rational normal curves is obtained already for $m = 2$, i.e. for conics.

REMARKS: (i) The canonical divisor of a complete intersection $X \subseteq \mathbf{P}^r$ of type $(n_1, \ldots, n_\nu)$ is given by:

$$K_X \sim (n-r-1)E, \tag{47}$$

where E is an hyperplane section. The condition $n \leq r$ of theorem 1 is therefore equivalent to $-K_X > 0$. Remembering that an algebraic variety X with $-K_X$ ample is called a *Fano variety*, theorem 1 may be restated by saying that the complete intersections covered by conics are exactly the ones which are Fano varieties. The Hodge (2, 2)-conjecture holds therefore for all Fano fourfolds which are complete intersections. More generally, S. Mori [3] has proved recently that on *every* Fano variety there is at least one rational curve. It seems very likely that his method would give also a proof of the fact that on every Fano variety there exists a family of rational curves covering it. This would give a proof of the Hodge (2, 2)-conjecture for all Fano fourfolds and not only for the complete intersections.

(ii) All Fano varieties have Kodaira dimension equal to $-\infty$ (i.e., all plurigenera vanish). Ueno made the conjecture (at least in dimension 3) that all varieties with Kodaira dimension $-\infty$ are uniruled. If this conjecture were true (also in dimension 4), it would give the Hodge conjecture for all fourfolds of Kodaira dimension $-\infty$.

REFERENCES

[1] A. Conte: Conics on complete intersections and the Hodge conjecture for some of them, to appear in Atti del Convegno Internazionale di Geometria Algebrica, Ist. Naz. di Alta Mat. "F. Severi" Roma, aprile 1979.
[2] A. Conte and J. P. Murre: The Hodge conjecture for fourfolds admitting a covering by rational curves, Math. Ann. *238* (1978) 79–88.
[3] S. Mori: Projective manifolds with ample tangent bundles, Ann. of Math. *110* (1979) 593–606.
[4] J.P. Murre: On the Hodge conjecture for unirational fourfolds, Indag. Math. *80* (1977) 230–32.
[5] A. Predonzan: Fibrazioni di varietà algebriche, Symposia Math. INDAM, vol. *V* (1970) 375–84.
[6] B. R. Tennison: On the quartic threefold, Proc. London Math. Soc., *29* (1974) 714–34.

Alberto Conte
Istituto di Geometria "C. Segre"
Università di Torino
Via Principe Amedeo 8,
10123 Torino, Italy

Jacob P. Murre
Mathematisch Instituut
der Rijksuniversiteit, Wassenaarseweg 80
2300 RA Leiden, The Netherlands.

THE DEGREE OF THE PRYM MAP ONTO THE MODULI SPACE OF FIVE DIMENSIONAL ABELIAN VARIETIES*

Ron Donagi** and Roy Smith***

1. Introduction

Prym varieties are the next more general class of abelian varieties after Jacobians and they have an analogous interpretation in terms of the theory of curves. They have played an important role in recent work on two classical problems: the "Lüroth problem" of distinguishing unirational from rational varieties [3], [4], [10], and the related "Schottky problem" of distinguishing Jacobian varieties among all principally polarized abelian varieties [2], [9, p.341]. Their usefulness in the study of 5-dimensional abelian varieties stems from the following result, due primarily to Wirtinger [3, p.386], [14, p.124]:

THEOREM: *There is a moduli space $\mathcal{R}_6$, parametrizing unramified double covers of smooth genus 6 curves, which is generically a finite covering both of the moduli space $\mathcal{M}_6$ of all smooth genus 6 curves, and of the moduli space $\mathcal{A}_5$ of all principally polarized 5-dimensional abelian varieties.*

This is illustrated by the following diagram of maps

$$\begin{array}{ccc} & \mathcal{R}_6 & \\ {}^{\pi}\swarrow & & \searrow^{P} \\ \mathcal{M}_6 & & \mathcal{A}_5 \end{array}$$

in which all three spaces have dimension 15, P assigns to a double cover its associated Prym variety, and π simply forgets the double cover. It is elementary that π is a finite map of degree $2^{12}-1$ to one. These maps provide a valuable tool for studying the subvarieties of the space $\mathcal{A}_5$ by comparing them with the relatively well-known subvarieties of $\mathcal{M}_6$. We undertake to sharpen this tool by showing that the degree of P is 27.

---

*These are notes from an expository lecture delivered in Angers by the second named author on joint work which we intend to publish in the near future.

**partially supported by NSF grant #MCS 77-03976
***partially supported by NSF grant #MCS 79-03717

2. The Definition of a Prym Variety

If $\eta:\tilde{C}\to C$ is an unramified double cover of a smooth curve, and $\eta_*:\tilde{J}\to J$ is the induced map of Jacobians, one defines the Prym variety $P(C,\eta)$ as follows:

$P(C,\eta)=(\text{Ker }\eta_*)^0=$ the connected component of the identity in Ker η_*.

It follows that $\dim P=\dim\tilde{J}-\dim J=g(\tilde{C})-g(C)=g(C)-1$ (by Hurwitz' formula). P has a natural polarizing divisor Ξ, equivalent to half the intersection divisor $(P.\Theta)$ [9, p. 342]. If $\lambda:\tilde{C}\to\tilde{C}$ is the involution interchanging the sheets of the double cover, and $\lambda_*:\tilde{J}\to\tilde{J}$ is the induced involution, then $P=\text{image }(1-\lambda_*)$ so that $\tilde{C}$ lies inside P via the composition

$$\tilde{C}\xrightarrow[\text{map}]{\text{Abel}}\tilde{J}\xrightarrow{1-\lambda_*}P.$$

3. Inverting the Prym map over the Jacobi locus in $\mathcal{A}_5$

In order to compute the degree of the map P, we need to know the fibre over some good point in $\mathcal{A}_5$, and we choose one we know most about, the Jacobian $J(X)$ of a generic genus 5 curve. The first step in describing $P^{-1}(J(X))$ was taken by Mumford, and we'll describe his results for our special case. Recall that an abelian variety (A,Θ) belongs to the "Andreotti-Mayer locus" $\mathcal{N}_1\subset\mathcal{A}_5$, if and only if $\dim\text{sing }\Theta\geq 1$, and that the Jacobi locus $\mathcal{J}_5$ is an irreducible component of $\mathcal{N}_1$ [1, p. 213]. Since the theta divisor Ξ on a Prym variety is obtained by intersecting P with $\tilde{\Theta}$, the singularities of Ξ come both from the singularities of $\tilde{\Theta}$ and from the points where the intersection is tangential. Mumford showed that singularities of the second type are responsible for the phenomenon $\dim(\text{sing }\Xi)\geq 1$ and that, when this occurs, they imply restrictions on the curve C. In fact he wrote down a list of loci in $\mathcal{R}_6$ which he proved contained all of $P^{-1}(\mathcal{N}_1)$.

To be precise:

THEOREM [9, p. 344]: *If $(C,\eta)\in\mathcal{R}_6$, and if $P(C,\eta)\in\mathcal{N}_1\subset\mathcal{A}_5$, then one of the following is true*:

(i) *C is hyperelliptic*
(ii) *C is a double cover of an elliptic curve*
(iii) *C is trigonal*
(iv) *C is a plane quintic and η is an "even" double cover.*

Now let X be a generic genus 5 curve and $J(X)$ its Jacobian. Since curves of types (i) and (ii) above have at most eleven moduli, while 5-dimensional Jacobians have twelve moduli, it follows that

$$P^{-1}(J(X)) \subset \{\text{trigonals}\} \cup \{\text{evenly covered plane quintics}\}.$$

Since trigonal curves of genus 6 have 13 moduli and plane quintics have 12 moduli, both these cases must be examined.

4. The plane quintic component of $P^{-1}(J(X))$

The last sentence of the proof of Mumford's theorem implies that the only double cover of a plane quintic whose Prym could be $J(X)$ is this one:

$$\eta : \tilde{C} = \text{sing}\,\Theta(X) \to \text{sing}\,\Theta(X)/_{q\sim -q} = C.$$

The proof that this Prym variety is in fact $J(X)$ is due independently to Tjurin and Masiewicki [8], [13, p.34]. First of all, since X is generic, sing $\Theta(X)$ is a curve [12, p. 162], and since Riemann's theta function is an even function, both Θ and sing Θ are invariant under the involution $q \mapsto -q$ on $J(X)$. Furthermore, C and $\tilde{C}$ are generically both smooth [8, p. 235]. But why is C a plane quintic?

Recall that the generic genus 5 curve X is a complete intersection of 3 quadrics in $\mathbf{P}^4$ [12, p.157]

$$X = \mathcal{Q}_0 \cap \mathcal{Q}_1 \cap \mathcal{Q}_2 \subset \mathbf{P}^4$$

so that

$$\begin{aligned} \Delta &= \{\det|t_0\mathcal{Q}_0 + t_1\mathcal{Q}_1 + t_2\mathcal{Q}_2| = 0\} \\ &= \text{set of singular quadrics containing } X \text{ in } \mathbf{P}^4 \end{aligned}$$

is a plane quintic. We claim $C \simeq \Delta$. For Riemann's singularities theorem [6, p. 348] and Clifford's theorem [12, p.158] imply that sing $\Theta \simeq W_4^1 =$ set of all g_4^1's on X = set of all 1-dimensional complete linear series $|D|$ of divisors of degree 4.
In particular, all singular points of Θ are double points. Then consider the map

$$\begin{aligned} \eta &: \text{Sing}\,\Theta \to \Delta \\ &|D| \mapsto \text{projective tangent cone to } \Theta \text{ at } |D| \end{aligned}$$

Since $|D|$ is a double point of Θ but smooth on the curve sing Θ, $\eta(|D|)$ is a rank 4 quadric cone in {the projectivized tangent space to $J(X)$ at $|D|\} \simeq \mathbf{P}^4$. It contains X because in fact

$$\eta(|D|) = \bigcup_{D \in |D|} \bar{D}$$

where $\bar{D}$ is the plane in $\mathbf{P}^4$ spanned by the points of the divisor D [6, p.348]. Since a rank 4 quadric has two rulings by planes, and two planes from opposite rulings span a hyperplane cutting a canonical divisor on X, the

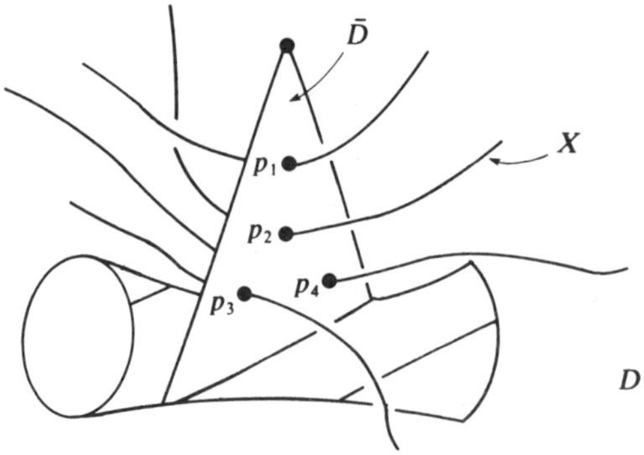

$D = p_1 + p_2 + p_3 + p_4$

map is 2 to 1, with fibres of form $\eta^{-1}(\eta(|D|)) = \{|D|, |K-D|\}$. The involution $|D| \mapsto |K-D|$ on W_4^1 corresponds to the minus map on sing Θ, so indeed,

$$C \simeq \text{sing } \Theta/_{q\sim -q} \simeq W_4^1/_{|D|\sim|K-D|} \simeq \Delta.$$

Probably the nicest way to see that $P(C, \eta) \simeq J(X)$ is to use Masiewicki's criterion:

THEOREM [8, p. 228]: *If a g-dimensional abelian variety (A, Θ) contains a symmetric curve $\tilde{C}$ such that the quotient curve $C = \tilde{C}/_{q\sim -q}$ is not hyperelliptic, then the following two statements are equivalent*:

(i) $(A, \Theta) \simeq P(\tilde{C} \to C)$

(ii) $[\tilde{C}] = \dfrac{2}{(g-1)!}[\Theta^{(g-1)}] \in H_2(A; \mathbf{Z})$.

In our case $(A, \Theta) = (J(X), \Theta(X))$ and $C \simeq W_4^1$, so Castelnuovo's formula shows that $\tilde{C}$ has the right homology class [6 p. 358].

So a generic 5-dimensional Jacobian is the Prym variety associated to a unique double cover of a plane quintic. In fact, Masiewicki generalizes everything to the case where $\tilde{C}, C$ are not necessarily smooth and proves the result for all 5-dimensional Jacobian varieties. The Prym varieties of singular curves will be very important to us, as we shall see in the following sections.

5. The trigonal component of $P^{-1}(J(X))$

Since the trigonal locus in $\mathcal{R}_6$ is 13-dimensional, if it maps onto the Jacobi locus the fibres will be curves. To each Jacobian we therefore seek to associate a natural one-parameter family of doubly covered trigonal curves. The following beautiful construction is due to Recillas [11 p. 9].

Choose a point $|D| \in W_4^1(X)$. We know there is a rank 4 quadric in $\mathbf{P}^4$ one of whose rulings cuts out the pencil of divisors in the linear series $|D|$. Since each divisor forms a quadrilateral in one of the ruling planes, the three pairs of diagonals define three intersection points which sweep out a trigonal curve C as the ruling plane moves. If $|D|$ is generic on W_4^1, C is smooth.

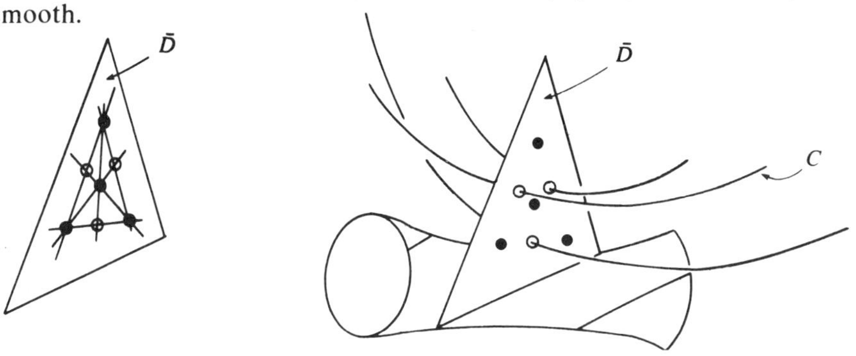

The double cover $\tilde{C}$ is the set of diagonals of the quadrilaterals. Since each diagonal determines a pair of points in one of the divisors $D \subset X$, there is a natural diagram of maps

$$\begin{array}{ccc} & X^{(2)} & \\ \nearrow & & \searrow \\ \tilde{C} & \longrightarrow & \mathrm{Pic}^{(2)}(X) \simeq J(X) \end{array}$$

in which $X^{(2)}$ denotes the second symmetric product of the curve X, and $X^{(2)} \to \mathrm{Pic}^{(2)}(X)$ is the map sending a divisor E to the associated line bundle $\{E\}$. It is easy to see that $f(\tilde{C})$ is symmetric for the involution $\{E\} \mapsto \{D-E\}$ on $\mathrm{Pic}^{(2)}(X)$. To see why $[f(\tilde{C})]$ is the right homology class, let X specialize momentarily to a trigonal curve, $X \to X_0$, and assume that $|D| = g_4^1 \to g_3^1 + x_0 = |D_0|$, for some fixed point $x_0 \in X_0$. Then the set of pairs of points in the divisors D_0 divides into two subsets, those pairs that contain x_0 and those that do not, and both subsets are isomorphic to X_0. That is:

$$\tilde{C} = \begin{cases} \{\langle a, b\rangle\text{: for some } c \in X_0,\ a+b+c+x_0 \in g_3^1 + x_0\} \simeq \{c \in X_0\} \\ \cup \{\langle a, x_0\rangle\text{: for some } b, c \in X_0,\ a+b+c+x_0 \in g_3^1 + x_0\} \simeq \{a \in X_0\} \end{cases}$$

Thus $[\tilde{C}_0] = 2[X_0] = 2/4!\ [\Theta_0^{(4)}]$, by Poincaré's formula [6, p.350]. Since the homology class of $\tilde{C}$ lies in a discrete group, it does not change during the deformation $X \to X_0$. Thus $[\tilde{C}] = 2/4!\ [\Theta^{(4)}]$.

Recillas also gives a (group theoretic) construction inverse to the one above, of which the following is a geometric version. If C is trigonal of genus 6 and $\eta: \tilde{C} \to C$ is a double cover, the induced map of symmetric products $\eta: \tilde{C}^{(3)} \to C^{(3)}$ is 8 to 1, hence $\eta: \tilde{C}^{(3)}/\lambda_* \to C^{(3)}$ is 4 to 1, where λ_* is the involution on $\tilde{C}^{(3)}$ induced by interchanging sheets of the double cover. Since C is trigonal, $C^{(3)}$ contains the unique linear series $|g_3^1| \simeq \mathbf{P}^1$, and restricting η to $X = \eta^{-1}(|g_3^1|)$ defines a g_4^1 on the curve X.

$$\eta: \tilde{C}^{(3)}/\lambda_* \to C^{(3)}$$

$$X = \eta^{-1}(|g_3^1|) \to |g_3^1| \simeq \mathbf{P}^1$$

The maps $X \to |g_3^1|$ and $C \to |g_3^1|$ have the same branch locus, so Hurwitz' formula implies that $g(X) = 5$. One can show that this inverts the previous construction.

6. Completing the fibre $P^{-1}(J(X))$

Recillas' result shows that even when $J(X)$ is generic, the trigonal component of $P^{-1}(J(X))$, is parametrized only by an open subset of W_4^1 (those $|D|$ for which C is smooth). This brings out two difficulties involved in calculating the degree of P: first, since the trigonal fibre is not complete the map is not proper, so we cannot be sure of getting the full degree of P by examining the components of $P^{-1}(J(X))$ consisting of smooth curves; second, the trigonal fibre has positive dimension, so that even determining its contribution to the degree is non-trivial.

The first problem was overcome by Beauville who gave the full generalization of Masiewicki's construction of Prym varieties of singular curves and proved that when the appropriate doubly-covered stable curves are included the Prym map is proper [2, p.177]. He also completed Mumford's list to include all doubly-covered stable curves whose Prym lies in the Andreotti-Mayer locus $\mathcal{N}_1$. Searching his list we find two further types of doubly-covered stable curves which occur in the completed fibre of P over a generic Jacobian. The first was already known to Wirtinger [14, p.118].

Let X be a smooth genus 5 curve, p, q any two distinct points of X, and let $C = X/_{p\sim q}$ be the singular genus 6 curve obtained by identifying p and q. If X_1 and X_2 are two isomorphic copies of X, then the normalization maps $X_i \to C$ induce a double cover of C by the reducible curve

$$\tilde{C} = X_1 \cup X_2/_{p_1\sim q_2, p_2\sim q_1}$$

and one can check directly that the Prym variety is isomorphic to $J(X)$ [2, p. 175]. This component of $P^{-1}(J(X))$ is parametrized by the possible choices of the two distinct points p, q, hence by $X^{(2)}$–diagonal.

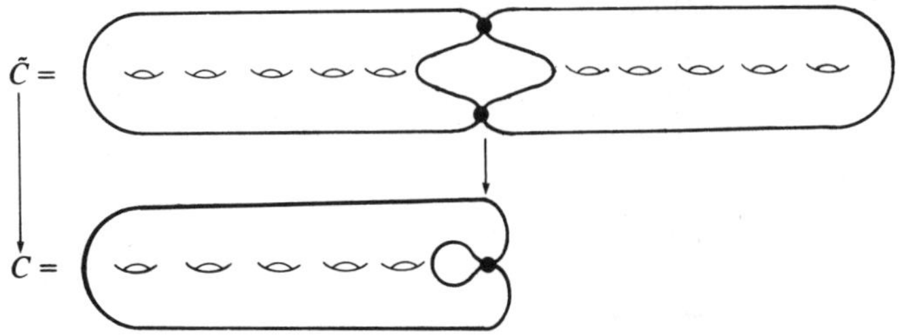

The second type of singular double cover from Beauville's list is the following [2, p.74]: Choose any point $p \in X$ and any elliptic curve with base point $(E, 0)$, and define C as follows:

$$C = X \cup E/_{p\sim 0}.$$

Now choose one of the three unramified double covers of E, $\tau : \tilde{E} \to E$, and let $\tau^{-1}(0) = \{\epsilon_1, \epsilon_2\} \subset \tilde{E}$. If X_1, X_2 are again isomorphic copies of X, then τ extends uniquely to a double cover of C by the curve $\tilde{C}$ where

$$\tilde{C} = X_1 \cup \tilde{E} \cup X_2/_{p_1\sim\epsilon_1, p_2\sim\epsilon_2}.$$

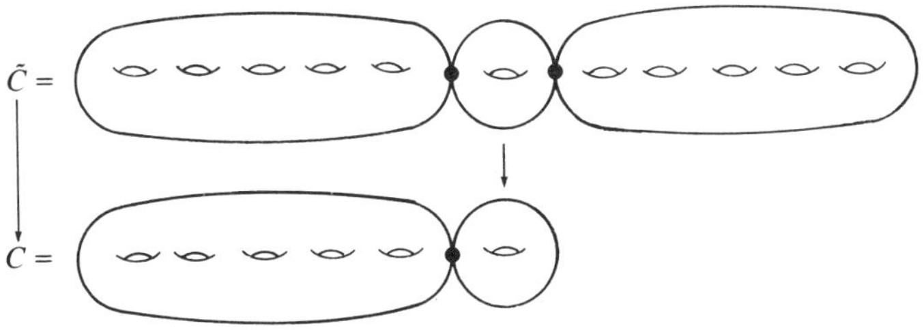

The Prym variety is again $J(X)$. The parameters for this construction are the point $p \in X$ and the double cover $\tau : \tilde{E} \to E$, so the parameter space for this component is $X \times \mathcal{R}_1$, ($\mathcal{R}_g$ = moduli space of connected, unramified double covers of smooth curves of genus g.)

Finally, the two types of singular double covers have a common degeneration

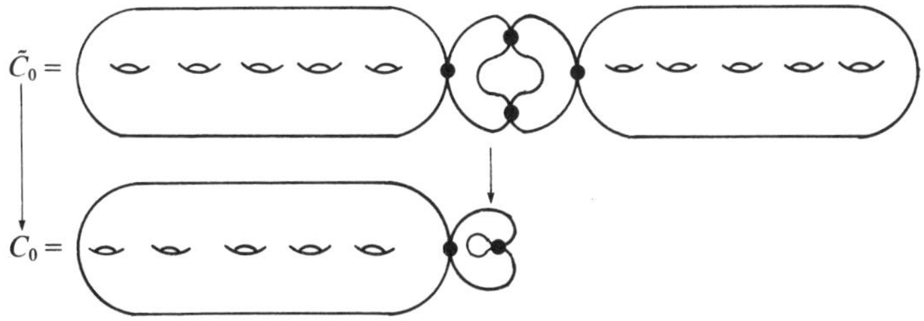

Therefore, when completed, the two irreducible components of singular curves in $P^{-1}(J(X))$ form one connected component parametrized by the reducible surface W defined as follows:

$$W = (X^{(2)} \cup X \times \bar{\mathcal{R}}_1) \Big/ \left(\text{diagonal of } X^{(2)} \sim X \times \left\{ \begin{matrix} \text{the étale, connected,} \\ \text{double cover of the} \\ \text{rational elliptic curve} \end{matrix} \right\} \right)$$

Moreover, since $\bar{\mathcal{R}}_1$ is a triple cover of $\mathbf{P}^1$,

$$\bar{\mathcal{R}}_1 \to \bar{\mathcal{M}}_1 \underset{j}{\overset{\simeq}{\to}} \mathbf{C} \cup \{\infty\} \simeq \mathbf{P}^1$$

branched simply over $j = 1728$ and $j = \infty$, and doubly over $j = 0$, [7, p. 321], Hurwitz' formula tells us that $\bar{\mathcal{R}}_1 \simeq \mathbf{P}^1$.

Thus the completed fibre $P^{-1}(J(X))$ has three connected components, we know their geometry, and we want to calculate their contributions to the degree of P.

7. The degree of P

Since P is proper and $J(X)$ is a smooth point on $\mathcal{A}_5$, the degree of P is the sum of the local degrees at each of the connected components of $P^{-1}(J(X))$. We will consider each of these individually.

(i) The plane quintic component is a single point at which Beauville proved the differential P_* is invertible; therefore the local degree is one [3, p.381, p.385].

(ii) The method we describe next for treating the component W of singular curves in $P^{-1}(J(X))$ was generously outlined for us by C.H. Clemens. A picture of a 3-dimensional cross section of the Prym map, transversal to the Jacobi locus at $J(X)$, and to the locus of singular curves at points of W, might appear as follows.

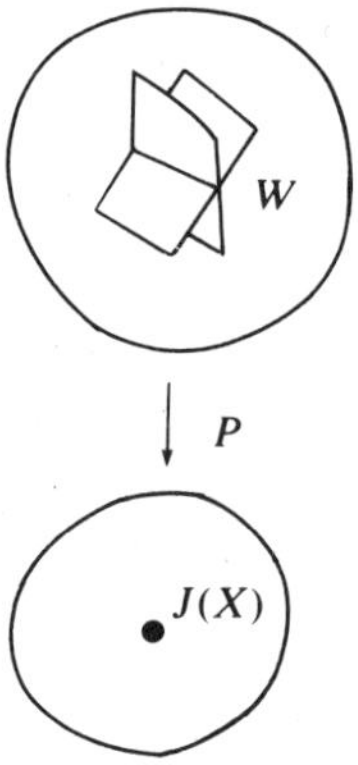

Let A be a generic point near $J(X)$ over which the fibre $P^{-1}(A)$ is finite, and let A approach $J(X)$ along a fixed direction normal to $\mathcal{J}_5$. The fundamental idea is that the points of the generic fibre $P^{-1}(A)$ will in general approach a finite set of limit points on the special fibre W.
This defines an equivalence relation on W whose quotient space is the set of directions normal to $\mathcal{J}_5$ at $J(X)$. If one can find this equivalence relation, the number of elements in one equivalence class will be the local degree of the map. Now observe two things:

(i) The quotient space of this equivalence relation is isomorphic to $\mathbf{P}^2$, since the Jacobi locus $\mathcal{J}_5$ has codimension three in $\mathcal{A}_5$.

(ii) Since $X = \mathcal{Q}_0 \cap \mathcal{Q}_1 \cap \mathcal{Q}_2 \subset \mathbf{P}^4$ is a complete intersection of three quadrics [12, p.157], the set of quadrics containing X in $\mathbf{P}^4$ is also isomorphic to $\mathbf{P}^2$:

$$t_0\mathcal{Q}_0 + t_1\mathcal{Q}_1 + t_2\mathcal{Q}_2 \leftrightarrow [t_0, t_1, t_2] \in \mathbf{P}^2$$

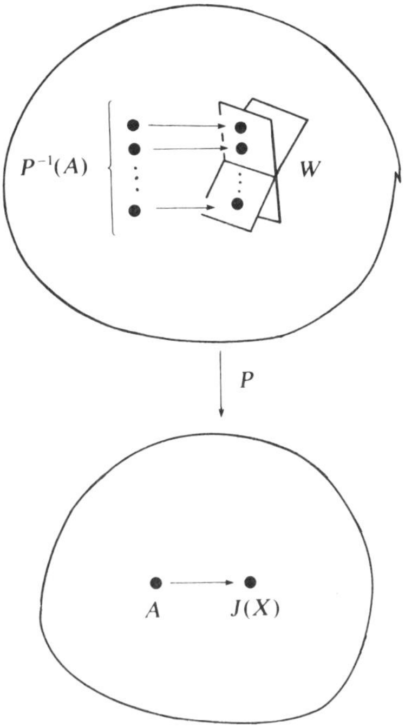

Then Clemens made the following surpassingly beautiful conjecture:

(1) The quotient space of the equivalence relation defined by the Prym map on $P^{-1}(J(X))$ is

$(\mathbf{P}^2)^*$ = the set of all pencils of quadrics containing X in $\mathbf{P}^4$.

(2) On the irreducible component parametrized by $X^{(2)}$,

$$(p, q) \sim (p', q') \Leftrightarrow \text{the secant lines } \overline{pq} \text{ and } \overline{p'q'} \text{ lie on the same pencil of quadrics through } X.$$

$$\Leftrightarrow \overline{pq} \text{ and } \overline{p'q'} \text{ lie on the same quartic del Pezzo surface through } X.$$

(3) Consequently, the contribution of $X^{(2)}$ to the degree of P is 16. (Recall that a generic pencil of quadrics in $\mathbf{P}^4$ intersects in a smooth quartic del Pezzo surface which contains exactly 16 lines [6, p. 550].

Let's check (1): Since $X = \mathcal{Q}_0 \cap \mathcal{Q}_1 \cap \mathcal{Q}_2 \subset \mathbf{P}^4$ is canonically embedded [7, p. 346], and since the codifferential J^* of the immersion $J : \mathcal{M}_5 \hookrightarrow \mathcal{J}_5 \subset \mathcal{A}_5$ is just the restriction map from quadratic polynomials on $\mathbf{P}^4 \simeq \mathbf{P}(H^\circ(X; \Omega^1)^*)$ to quadratic differentials on the canonical curve [5, p.806, p.843], it follows that:

$$\{\text{pencils of quadrics containing } X \text{ in } \mathbf{P}^4\}$$
$$\simeq \{\text{lines in } \mathbf{P}(\mathrm{Ker}\ J^*)\}$$
$$\simeq \{\text{pencils of tangent hyperplanes at } J(X) \text{ in } \mathcal{A}_5 \text{ which contain the tangent space along } \mathcal{J}_5\}$$

$\simeq$ {normal directions to $\mathcal{J}_5$ at $J(X)$}

$\simeq$ the quotient space of the equivalence relation on $P^{-1}(J(X))$.

Q.E.D.

The idea of the proof of (2) is to formulate the equivalence relation on $X^{(2)}$ in terms of the codifferential P^*, and then to compute the kernel of P^* analogously to the computation of the kernel of J^*[3, p. 381]. Looking at our three dimensional cross-section of $\bar{\mathcal{M}}_6$ containing $X^{(2)}$, we note that $X^{(2)}$ has codimension one there and P is constant along $X^{(2)}$, so that P^* is expected to have rank one. Assume this is true at the points (p, q) and (p', q') in $X^{(2)}$. Then we claim that $(p, q) \sim (p', q')$ if and only if P_* maps the tangent spaces at (p, q) and (p', q') both onto the same subspace of tangent vectors at $J(X)$. For $(p, q) \sim (p', q')$ means there is some curve σ approaching $J(X)$ along a fixed direction in $\mathcal{A}_5$ and that $P^{-1}(\sigma)$ contains at least one curve approaching (p, q) and one approaching (p', q'). Then the tangent directions determined by these two curves at (p, q) and (p', q') respectively are both mapped by P_* to the direction at $J(X)$ determined by σ. Thus image $P_{*,(p,q)}$ = image $P_{*,(p',q')}$. Since image $P_* = (\text{Ker } P^*)^{\perp}$, this is equivalent to saying $\text{Ker } P^*_{(p,q)} = \text{Ker } P^*_{(p',q')}$.
But just as $\mathbf{P}(\text{Ker } J^*_X) \simeq$ {quadrics containing the canonical curve $X \subset \mathbf{P}^4$}, one can prove that

$$\mathbf{P}(\text{Ker } P^*_{(p,q)}) \simeq \{\text{quadrics containing the "Prym-canonical" curve } X \cup \overline{pq} \subset \mathbf{P}^4\},$$

and this gives (2).

(iii) Now consider the trigonal component of $P^{-1}(J(X))$, parametrized by $W^1_4(X) \simeq \text{sing } \Theta(X)$. This fibre is one-dimensional, whereas the set of directions normal to $\mathcal{J}_5$ at $J(X)$ is two-dimensional. Therefore as generic points A_1, A_2 approach $J(X)$ from different directions it will sometimes happen that the two fibres $P^{-1}(A_1)$ and $P^{-1}(A_2)$ will converge to some common points on W^1_4, although from different directions.

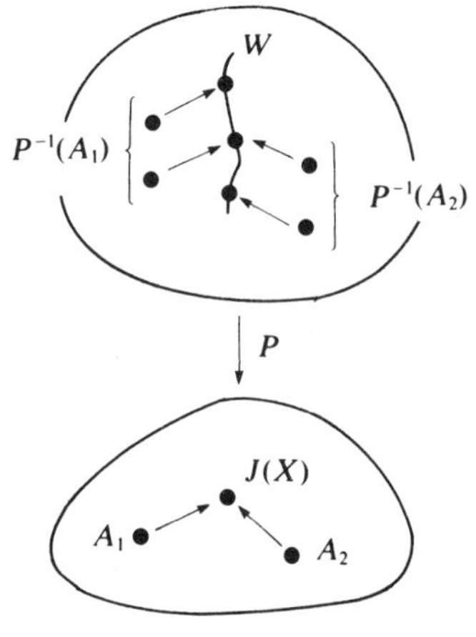

Therefore the equivalence relation in this case is defined not on W_4^1 itself but on the $\mathbf{P}^1$ bundle of directions normal to W_4^1, with the same quotient space as before.

So if $N =$ the projective normal bundle to W_4^1, in ($\bar{\mathscr{R}}_6$, it's the bundle of directions normal to the trigonal locus at points of $P^{-1}(J(X)) \cap \{\text{trigonals}\}$), and if $(\mathbf{P}^2)^* = \{$pencils of quadrics through X in $\mathbf{P}^4\}$, then the equivalence relation should be given by a natural map

$$f: N \to (\mathbf{P}^2)^*.$$

Since f is induced by the differential of P, the technical condition needed here is that rank $P_* =$ one at every point of W_4^1. Assuming that, the map f exists and is linear and injective on the fibers of the $\mathbf{P}^1$–bundle N. Accordingly, f represents N as a family of lines in $(\mathbf{P}^2)^*$ parametrized by W_4^1. Since there is a universal family of lines in $(\mathbf{P}^2)^*$, the family $N \xrightarrow{\sigma} W_4^1$ is induced by some natural classifying map $W_4^1 \to \mathbf{P}^2$. In fact, if Δ is the plane quintic parametrizing singular quadrics through X in $\mathbf{P}^4$, then the classifying map is just the double cover we described in section §4, of Δ by W_4^1, followed by the inclusion $\Delta \subset \mathbf{P}^2$.

Let's recall the definition of the double cover $\eta: W_4^1 \to \Delta$. If $|D| \in W_4^1$ is a pencil of divisors of degree 4 on X, then

$$\eta(|D|) = \bigcup_{D \in |D|} \bar{D} = \text{the singular quadric in } \mathbf{P}^4 \text{ ruled by the planes spanned by the divisors } D \in |D|.$$

Therefore if $E = \{\langle x, l \rangle : x \in l\} \subset \Delta \times (\mathbf{P}^2)^*$ is the restriction to Δ of the universal family of lines in $(\mathbf{P}^2)^*$, and $\pi_1: E \to \Delta$, $\pi_2: E \to (\mathbf{P}^2)^*$ are the two projections, then there is a pull-back diagram

$$\begin{array}{ccc} N & \xrightarrow{\tilde{\eta}} & E \\ {\scriptstyle \sigma}\downarrow & & \downarrow{\scriptstyle \pi_1} \\ W_4^1 & \xrightarrow{\eta} & \Delta \end{array}$$

such that $f = \pi_2 \circ \tilde{\eta}$. Consequently, degree $(f) =$ degree $(\pi_2) \cdot \text{degree}(\tilde{\eta}) = 5 \cdot 2 = 10$.

We want to verify that the classifying map η is really the double cover of Δ given above. Let $|D| \in W_4^1$. Since the diagram commutes, we have

$$\begin{aligned} \eta(|D|) &= \pi_1(\tilde{\eta}(\sigma^{-1}(|D|))) \\ &= \pi_1(\{\langle \eta(|D|), \ell \rangle : \ell \in (\mathbf{P}^2)^* \quad \text{and} \quad \eta(|D|) \in \ell\}) \\ &= \bigcap_{\ell \in \pi_2(\tilde{\eta}(\sigma^{-1}(|D|)))} \in \mathbf{P}^2 \end{aligned}$$

$$= \bigcap_{n \in \sigma^{-1}(|D|)} f(n), \text{ (since } \pi_2 \circ \tilde{\eta} = f),$$

$$= \bigcap_{n \in \sigma^{-1}(|D|)} P^{*-1}_{|D|}(n),$$

(when directions normal to a given locus are identified with hyperplanes in the space of covectors orthogonal to that locus, f is identified with P^{*-1})

$$= \text{Ker } P^{*}_{|D|}$$

= the unique quadric in $\mathbf{P}^4$ containing the (Prym-canonical) trigonal curve C defined by the points of intersection of the pairs of diagonals of the divisors $D \in |D|$.

$$= \bigcup_{D \in |D|} \bar{D}, \qquad \text{Q.E.D.}$$

(iv) It can be shown that the other irreducible piece of the connected component of singular curves in $P^{-1}(J(X))$ (the set of curves with "elliptic tails", parametrized by $X \times \bar{\mathcal{R}}_1$), is blown down by P onto the diagonal of $X^{(2)}$ and thus contributes zero to the local degree of P.

Consequently, degree $P = 16 + 10 + 1 = 27$.

Remark: The first named author has recently shown that the points of a generic fibre of P carry a natural relation isomorphic to the incidence relation of lines on a non-singular cubic surface!

REFERENCES

[1] A. ANDREOTTI and A. MAYER: On period relations for abelian integrals on algebraic curves, Ann. Scuola Norm. Sup. Pisa *21* (1967) 189–238.

[2] A. BEAUVILLE: Prym varieties and Schottky problem, Invent. Math. *41* (1977) 149–196.

[3] A. BEAUVILLE: Variétés de Prym et Jacobiennes intermédiaires, Ann. Sci. Ecole Norm. Sup. *10* (1977) 309–391.

[4] H. CLEMENS and P. GRIFFITHS: The intermediate Jacobian of the cubic threefold, Annals of Math. *95* (1972) 281–356.

[5] P. GRIFFITHS: Periods of integrals on algebraic manifolds II, Am. J. Math. *90* (1968) 805–865.

[6] P. GRIFFITHS and J. HARRIS: Principles of Algebraic Geometry, John Wiley and Sons, New York, 1978.

[7] R. HARTSHORNE: Algebraic Geometry, Springer-Verlag, New York, 1977.

[8] L. MASIEWICKI: Universal properties of Prym varieties with an application to algebraic curves of genus five, Trans. Am. Math. Soc., *222* (1976) 221–240.

[9] D. MUMFORD: Prym Varieties I, in Contributions to Analysis, Academic Press, New York, 1974.

[10] J.P. MURRE: Reduction of the proof of non-rationality of a non-singular cubic threefold to a result of Mumford, Compositio Math., *27* (1973) 63–82.

[11] S. RECILLAS: Jacobians of curves with a g_4^1 are Prym varieties of trigonal curves, Bol. Soc. Mat. Mexicana, *19* (1974) 9–13.
[12] B. SAINT-DONAT: On Petri's analysis of the linear system of quadrics through a canonical curve, Math. Ann. *206* (1973) 157–175.
[13] A. TJURIN: Five lectures on three-dimensional varieties, Russian Math. Surveys, vol. *27* (1972).
[14] W. WIRTINGER: Untersuchungen Über Theta Funktionen, Teubner, Berlin, 1895.

R. Smith
Department of Mathematics
University of Georgia
Athens, Georgia 30602 (USA)

R. Donagi
Department of Mathematics
University of Utah
Salt Lake City, Utah 84112 (USA)

THE LOCAL TORELLI PROBLEM FOR ALGEBRAIC CURVES

Frans Oort and Joseph Steenbrink

Let C be an algebraic curve. Its Jacobian variety Jac(C) has a canonical polarization Θ_C. In this way

$$C \longmapsto (\mathrm{Jac}(C), \Theta_C)$$

defines a morphism

$$j: M_g \longrightarrow A_{g,1}$$

the *Torelli mapping* (we work over a field, M_g is the coarse moduli scheme for curves of genus g, and $A_{g,1}$ is the coarse moduli scheme for principally polarized abelian varieties). It is Torelli's theorem that j is injective on geometric points.

Local Torelli problem for curves: is j a (locally closed) *immersion*?

In this paper we prove that j is an immersion if char(k) = 0, that it is an immersion at almost all points if char(k) = $p > 2$ (cf. Corollary (3.2)), and that j is not an immersion if char(k) = 2 and $g \geq 5$ (cf. Corollary (5.3)).

We give a brief description of the proofs we are going to present. For more details, and for a discussion of related results we refer to detailed description in the paper.

The coarse moduli scheme M_g does not represent the moduli functor; in particular tangent vectors to M_g are not directly associated with infinitesimal deformations of curves. In order to improve this, we study the fine moduli schemes

$$j^{(n)}: M_g^{(n)} \longrightarrow A_{g,1}^{(n)}.$$

However, $j^{(n)}$ is a 2–1 mapping outside the hyperelliptic locus. Thus we factor this map

$$M_g^{(n)} \longrightarrow M_g^{(n)}/\Sigma = V^{(n)} \stackrel{\iota}{\longrightarrow} A_{g,1}^{(n)}.$$

Then ι is injective, and the main theorem of this paper asserts that ι is an immersion (in case char(k) = 0, or $p > 2$).

For a proof, first we study the tangent space at $V^{(n)}$ at a "hyperelliptic point" $y \in V^{(n)}$. It turns out that(!)

$$\dim(T_{V^{(n)},y}) = \tfrac{1}{2}g(g+1) = \dim(A_{g,1}^{(n)}).$$

The injectivity (and hence the surjectivity) of $d\iota : T_{V^{(n)},y} \to T_{A_{g,1}^{(n)},z}$ we prove in two ways. Our first proof (Section 3) is *inspired by a result of Griffiths*: for a curve C, with $x \in M_g^{(n)}$, $z \in A_{g,1}^{(n)}$ the corresponding points, the co-tangent space at z is the space of symmetric tensors in $\bigotimes^2 H^0(C, \Omega_C^1)$; with this identification the period mapping can be described explicitly (cf. Theorem (2.6) and (3.6)), which enables us to compute $d\iota$ on a basis, and the result follows.

Our second proof uses the methods in the *proof by Andreotti of Torelli's theorem.* The "dual of $\phi_K(C)$", where

$$\phi_K : C \longrightarrow \mathbf{P}(H^0(C, \Omega_C^1)^D)$$

is the mapping defined by the canonical series K on C, is a hypersurface

$$h \subset \mathbf{P}(H^0(\Omega_C^1))$$

of degree $6g-6$; this h is canonically the branch locus of the Gauss mapping defined by $(\mathrm{Jac}(C), \Theta_C)$. This method reduces computations of the period mapping $j^{(n)}$ to a computation of the discriminant locus h; this can be done locally (cf. Section 4),

$$h \in H = \text{Hilbert scheme of hypersurfaces of degree } 6g-6 \text{ in } \mathbf{P},$$

$$\begin{array}{ccc} V^{(n)} & \xrightarrow{\ \iota\ } & A_{g,1}^{(n)} \\ & {\scriptstyle\beta}\searrow \quad \swarrow{\scriptstyle\gamma} & \\ & H & \end{array}$$

then $d\beta$ is shown to be injective, and injectivity of $d\iota$ follows.

We have chosen to present both proofs. Perhaps the first method can be applied to other situations, it is short and direct. The second proof is somewhat more complicated, but the underlying geometric idea is very transparent, and the only difficulties are of a trivial algebraic nature.

We hope that the terminology "local Torelli problem" does not cause confusion (cf. Section 2).

We could not find a proof for our main theorem in the literature; probably it was not proved in the classical literature.

We thank A. Andreotti, W.L. Baily, W. van der Kallen, D. Lieberman, D. Mumford, C. Peters and K. Ueno for information and for stimulating discussions.

Contents

§1. Moduli of curves and abelian varieties 159
§2. Deformation theory and the infinitesimal Torelli problem 165
§3. The local Torelli problem, char(k) $\neq 2$, first proof 176
§4. The local Torelli problem, char(k) $\neq 2$, second proof 186
§5. Some remarks, and the case char(k) $= 2$ 197
References 203

§1. Moduli of curves and abelian varieties

In the sequel k is an algebraically closed field of arbitrary characteristic $p \geq 0$. All schemes considered are k-schemes, and "curve" always means: irreducible smooth projective algebraic curve.

(1.1) Definition: Let S be a noetherian scheme. A *curve of genus g over S* is a morphism

$$\pi : C \longrightarrow S$$

which is proper and smooth and whose geometric fibres are irreducible curves of genus g (cf. [20], page 98, Definition 5.3).

We define a functor $\mathcal{M}_g$ from the category of schemes to the category of sets by

$$\mathcal{M}_g(S) = \text{the set of } S\text{-isomorphism classes of curves of genus } g \text{ over } S.$$

(1.2) Definition: A *coarse moduli scheme* for the functor $\mathcal{M}$: Schemes $\rightarrow$ Sets is a scheme M together with a morphism of functors

$$\phi : \mathcal{M} \longrightarrow h_M = \mathrm{Hom}(-, M)$$

such that

(1) for any algebraically closed field Ω, the map

$$\phi_\Omega : \mathcal{M}(\mathrm{Spec}(\Omega)) \longrightarrow h_M(\mathrm{Spec}(\Omega))$$

is bijective;

(2) for any scheme N and any morphism of functors

$$\theta : \mathcal{M} \longrightarrow h_N$$

there exists a unique morphism of schemes

$$\xi : M \to N$$

such that

$$\theta = h_\xi \circ \phi$$

(cf. [20], page 99).

(1.3) THEOREM: *A coarse moduli scheme M_g for curves of genus g exists; M_g is a normal quasi-projective variety of dimension* $3g-3$ (*if* $g \geq 2$).

For $g=0, 1$ this is easy: $M_0 = \mathrm{Spec}(k)$ and $M_1 =$ the affine line over k, with ϕ given by the j-invariant.

For $g \geq 2$, Mumford has constructed M_g as the geometric quotient of the locally closed subscheme Z of the Hilbert scheme of curves in $\mathbf{P}^{5g-6}$ with genus g and degree $6g-6$, consisting of tricanonically embedded curves, under the action of $PGL(5g-6)$ (cf. [20], p. 103, Proposition 5.3 and 5.4, and page 143, Theorem 7.13). Mumford's proof uses the coarse moduli space for principally polarized abelian varieties; his notion of reductive is, what nowadays is called linearly reductive, and very few groups in positive characteristic have this property. However, by a theorem of Haboush (cf. [11], Theorem 5.2) every reductive group is geometrically reductive, and Seshadri showed that geometric invariant theory works for geometrically reductive groups (cf. [26] and [27] for more details). Hence the geometric quotient $Z/PGL(5g-6)$ exists. The scheme M_2 over Spec($\mathbf{Z}$) has been described explicitly by Igusa [13].

(1.4) DEFINITION: Let S be a noetherian scheme. An *abelian scheme over* S is a group scheme

$$\pi : X \longrightarrow S$$

for which π is smooth and proper and has connected geometric fibres.

For any abelian scheme X over S, there is a dual abelian scheme, denoted by X^t (cf. [20], p. 118 for π projective). If L is an invertible sheaf on X, it induces a map

$$\Lambda(L) : X \longrightarrow X^t.$$

(1.5) DEFINITION: Let $\pi : X \to S$ be an abelian scheme. A *polarization* of X is an S-homomorphism

$$\lambda : X \longrightarrow X^t$$

such that for every geometric point s of S, the induced homomorphism

$$\lambda_s : X_s \longrightarrow X_s^t$$

is of the form $\Lambda(L)$ for some ample invertible sheaf L on X_s; we call λ a *principal polarization* if λ is an isomorphism.

We define a functor

$$\mathcal{A}_{g,1} : \text{Schemes} \longrightarrow \text{Sets}$$

by

$$\mathcal{A}_{g,1}(S) = \text{the set of } S\text{-isomorphism classes of principally polarized abelian schemes over } S \text{ of relative dimension } g.$$

(1.6) THEOREM: *A coarse moduli scheme $A_{g,1}$ for principally polarized abelian varieties of dimension g (i.e. for the functor $\mathcal{A}_{g,1}$) exists; $A_{g,1}$ is a normal quasi-projective variety of dimension $g(g+1)/2$.*
Cf. [20], Theorem 7.10, page 139.

If S is a noetherian scheme and $\pi: C \to S$ a curve over S, then the Jacobian $J(C/S) = J$ is a principally polarized abelian scheme over S: there exists a canonical isomorphism

$$\theta : J \longrightarrow J^t$$

(cf. Mumford [20], proposition 6.9). Hence there exists a unique morphism

$$j : M_g \longrightarrow A_{g,1}$$

such that the diagram of functors

$$\begin{array}{ccc} \mathcal{M}_g & \xrightarrow{j} & \mathcal{A}_{g,1} \\ \downarrow & & \downarrow \\ h_{M_g} & \xrightarrow{h_j} & h_{A_{g,1}} \end{array}$$

commutes. Remark that the vertical arrows are not isomorphisms: due to the existence of non-trivial automorphisms on special curves and abelian varieties the functors $\mathcal{M}_g$ and $\mathcal{A}_{g,1}$ are not representable.

Let X be an abelian scheme over S of relative dimension g. Let n be a positive integer, not divisible by p. Then the kernel ${}_nX$ of multiplication by n is an étale group scheme over S, locally isomorphic in the étale topology to the constant group scheme $(\mathbf{Z}/n)^{2g}$. Denote by μ_n the group scheme of nth roots of unity over S. Then a principal polarization λ of X determines a skew symmetric bilinear form

$$E_n : {}_nX \times_S {}_nX \longrightarrow \mu_n$$

as follows.

First remark that ${}_n(X^t)$ is the Cartier dual group scheme $({}_nX)^D$ of ${}_nX$ (cf. [21], §15, Theorem 1 and [23], Corollary 19.2) and that one has a canonical map

$$e_n : {}_nX \times ({}_nX)^D \longrightarrow \mu_n.$$

One defines

$$E_n = e_n \circ (1 \times \lambda).$$

We provide $(\mathbf{Z}/n)^{2g}$ with the standard symplectic structure, described by the $2g \times 2g$ matrix

$$\begin{pmatrix} 0 & I_g \\ -I_g & 0 \end{pmatrix}.$$

(1.7) Definition: Let (X, λ) be a principally polarized abelian scheme over S, $n \in \mathbf{N}$, $p \nmid n$. A *level n structure* on X consists of a symplectic isomorphism

$$\alpha : ({}_nX, E_n) \longrightarrow (\mathbf{Z}/n)^{2g}.$$

If C is a curve of genus g over S, a level n structure on C is a level n structure on $J(C/S)$.

This definition is different from the one in Mumford's book [20], because Mumford does not consider the symplectic structure. Our notion is equivalent with Popp's (cf. [26]). For the local Torelli problem the symplectic structure is not important.

One defines functors $\mathcal{M}_g^{(n)}, \mathcal{A}_{g,1}^{(n)}$: Schemes → Sets by

$\mathcal{M}_g^{(n)}(S)$ = the set of S-isomorphism classes of curves of genus g over S with level n structure;

$\mathcal{A}_{g,1}^{(n)}(S)$ = the set of S-isomorphism classes of principally polarized abelian schemes over S of relative dimension g with level n structure.

(1.8) Theorem: *If $n \geq 3$ with $(p, n) = 1$, the functor $\mathcal{M}_g^{(n)}$ is represented by a smooth quasi-projective variety $M_g^{(n)}$, which is a Galois covering of M_g with group* $\mathrm{Sp}(g, \mathbf{Z}/n)$ if $g \geq 3$.

(1.9) THEOREM: *If $n \geq 3$ with $(p, n) = 1$, the functor $\mathscr{A}^{(n)}_{g,1}$ is represented by a smooth quasi-projective variety $A^{(n)}_{g,1}$, which is a Galois covering of $A_{g,1}$ with group* $\mathrm{Sp}(g, \mathbf{Z}/n)/\{\pm I\}$.

Theorem (1.9) can be found in Mumford's book ([20], Theorem 7.9). We sketch a proof of theorem (1.8), following Popp [26].

One considers the scheme Z (cf. (1.3)) which parametrizes tricanonically embedded curves of genus g. Denote by

$$\pi : \Gamma \longrightarrow Z$$

the corresponding family of curves and by

$$q : J(\Gamma/Z) \longrightarrow Z$$

the Jacobian of Γ. Then ${}_nJ(\Gamma/Z)$ is an étale group scheme over Z, on which $\pi_1(Z)$ acts by symplectic automorphisms. Denote by $Z^{(n)}$ the finite unramified covering of Z with this monodromy group as a Galois group. Then ${}_nJ(\Gamma/Z)$ becomes a constant group scheme over $Z^{(n)}$, hence

$$\Gamma^{(n)} = \Gamma \times_Z Z^{(n)}$$

carries a universal level n structure.

Moreover $PGL(5g-6)$ acts on $\Gamma^{(n)} \to Z^{(n)}$ without fixed points. Hence the quotient

$$M^{(n)}_g = Z^{(n)}/PGL(5g-6)$$

is a smooth algebraic space, and according to work of Mumford and Knudsen [14, 22] even a quasi-projective variety. The universal family over $M^{(n)}_g$ is $\Gamma^{(n)}/PGL(5g-6)$.

Clearly one has a morphism

$$j^{(n)} : M^{(n)}_g \longrightarrow A^{(n)}_{g,1}$$

representing the morphism of functors

$$J : \mathscr{M}^{(n)}_g \longrightarrow \mathscr{A}^{(n)}_{g,1}.$$

Remark that $j^{(n)}$ has generically degree two onto its image, if $g \geq 3$. Namely, if C is a curve of genus g with $\mathrm{Aut}(C) = \{id_C\}$, then for a level n structure α on C, (C, α) and $(C, -\alpha)$ are not isomorphic, whereas $(J(C), \alpha)$ and $(J(C), -\alpha)$ are isomorphic, because $-id \in \mathrm{Aut}(J(C))$. This implies that j and $j^{(n)}$ are not very well comparable. To circumvent this fact, we define an automorphism Σ on $M^{(n)}_g$ by

$$\Sigma(C, \alpha) = (C, -\alpha).$$

If $g \leq 2$ then $\Sigma = id$; Σ is an involution if $g \geq 3$. The geometric quotient

$$V^{(n)} = M_g^{(n)}/\Sigma$$

exists and is a Galois covering of M_g with Galois group $\mathrm{Sp}(g, \mathbf{Z}/n)$ (cf. Mumford [21], section II.7). One has the following commutative diagram, on which $\mathrm{Sp}(g, \mathbf{Z}/n)$ acts:

(1.10)

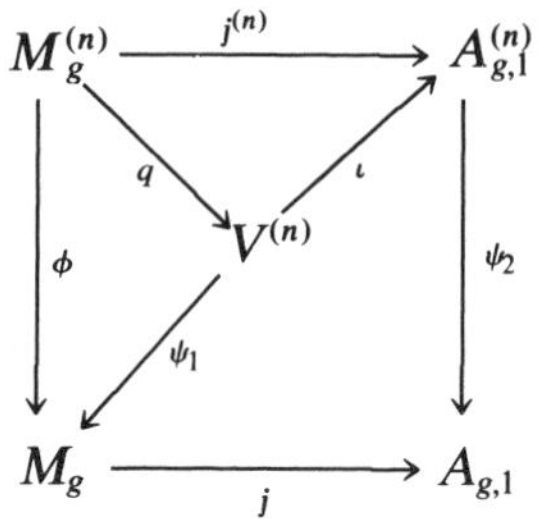

In this diagram, j is injective (Torelli's theorem, cf. [1, 18, 19, 30, 31]), ψ_1 and ψ_2 are Galois with the same group.

(1.11) LEMMA: *The map* $\iota: V^{(n)} \to A_{g,1}^{(n)}$ *is injective (on geometric points).*

PROOF: Let C and C' be curves over $\mathrm{Spec}(\Omega)$ where Ω is an algebraically closed field, and let α and α' be level n structures on C and C' respectively.

For an isomorphism $\sigma: C \to C'$ we denote the induced isomorphism $J(C) \to J(C')$ by $J(\sigma)$. Suppose we have an isomorphism

$$\tau: (J(C), \Theta_C, \alpha) \longrightarrow (J(C'), \Theta_{C'}, \alpha')$$

of principally polarized abelian varieties with level n structure. Then according to Matsusaka [19] there exists an isomorphism σ of C with C' such that

$$J(\sigma) = \pm\tau.$$

If $J(\sigma) = \tau$, then $\tau: (C, \alpha) \xrightarrow{\sim} (C', \alpha')$, and if $J(\sigma) = -\tau$, then $\tau: (C, \alpha) \xrightarrow{\sim} (C', -\alpha')$. Hence

$$(C, \alpha) \xrightarrow{\sim} (C', \alpha') \quad \text{or} \quad (C, \alpha) \xrightarrow{\sim} \Sigma(C', \alpha').$$

§2. Deformation theory and the infinitesimal Torelli problem

We describe the deformation theory of curves and abelian varieties over a field k, $k = \bar{k}$.

(2.1) DEFINITION: We let $\mathscr{C}_k$ denote the category of local artinian k-algebras. If X is a k-scheme and $R \in \mathrm{Ob}(\mathscr{C}_k)$, a *deformation of X over R* is a flat R-scheme $\mathscr{X}_R$ together with an isomorphism

$$\mathscr{X}_R \otimes_R k \xrightarrow{\sim} X.$$

Two deformations $\mathscr{X}_R$ and $\mathscr{X}'_R$ of X over R are *isomorphic* if there exists an R-isomorphism

$$\phi : \mathscr{X}_R \longrightarrow \mathscr{X}'_R$$

which induces the identity on X, i.e. $\phi \otimes_R k = id_X$.

We define the *deformation functor* Def_X of X by

$$\mathrm{Def}_X(R) = \text{the set of isomorphism classes of deformations of } X \text{ over } R.$$

(2.2) THEOREM: *Suppose C is a curve of genus g, which is smooth and projective. Then* Def_C *is pro-represented by the formal power series ring*

$$A = k[[s_1, \ldots, s_{3g-3}]],$$

i.e. there is an isomorphism of functors $\mathscr{C}_k \to \mathrm{Sets}$:

$$\mathrm{Hom}_{k-alg}(A, R) \longrightarrow \mathrm{Def}_C(R).$$

(2.3) THEOREM: *Suppose X is an abelian variety of dimension g. Then* Def_X *is pro-represented by the formal power series ring*

$$B = k[[t_{ij};\, i, j = 1, \ldots, g]].$$

If (X, λ) is a principally polarized abelian variety, we denote by $\mathrm{Def}_{(X,\lambda)}$ the deformation functor of the pair (X, λ), where a deformation of (X, λ) over R is a principally polarized abelian scheme $(\mathscr{X}_R, \lambda_R)$ over R, with fibre (X, λ) over k.

(2.4) THEOREM: *If (X, λ) is a principally polarized abelian variety of dimension g, then* $\mathrm{Def}_{(X,\lambda)}$ *is pro-represented by the complete local ring*

$$B/(t_{ij} - t_{ji};\, i \neq j),$$

where B is the ring mentioned in Theorem (2.3).

Theorem (2.2) is a consequence of [9], Théorème 10, and the fact that

$$\mathrm{Def}_C(k[\epsilon]) \cong H^1(C, \mathcal{T}_C)$$

has dimension $3g-3$. Here $\epsilon^2 = 0$ and $\mathcal{T}_C$ is the tangent sheaf of C. Theorems (2.3) and (2.4) are due to Grothendieck and Mumford (see [24] for a more detailed discussion, and see Theorem (2.6) below).

If $k = \mathbf{C}$, a principally polarized abelian variety can be considered as a complex torus $X = \mathbf{C}^g/\Gamma$, where $\Gamma \cong \mathbf{Z}^{2g}$ can be normalized in such a way that it has a $\mathbf{Z}$-basis consisting of the columns of a matrix

$$\Omega = (I_g, Z)$$

with Z is the Siegel upper half space

$$\mathcal{S}_g = \{Z \in M(g, \mathbf{C}),\, {}^tZ = Z \text{ and } \mathrm{Im}(Z) > 0\}.$$

Small deformations of X arise when Z varies in $M(g, \mathbf{C})$, but the resulting complex tori are abelian varieties iff Z stays in $\mathcal{S}_g$. Hence, with notations as above, we obtain

$$B \cong \hat{\mathcal{O}}_{M(g,\mathbf{C}),Z} \cong \mathbf{C}[[z_{ij};\, i, j = 1, \ldots, g]]$$

and

$$\hat{\mathcal{O}}_{\mathcal{S}_g,Z} \cong \mathbf{C}[[z_{ij}]]/[z_{ij} - z_{ji};\, i \neq j].$$

One sees that $\dim \mathrm{Def}_X(\mathbf{C}[\epsilon]) = g^2$ and $\dim \mathrm{Def}_{(X,\lambda)}(\mathbf{C}[\epsilon]) = g(g+1)/2$.

Let (X, λ, α) be a principally polarized abelian variety over k with level n structure, where $p \nmid n$ and $n \geq 3$. If $R \in \mathrm{Ob}(\mathcal{C}_k)$ and $(\mathcal{X}_R, \lambda_R)$ is a deformation of (X, λ) over R, then there exists a unique extension α_R of the level n structure α to a level n structure of $\mathcal{X}_R$ over R. This is due to the fact that every étale group scheme over $\mathrm{Spec}(R)$ is constant. Hence $(\mathcal{X}_R, \lambda_R, \alpha_R)$ is induced by a unique morphism

$$\eta : \mathrm{Spec}(R) \longrightarrow A^{(n)}_{g,1}.$$

In this way one can show

(2.5) PROPOSITION: *Let (C, α) be a curve of genus g, $g \geq 1$, with level n structure, where $n \geq 3$. Denote by $x \in M^{(n)}_g$ the corresponding moduli point and let $z = j^{(n)}(x)$. Then there are natural isomorphisms*

$$A \xrightarrow{\sim} \hat{\mathcal{O}}_{M_g^{(n)},x}$$

$$B/(t_{ij} - t_{ji}) \xrightarrow{\sim} \hat{\mathcal{O}}_{A_{g,1}^{(n)},z}.$$

In particular $M_g^{(n)}$ *and* $A_{g,1}^{(n)}$ *are smooth of dimension* $3g-3$ *and* $g(g+1)/2$ *respectively.*

We can now formulate the main problem in this paper. To avoid confusion with respect to terminology, let us make a distinction between two different problems:

(a) *The infinitesimal Torelli problem*

This is the question whether $j^{(n)}$ is an immersion. It follows from Proposition (2.5) that one has an equivalent formulation as follows. Let C_0 be an algebraic curve over a field k, let R be a local artinian k-algebra, and let

$$\mathscr{C} \longrightarrow \mathrm{Spec}(R)$$

be an algebraic curve with $\mathscr{C} \otimes_R k \cong C_0$. Suppose

$$J(C_0) \otimes R \cong J(\mathscr{C})$$

(i.e., the Jacobian of C is constant over R). Does this imply that C is constant over R? Another equivalent formulation is this: does $j^{(n)}$ induce a surjection $B/(t_{ij} - t_{ji}) \to A$?

Answer to the infinitesimal Torelli problem:
"yes" if $g = 1$, $g = 2$ or $g \geq 3$ and C_0 not hyperelliptic,
"no" if $g \geq 3$ and C_0 hyperelliptic.
The proof is a combination of (2.6) and (2.7) below.

(b) *The local Torelli problem*

This is the question, whether the map $j: M_g \to A_{g,1}$ is an immersion, and in this paper we give a partial answer to it (Corollaries (2.8), (3.2) and (5.3)).

It should be mentioned that the infinitesimal Torelli problem is usually called "local Torelli problem" in the literature.

To treat the infinitesimal Torelli problem, we compute the tangent map

$$dj_x^{(n)}: T_{M,x} \longrightarrow T_{A,z}$$

where for brevity in the sequel we denote $\mathrm{Spec}(\mathcal{O}_{M_g^{(n)},x})$ by M and

$\mathrm{Spec}(\mathcal{O}_{A_{g,1,z}^{(n)}})$ by A. Our treatment is inspired by Beauville (cf. [4], Chapter VII). See Griffiths ([8], Section 1) for a transcendental approach.

We first discuss the Kodaira–Spencer map associated with a deformation. Let S be a k-scheme, $0 \in S$ a k-rational point and $\pi : \mathscr{X} \to S$ a smooth and proper morphism. Denote by X_0 the fibre over $0 \in S$. One has an exact sequence of sheaves on $\mathscr{X}$

$$0 \longrightarrow \mathscr{T}_{\mathscr{X}|S} \longrightarrow \mathscr{T}_{\mathscr{X}} \longrightarrow \pi^* \mathscr{T}_S \longrightarrow 0$$

defining the relative tangent sheaf $\mathscr{T}_{\mathscr{X}|S}$. By restriction to X_0 one obtains an exact sequence

$$0 \longrightarrow \mathscr{T}_{X_0} \longrightarrow \mathscr{T}_{\mathscr{X}} \otimes \mathcal{O}_{X_0} \longrightarrow T_{S,0} \otimes_k \mathcal{O}_{X_0} \longrightarrow 0.$$

From the exact sequence of cohomology we obtain a map

$$\kappa_{\pi,0} : T_{S,0} \longrightarrow H^1(X_0, \mathscr{T}_{X_0})$$

called the Kodaira–Spencer map. In the case $S = \mathrm{Spec}(k[\epsilon])$, where $\mathscr{X}$ is an infinitesimal deformation, we obtain a class

$$\kappa_{\pi,0}(\partial/\partial\epsilon) \text{ in } H^1(X_0, \mathscr{T}_{X_0}),$$

called the Kodaira–Spencer deformation class.

Remark that $\kappa_{\pi,0}$ is the fibre over 0 of a sheaf homomorphism

$$\kappa_\pi : \mathscr{T}_S \longrightarrow R^1\pi_* \mathscr{T}_{\mathscr{X}|S}.$$

Denote by C_0 the curve corresponding to the closed point $x \in M$ and let $\pi : \mathscr{C} \to M$ be the universal family. Likewise $\nu : \mathscr{X} \to A$ denotes the universal family of principally polarized abelian varieties. Associated with π is the universal family of Jacobians $\pi' : \mathscr{J}(\mathscr{C}/M) \to M$. Up to a finite unramified covering of M we may assume that π has a section. Let $u : \mathscr{C} \to \mathscr{J}(\mathscr{C}/M)$ be the corresponding natural map. We have a commutative diagram

$$\begin{array}{ccccc}
\mathscr{C} & \longrightarrow & \mathscr{J}(\mathscr{C}/M) & \longrightarrow & \mathscr{X} \\
\downarrow{\scriptstyle \pi} & & \downarrow{\scriptstyle \pi'} & & \downarrow{\scriptstyle \nu} \\
M & = & M & \xrightarrow{j(n)} & A
\end{array}$$

The induced map $u_x : C_0 \to J(C_0))$ gives an isomorphism

$$u_x^* : H^1(J(C_0), \mathcal{O}_{J(C_0)}) \longrightarrow H^1(C_0, \mathcal{O}_{C_0}).$$

Denote by T_0 the tangent space of $J(C_0)$ at zero. Because $J(C_0)$ is a group variety, one has a canonical isomorphism

$$\mathscr{T}_{J(C_0)} \xrightarrow{\sim} \mathscr{O}_{J(C_0)} \otimes_k T_0,$$

so

$$u_x^* \mathscr{T}_{J(C_0)} \cong \mathscr{O}_{C_0} \otimes_k T_0.$$

We have a commutative diagram

$$\begin{array}{ccc}
T_{M,x} & \xrightarrow{\kappa_{\pi,x}} & H^1(C_0, \mathscr{T}_{C_0}) \\
 & & \downarrow{\scriptstyle du_x} \\
\downarrow{\scriptstyle dj_x^{(n)}} & & H^1(C_0, u_x^* \mathscr{T}_{J(C_0)}) \\
 & \nearrow{\scriptstyle \beta} & \uparrow{\scriptstyle u_x^*} \\
T_{A,z} & & \\
 & \searrow{\scriptstyle \kappa_{\nu,z}} & H^1(J(C_0), \mathscr{T}_{J(C_0)})
\end{array}$$

in which u_x^* is just the natural isomorphism

$$u_x^* \otimes 1 : H^1(J(C_0), \mathscr{O}) \otimes_k T_0 \xrightarrow{\sim} H^1(C_0, \mathscr{O}) \otimes_k T_0.$$

If we denote by V^D the linear dual of a coherent sheaf or a k-vector space V, we obtain

$$(u_x^* \mathscr{T}_{J(C_0)})^D \cong u_x^* \Omega^1_{J(C_0)} \cong \mathscr{O}_{C_0} \otimes_k H^0(J(C_0), \Omega^1_{J(C_0)}) \cong \mathscr{O}_{C_0} \otimes_k H^0(C_0, \omega_{C_0})$$

where $\omega_{C_0} = \Omega^1_{C_0}$ is the dualizing sheaf of C_0, so by Serre duality

$$H^1(C_0, u_x^* \mathscr{T}_{J(C_0)})^D \cong H^0(C_0, \omega_{C_0}) \otimes_k H^0(C_0, \omega_{C_0}).$$

In the same way

$$H^1(C_0, \mathscr{T}_{C_0})^D \cong H^0(C_0, \omega_{C_0}^{\otimes 2}).$$

(2.6) THEOREM: *In the above diagram $\kappa_{\pi,x}$ is an isomorphism and $\kappa_{\nu,z}$ is injective. Moreover, under the above identifications, the map du_x is equal to the transpose of the pointwise multiplication map*

$$\mu : H^0(C_0, \omega_{C_0}) \bar{\otimes}_k H^0(C_0, \omega_{C_0}) \longrightarrow H^0(C_0, \omega_{C_0}^{\otimes 2})$$

and β can be considered as the transpose of the natural projection

$$H^0(C_0, \omega_{C_0}) \otimes_k H^0(C_0, \omega_{C_0}) \longrightarrow S^2 H^0(C_0, \omega_{C_0}).$$

PROOF: Because a level n structure on C_0 can be extended in a unique way to any infinitesimal deformation of C_0, it follows that $\kappa_{\pi,x}$ is an isomorphism. Similarly $\kappa_{\nu,z}$ identifies $T_{A,z}$ with those infinitesimal deformations of $J(C_0)$ to which the principal polarization can be extended. The statement about du_x is obvious from the preceding arguments. To compute β, we need some general facts on polarized abelian varieties. Suppose X is a smooth variety, $\mathscr{L}$ a line bundle on X and W an infinitesimal deformation of X. Denote by $\lambda \in \mathrm{Pic}(X)$ the class of $\mathscr{L}$ and by $w \in H^1(X, \mathscr{T}_X)$ the class of W. The map $\mathscr{O}_X^* \to \Omega_X^1$, defined by $f \mapsto f^{-1}df$, gives a homomorphism

$$\alpha : \mathrm{Pic}(X) \longrightarrow H^1(X, \Omega_X^1).$$

From the exact sequence

$$0 \longrightarrow \mathscr{O}_X \xrightarrow{\gamma} \mathscr{O}_W^* \longrightarrow \mathscr{O}_X^* \longrightarrow 1$$

where $\gamma(f) = 1 + f\epsilon$, we get a map

$$\delta : \mathrm{Pic}(X) \longrightarrow H^2(X, \mathscr{O}_X).$$

Claim. Under cup product

$$H^1(X, \mathscr{T}_X) \otimes H^1(X, \Omega_X^1) \xrightarrow{\cup} H^2(X, \mathscr{O}_X)$$

one has

$$\delta(\lambda) = w \cup \alpha(\lambda).$$

PROOF: Use Čech-cohomology with respect to an affine covering $\mathscr{U}$ of X. Represent λ by a 1-cocycle (f_{ij}), and w by (w_{ij}). If $X = \cup U_i$ then $W = \cup W_i$ where we have isomorphisms

$$\phi_i : U_i \times \mathrm{Spec}(k[\epsilon]) \longrightarrow W_i$$

such that for $a + b\epsilon \in \Gamma((U_i \cap U_j) \otimes k[\epsilon], \mathscr{O})$ one has

$$(\phi_i^{-1}\phi_j)^*(a + b\epsilon) = a + (b + w_{ij}(a))\epsilon.$$

Lift the cocycle f to the cochain $\tilde{f} \in C^1(\mathcal{U}, \mathcal{O}_W^*)$, which satisfies

$$\phi_i^*(\tilde{f}_{ij}) = f_{ij} + 0 \cdot \epsilon.$$

Then its Čech coboundary satisfies

$$\begin{aligned}\phi_i^*(\check{\partial}\tilde{f})_{ijk} &= \phi_i^*(\tilde{f}_{ij})\phi_i^*(\tilde{f}_{ik})^{-1}\phi_i^*(\tilde{f}_{jk}) = \\ &= f_{ij}f_{ik}^{-1}(\phi_i^{-1}\phi_j)^*(f_{jk}) = \\ &= 1 + \epsilon f_{jk}^{-1} w_{ij}(f_{jk}).\end{aligned}$$

So $\delta(\lambda)$ is represented by the 2-cocycle c with

$$c_{ijk} = f_{jk}^{-1} w_{ij}(f_{jk}) = \langle w_{ij}, f_{jk}^{-1} df_{jk}\rangle$$

while $\alpha(\lambda)$ is represented by the cocycle $(f_{ij}^{-1} df_{ij})$. Because on the cochain level the cup product is given by the formula

$$(a \cup b)_{i_0, \dots, i_{p+q}} = a_{i_0, \dots, i_p} \otimes b_{i_p, \dots, i_{p+q}}$$

(cf. [7], Chapitre II, Section 6.6), we have proved our claim.

Next we assume that X is an abelian variety, $\mathcal{L}$ a line bundle and W as before. Then clearly $\mathcal{L}$ can be lifted to a line bundle on W if and only if λ comes from $H^1(W, \mathcal{O}_W^*)$, or equivalently $\delta(\lambda) = 0$ or $w \cup \alpha(\lambda) = 0$.

The line bundle $\mathcal{L}$ defines a morphism $\phi_{\mathcal{L}}: X \to X^t$ as follows. Denote by $s: X \times X \to X$ the addition. Any morphism $a: S \to X$ gives a line bundle $\mathcal{L}_a$ on $S \times X$, given by

$$\mathcal{L}_a = (a \times 1)^* s^* \mathcal{L}.$$

Then

$$\phi_{\mathcal{L}}(a) = \mathrm{cl}(\mathcal{L}_a \otimes \mathcal{L}^{-1}) \in X^t(S).$$

This enables one in particular to compute the tangent map of $\phi_{\mathcal{L}}$ explicitly. To do this, we observe that one has canonical isomorphisms

$$T_{X,e} = T_0 \xrightarrow{\sim} H^0(X, \mathcal{T}_X)$$

and

$$T_{X^t,e} = T_0^t \xrightarrow{\sim} H^1(X, \mathcal{O}_X).$$

The latter is obtained in the following way. Any morphism

$$\mathrm{Spec}(k[\epsilon]) \longrightarrow X^t$$

mapping the closed point to e, corresponds to a line bundle on $\mathrm{Spec}(k[\epsilon])\times X$ which is trivial on $\mathrm{Spec}(k)\times X$; hence its cohomology class is represented by a cocycle of the form $(1+\epsilon g_{ij})$ with $[g_{ij}]\in H^1(X,\mathcal{O}_X)$.

Claim: *The tangent map $d(\phi_{\mathcal{L}})_e$ is equal to the map*

$$\xi \longmapsto \xi\cup\alpha(\lambda): H^0(X,\mathcal{T}_X)\longrightarrow H^1(X,\mathcal{O}_X).$$

PROOF: Consider $\xi\in T_0$ as a map $\mathrm{Spec}(k[\epsilon])\to X$; then the line bundle $\mathcal{L}_\xi$ on $\mathrm{Spec}(k[\epsilon])\times X$ corresponds to the cocycle $(f_{ij}+v(f_{ij})\epsilon)$, hence $\mathcal{L}_\xi\otimes\mathcal{L}^{-1}$ corresponds to $(1+f_{ij}^{-1}v(f_{ij})\epsilon)$, mapping to $[f_{ij}^{-1}v(f_{ij})]=\xi\cup\alpha(\lambda)\in H^1(X,\mathcal{O}_X)$.

Applying these results to the case where $X=J(C_0)$ and $\mathcal{L}=\mathcal{O}_X(\Theta)$, so

$$\phi_{\mathcal{L}}=\theta: X\longrightarrow X^t,$$

we see that

$$\langle * \rangle \quad \left[\begin{array}{l} \text{if } \xi\in H^1(J(C_0),\mathcal{T}_{J(C_0)})=T_0\otimes T_0^t, \text{ then} \\ \xi\in\kappa_{\nu,z}(T_{A,z})\Leftrightarrow (d\theta_e\otimes 1)(\xi) \text{ has zero image in } \Lambda^2T_0^t\Leftrightarrow \\ \qquad\qquad \Leftrightarrow (d\theta_e\otimes 1)(\xi) \text{ is symmetric.}\end{array}\right.$$

If one dualizes statement $\langle * \rangle$ one has the last part of the theorem.

(2.7) THEOREM (M. Noether): *The map*

$$\bar\mu: S^2H^0(C_0,\omega_{C_0})\longrightarrow H^0(C_0,\omega_{C_0}^{\otimes 2})$$

is onto if $g=2$ or if $g>2$ and C_0 is non-hyperelliptic.

For a proof, see Griffiths [8], page 843 and Andreotti [1], Proposition 8, or Saint-Donat [28], Theorem 2.10.

(2.8) COROLLARY: *If (C,α) is a curve of genus $g\geq 2$ with level n structure, $p\nmid n$, $n\geq 3$, and if C is not hyperelliptic, then the map*

$$\iota: V^{(n)}\longrightarrow A_{g,1}^{(n)}$$

is an immersion at the corresponding point $q(x)=y$ of $V^{(n)}$. If moreover $|\mathrm{Aut}(C)|$ is not divisible by p, then j is an immersion at the point $\phi(x)$ of M_g, corresponding to C.

PROOF: If C is not hyperelliptic and (C,α) corresponds to the point

$x \in M_g^{(n)}$, then $\Sigma(x) \neq x$, hence q is unramified at x. So dq_x is an isomorphism. This implies that

$$d\iota_y = dj_x^{(n)} \circ (dq_x)^{-1}$$

is injective. Hence ι is an immersion, so $V^{(n)}$ is a locally closed subvariety of $A_{g,1}^{(n)}$ near y; denote $z = \iota(y) \in A_{g,1}^{(n)}$. We must show that the map

$$j^* : \hat{\mathcal{O}}_{A_{g,1},\psi_2(z)} \longrightarrow \hat{\mathcal{O}}_{M_g,\phi(x)}$$

is onto. Denote $G = \mathrm{Aut}(C)$. Then G is isomorphic to the isotropy subgroup of the points y and z in $\mathrm{Sp}(g, \mathbf{Z}/n)/\{\pm I\}$. We get:

$$\hat{\mathcal{O}}_{A_{g,1},\psi_2(z)} \cong [\hat{\mathcal{O}}_{A_{g,1}^{(n)},z}]^G$$
$$\hat{\mathcal{O}}_{M_g,\phi(x)} \cong [\hat{\mathcal{O}}_{V^{(n)},y}]^G.$$

Because $|G|$ is not divisible by p, the surjectivity of j^* follows from the surjectivity of the map

$$\iota^* : \hat{\mathcal{O}}_{A_{g,1}^{(n)},z} \longrightarrow \hat{\mathcal{O}}_{V^{(n)},y}.$$

(2.9) Remark: Suppose $p \neq 2$. There is an obvious reason for the fact that μ is *not* onto if $g \geq 3$ and C is hyperelliptic. For then the hyperelliptic involution acts on $H^0(\omega_C)$ by multiplication with -1, so its acts trivially on $S^2H^0(\omega_C)$; however C carries in that case quadratic differentials $\xi \neq 0$ with $\sigma^*\xi = -\xi$, as we will show in the sequel; hence the map μ cannot be surjective.

(2.10): We study all deformations of hyperelliptic curves C in more detail. We first derive a normal form, then construct bases for $H^0(\omega_C)$, $H^0(\omega_C^{\otimes 2})$ and $H^1(\mathcal{T}_C)$. Cf. also Laudal and Lønsted [15], or Lønsted and Kleiman [16, 17].

We call a curve C of genus $g \geq 2$ hyperelliptic if the canonical image of C is a rational curve, or equivalently, if C carries an involution σ such that

$$C/\langle\sigma\rangle \cong \mathbf{P}^1,$$

or if C is a double covering of $\mathbf{P}^1$. If we choose an inhomogeneous coordinate X on $\mathbf{P}^1$ such that ∞ is not a branch point of $C \to \mathbf{P}^1$, then C is given by an affine equation

$$Y^2 + a(X)Y + b(X) = 0.$$

If $p \neq 2$ we can change coordinates, replacing Y by $Y + \frac{1}{2}a(X)$, to obtain

$$Y^2 - F(X) = 0$$

where F has no multiple factor. Riemann–Hurwitz formula shows that $\deg(F) = 2g + 2$.

If $p = 2$ we have a wild ramification, and C can be given by an equation

$$Y^2 + Q_1(X)Y = Q_2(X)P(X)$$

where $Q_1(X) = \prod_1^r (X - \alpha_i)^{a_i}$ with $\sum_1^r a_i = g + 1$, $Q_2(X) = \prod_1^r (X - \alpha_i)$ and $\deg(P) = 2g + 2 - r$, $P(\alpha_i) \neq 0$ for $i = 1, \ldots, r$.

Then C is ramified over $\alpha_1, \ldots, \alpha_r$ with ramification index $e_i = 2a_i$, and σ is given by

$$\sigma(X, Y) = (X, Y + Q_1(X)).$$

Cf. Hasse [12].

(2.11): From now on we assume that $p \neq 2$. The projective nonsingular model of C is then a union $C = U_0 \cup U_1$, with

$$U_0 = \mathrm{Spec}(k[X, Y]/(Y^2 - F(X))) = \mathrm{Spec}(k[x, y]);$$
$$U_1 = \mathrm{Spec}(k[U, V]/(V^2 - \tilde{F}(U))) = \mathrm{Spec}(k[u, v]);$$

and $U_0 \cap U_1$ given by $x \neq 0$ respectively $u \neq 0$, with the identifications on $U_0 \cap U_1$:

$$\begin{cases} x = u^{-1}; & \tilde{F}(U) = U^{2g+2}F(U^{-1}) \\ y = u^{-g-1}v; & \end{cases}$$

and

$$\sigma(x, y) = (x, -y)$$
$$\sigma(u, v) = (u, -v).$$

(2.12) LEMMA: *A basis for $H^0(C, \Omega_C^1)$ is given by the forms*

$$\omega_i = x^{i-1}\,\mathrm{d}x/y, \quad i = 1, \ldots, g.$$

PROOF: As $\omega_1 = \mathrm{d}x/y$ generates Ω_C^1 on U_0, every $\omega \in H^0(C, \Omega_C^1)$ has the form

$$\omega = [a(x) + b(x)y]\omega_1, \quad a, b \in k[x].$$

Substituting $x = u^{-1}$, $y = u^{-g-1}v$, $\omega_1 = -u^{g-1}\,\mathrm{d}u/v$ and because $\mathrm{d}u/v$ generates Ω^1_C on U_1, we get conditions $\deg(a) \leq g-1$, $b = 0$.

In an analogous way one shows:

(2.13) LEMMA: *A basis for $H^0(C, (\Omega^1_C)^{\otimes 2})$ is given by the forms*

$$\begin{cases} \rho_i = x^{i-1}(\mathrm{d}x/y)^2, & i = 1, \ldots, 2g-1; \\ \sigma_i = x^{i-1}(\mathrm{d}x)^2/y, & i = 1, \ldots, g-2. \end{cases}$$

Moreover: $\sigma^*(\rho_i) = \rho_i$, $\sigma^*(\sigma_i) = -\sigma_i$.

The cohomology groups of any coherent sheaf on C can be computed using the covering $\{U_0, U_1\}$ of C. In this manner every k-derivation D of $\Gamma(U_0 \cap U_1, \mathcal{O}_C)$ corresponds to a class $[D] \in H^1(C, \mathcal{T}_C)$. Denote by $y\partial_x$ the k-derivation given by

$$\begin{cases} y\partial_x(x) = y; \\ y\partial_x(y) = \tfrac{1}{2}F'(x). \end{cases}$$

Because $y\partial_x(y^2) = 2y \cdot \frac{1}{2}F'(x) = yF'(x) = y\partial_x(F)$ this indeed gives a derivation of $\Gamma(U_0 \cap U_1, \mathcal{O}_C)$.

(2.14) LEMMA: *A basis for $H^1(C, \mathcal{T}_C)$ is given by*

$$\begin{cases} \theta_i = [y^2x^{-i}\partial_x], & i = 1, \ldots, 2g-1 \\ \eta_j = [yx^{-j}\partial_x], & j = 1, \ldots, g-2. \end{cases}$$

Moreover: $\sigma(\theta_i) = \theta_i$, $\sigma(\eta_j) = -\eta_j$.

(2.15) LEMMA: *A basis for $H^1(C, \mathcal{O}_C)$ is given by*

$$z_i = [yx^{-i}], \quad i = 1, \ldots, g$$

where $[yx^{-i}]$ is the class of $yx^{-i} \in \Gamma(U_0 \cap U_1, \mathcal{O}_C)$ in $H^1(C, \mathcal{O}_C)$.

(2.16) REMARK: Recall from Theorem (2.6) that we have isomorphisms

$$T_{A^{(n)}_{g,1},z} \xrightarrow{\sim} [S^2H^0(C, \omega_C)]^D \cong I^2H^1(C, \mathcal{O}_C)$$

where $I^2V \subset V \otimes V$ is the fixed subspace under the twist $x \otimes y \mapsto y \otimes x$. As $\mathrm{char}(k) \neq 2$, we have $I^2 \cong S^2$. Hence the elements z_iz_j, $1 \leq i \leq j \leq g$ form a basis for $T_{A^{(n)}_{g,1},y}$. Moreover the map

$$\mathrm{d}j^{(n)}_x : T_{M^{(n)}_g,x} \longrightarrow T_{A^{(n)}_{g,1},y}$$

is given by

$$\mathrm{d}j_x^{(n)}(\theta_k) = \sum_{\substack{i+j=k+1\\ i\le j}} z_i z_j;$$

$$\mathrm{d}j_x^{(n)}(\eta_j) = 0$$

(here and in the sequel we let $z_i = 0$ if $i<0$ or $i>g$). In particular the restriction of $j^{(n)}$ to the hyperelliptic locus is an immersion.

§3. The local Torelli problem: case $p \neq 2$, first proof

In this chapter we give a proof of

(3.1) THEOREM: *Suppose $g \geq 2$ and* char$(k) \neq 2$. *Then the map*

$$\iota : V^{(n)} \longrightarrow A_{g,1}^{(n)}$$

is an immersion.

(3.2) COROLLARY: *If C is a curve of genus $g \geq 2$ such that* Aut(C) *has no elements of order p, where $p =$* char$(k) \neq 2$, *then the map*

$$j : M_g \longrightarrow A_{g,1}$$

is an immersion at the point $c \in M_g$ corresponding to the isomorphism class of C.

PROOF OF THE COROLLARY: One uses the same arguments as in the proof of Corollary (2.8).

In view of Corollary (2.8) it is sufficient to prove that ι is an immersion at all hyperelliptic points. So in the sequel, we fix a hyperelliptic curve C_0 of genus $g \geq 2$; let $x \in M_g^{(n)}$ correspond to C_0 (with a certain level n structure, $n \geq 3$), and put

$$y = q(x) \in V^{(n)},$$

$$z = j^{(n)}(x) \in A_{g,1}^{(n)}.$$

Write

$$M = \mathrm{Spec}(\mathcal{O}_{M_g^{(n)},x}),$$

$$V = \mathrm{Spec}(\mathcal{O}_{V^{(n)},y}),$$

$$A = \mathrm{Spec}(\mathcal{O}_{A_{g,1}^{(n)},z}).$$

The proof of Theorem (3.1) has five steps.

(a) From (2.16) we get an explicit basis

$$\{z_i z_j;\ 1 \le i \le j \le g\}$$

for $T_{A,z}$;
(b) We determine an explicit basis for $T_{V,y}$ of the form

$$\{A_k, B_{ij};\ 1 \le k \le 2g-1,\ 1 \le i \le j \le g-2\}$$

with $A_k = \mathrm{d}q_x(\theta_k)$ (Notations as in (1.10) and (2.14)).
(c) We prove a relative version of Theorem (2.6).
(d) We use this to compute $\mathrm{d}\iota_y(A_k)$ and $\mathrm{d}\iota_y(B_{ij})$ and
(e) finally show that these are linearly independent.

(3.3): We perform step (b) of the proof.

Because C_0 is hyperelliptic, x is a fixpoint of $\Sigma: M_g^{(n)} \to M_g^{(n)}$, hence we have $\Sigma: M \to M$, and $V \cong M/\Sigma$. Observe that one may choose local coordinates $t_1, \dots, t_{3g-3} \in \hat{\mathcal{O}}_{M,x}$ on M at x with the following properties:

(i) $\Sigma^* t_i = t_i$ if $1 \le i \le 2g-1$,

$\Sigma^* t_i = -t_i$ if $2g \le i \le 3g-3$;

(ii) if θ_i, η_j are as in lemma (2.14), then

$\partial/\partial t_i = \theta_i$ for $i = 1, \dots, 2g-1$;

$\partial/\partial t_i = \eta_{i+1-2g}$ for $i = 2g, \dots, 3g-3$;

(iii) $\hat{\mathcal{O}}_{M,x} \cong k[[t_1, \dots, t_{3g-3}]]$.

To achieve this, first choose arbitrary coordinates such that (iii) holds. Then apply the automorphism

$$t \longmapsto \tfrac{1}{2}(t + (\mathrm{d}\Sigma_x)^{-1}\Sigma t)$$

to get new coordinates such that the action of Σ becomes linear. Then apply a linear change of coordinates to let (i) and (ii) hold.

Clearly we have an isomorphism

$$\hat{\mathcal{O}}_{V,y} = (\hat{\mathcal{O}}_{M,x})^{\Sigma} \cong k[[t_1, \dots, t_{2g-1}, t_{2g}^2, t_{2g}t_{2g+1}, \dots, t_{3g-3}^2]].$$

We define k-derivations $A_r, B_{ij}: \hat{\mathcal{O}}_{V,y} \to k$ $(r = 1, \dots, 2g-1,\ 1 \le i \le j \le g-2)$

as follows:

$$A_r(t_s) = \delta_{rs} \quad s = 1, \ldots, 2g-1;$$
$$A_r(t_u t_v) = 0 \quad 2g \leq u, v \leq 3g-3;$$
$$B_{ij}(t_r) = 0 \quad r = 1, \ldots, 2g-1;$$
$$B_{ij}(t_u t_v) = 1 \quad \text{if } u, v \in \{2g-1+i, 2g-1+j\};$$
$$B_{ij}(t_u t_v) = 0 \quad \text{else.}$$

(3.4) LEMMA: *$\{A_k, B_{ij} \mid 1 \leq k \leq 2g-1,\ 1 \leq i \leq j \leq g-2\}$ is a basis for $T_{V,y}$ and $A_k = dq_x(\theta_k)$, $k = 1, \ldots, 2g-1$.*

PROOF: Easy.

The basis of $T_{V,y}$, constructed above, is a very special one: all of its elements lie in the tangent cone $C_{V,y}$ of V at y; this means that they can be considered as tangent vectors to smooth curves on V which pass through y. Of course it is easy to compute $\mathrm{d}\iota_y(A_k)$, as

$$\mathrm{d}\iota_y(A_k) = \mathrm{d}\iota_y(\mathrm{d}q_x(\theta_k)) = \mathrm{d}j_x^{(n)}(\theta_k)$$

which has been computed in (2.16). This method fails for the B_{ij}'s, as these do not come from tangent vectors to M at x. However, they correspond to higher order "tangent vectors" in the following way.

Denote

$$S = \mathrm{Spec}(k[[t]]),$$
$$S_m = \mathrm{Spec}(k[[t]]/(t^{m+1}))$$

for $m \in \mathbf{N}$, and denote by σ the automorphism of S or S_m with $\sigma^*(t) = -t$. Suppose $w: S \to M$ is an equivariant morphism, with $w(0) = x$, so $\Sigma w = w\sigma$. Then we have a factorization

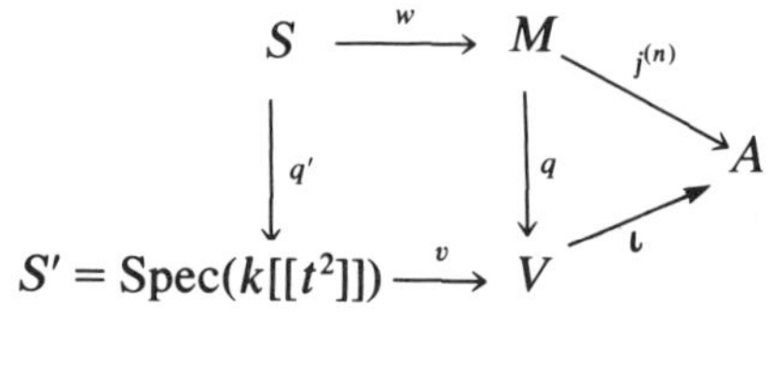

and the restriction of v to $\mathrm{Spec}(k[[t^2]]/(t^4))$ defines a tangent vector $\rho(w)$ to V at y. Remark that $\rho(w)$ only depends on $w|_{S_2}$, so it comes from a second order tangent vector to M at x. The diagram above gives rise to a sheaf map

$$d(j^{(n)}w): \mathcal{T}_S \longrightarrow (j^{(n)}w)^*\mathcal{T}_A.$$

Applying this to the section $\partial/\partial t$ of $\mathcal{T}_S$ we obtain a section

$$\lambda \in \Gamma(S, (j^{(n)}w)^*\mathcal{T}_A)$$

which satisfies $\sigma^*(\lambda) = -\lambda$; hence

$$\lambda = 2t\lambda'$$

for some section $\lambda' \in \Gamma(S, (j^{(n)}w^*\mathcal{T}_A)$.

Remark that

$$\lambda'(0) \in \Gamma(S, (j^{(n)}w)^*\mathcal{T}_A) \otimes_{k[[t]]} k \cong T_{A,z}.$$

(3.5) LEMMA: $d\iota_y(\rho(w)) = \lambda'(0)$.

PROOF: Put $\tau = t^2$, then $dq'(\partial/\partial t) = 2t\, \partial/\partial\tau$ (as sections of $q'^*\mathcal{T}_{S'}$). Hence as sections of

$$(j^{(n)}w)^*\mathcal{T}_A = (\iota v q')^*\mathcal{T}_A$$

one has

$$d(j^{(n)}w)(\partial/\partial t) = 2t\, d(\iota v)(\partial/\partial\tau).$$

Hence

$$\lambda' = d(\iota v)(\partial/\partial\tau)$$

so

$$\lambda'(0) = d(\iota v)_0(\partial/\partial\tau) = d\iota_y(\rho(w)).$$

(3.6): Next we show how one can compute the section $\lambda \in \Gamma(S, (j^{(n)}w)^*\mathcal{T}_A)$. Recall that the maps $w: S \to M$ and $j^{(n)}: S \to A$ give rise to a diagram

$$\begin{array}{ccc} \mathcal{C}_w & \xrightarrow{u_w} & \mathcal{J}_w \\ & {\scriptstyle \pi_w}\searrow \quad \swarrow{\scriptstyle v_w} & \\ & S & \end{array}$$

where $\mathscr{C}_w$ is the pull-back of the universal curve $\pi:\mathscr{C}\to M$ and J_w the pull-back of the universal Jacobian $\mathscr{J}(\mathscr{C}/M)$ by w (or equivalently: the pull-back of $\nu:\mathscr{X}\to A$ by the map $j^{(n)}w$). We have Kodaira–Spencer maps

$$\kappa_{\pi_w}:\mathscr{T}_S\longrightarrow R^1\pi_{w*}\mathscr{T}_{\mathscr{C}_w/S}$$

and

$$\kappa_{\nu_w}:\mathscr{T}_S\longrightarrow R^1\nu_{w*}\mathscr{T}_{\mathscr{J}_w/S}.$$

Moreover we have a natural isomorphism

$$w^*\mathscr{T}_M\longrightarrow R^1\pi_{w*}\mathscr{T}_{\mathscr{C}_w/S}$$

because $\mathscr{T}_{\mathscr{C}_w/S}\cong w^*\mathscr{T}_{\mathscr{C}/M}$.

Under these identifications the sections $\kappa_{\pi_w}(\partial/\partial t)$ and $\mathrm{d}w(\partial/\partial t)$ are the same element of

$$\Gamma(S,R^1\pi_{w*}\mathscr{T}_{\mathscr{C}_w/S})=H^1(\mathscr{C}_w,\mathscr{T}_{\mathscr{C}_w/S}).$$

Next observe that cup product induces a map

$$c:H^1(\mathscr{C}_w,\mathscr{T}_{\mathscr{C}_w/S})\longrightarrow \mathrm{Hom}_{k[[t]]}(H^0(\mathscr{C}_w,\omega_{\mathscr{C}_w/S}),H^1(\mathscr{C}_w,\mathscr{O}_{\mathscr{C}_w}))\cong$$

$$\cong H^1(\mathscr{C}_w,\mathscr{O}_{\mathscr{C}_w})\otimes H^1(\mathscr{C}_w,\mathscr{O}_{\mathscr{C}_w})\cong H^1(\mathscr{J}_w,\mathscr{T}_{\mathscr{J}_w/S})$$

given by

$$c(\xi)(\omega)=\xi\cup\omega.$$

Moreover it follows from the reasoning in (2.5) that in this way

$$c(\kappa_{\pi_w}(\partial/\partial t))=\kappa_{\nu_w}(\partial/\partial t)$$

and these sections of $R^1\nu_{w*}\mathscr{T}_{\mathscr{J}_w/S}$ coincide with the image of the section λ under the inclusion map (determined by the principal polarization of $\mathscr{J}_w$):

$$(j^{(n)}w)^*\mathscr{T}_A\hookrightarrow R^1\nu_{w*}\mathscr{T}_{\mathscr{J}_w/S}.$$

So to compute λ we first construct $\mathscr{C}_w$ and a cocycle representing $\kappa_{\pi_w}(\partial/\partial t)$, then calculate $c(\kappa_{\pi_w}(\partial/\partial t))$ on an explicit basis. If w satisfies $w\sigma=\Sigma w$, then we only need to construct $\mathscr{C}_w$ over S_2 to know $\lambda'(0)$.

(3.7): Remark that the elements $B_{ij}\in T_{V,y}$ for $1\le i\le j\le g-2$ are all of the form $\rho(w)$ for some $w:S\to M$. Namely, take w_{ij} such that the dual map

$$w_{ij}^{*}: k[[t_1, \ldots, t_{3g-3}]] \longrightarrow k[[t]]$$

satisfies

$$\begin{cases} w_{ij}^{*}(t_r) = 0 & \text{if } r \neq 2g-1+i, 2g-1+j; \\ w_{ij}^{*}(t_{2g-1+i}) = w^{*}(t_{2g-1+j}) = t; \end{cases}$$

then clearly

$$B_{ij} = \rho(w_{ij}).$$

(3.8): Take a tangent vector

$$\eta = \sum_{i=1}^{g-2} (b_i \eta_i) \in T_{M,x}.$$

We consider η as an element of $H^1(C_0, \mathcal{T}_{C_0})$. Recall that C_0 has the normal form as described in (2.11) and that η can be represented by the derivation

$$D = \sum_{j=1}^{g-2} (b_j y x^{-j} \partial_x) \in \Gamma(U_0 \cap U_1, \mathcal{T}_{C_0}).$$

We construct a deformation $\pi_2 : \mathcal{C}_2 \to S_2$ of C_0 as follows. We let

$$\mathcal{C}_2 = U_0^{(2)} \cup U_1^{(2)}$$

where $U_i^{(2)} = U_i \otimes_k k[[t]]/(t^3) = U_i \times_{\mathrm{Spec}(k)} S_2$, and $U_0^{(2)}$ and $U_1^{(2)}$ are glued by the isomorphism

$$\Phi_\eta : A_{01}[[t]]/(t^3) \longrightarrow A_{01}[[t]]/(t^3)$$
$$\Phi_\eta(f_0 + f_1 t + f_2 t^2) = f_0 + (f_1 + Df_0)t + (f_2 + Df_1 + D^2 f_2/2)t^2.$$

The fibre $\pi_1 : \mathcal{C}_1 \to S_1$ of $\mathcal{C}_2$ over S_1 corresponds exactly to the tangent vector η.

Recall that $U_0 \cap U_1 = \mathrm{Spec}(A_{01})$ where

$$A_{01} = k[x, y, x^{-1}] = k[u, v, u^{-1}]$$

with $x = u^{-1}$, $y = u^{-g-1}v$, $y^2 = F(x)$.

(3.9) LEMMA: *A basis for $\Gamma(\mathcal{C}_1, \omega_{\mathcal{C}_1/S_1})$ is given by the forms*

$$\tilde{\omega}_1, \ldots, \tilde{\omega}_g$$

where

$$\tilde{\omega}_k = \left\{x^{k-1} - ty\sum_{i=1}^{k-2}(i-k+1)b_k x^{k-i-2}\right\} dx/y.$$

PROOF: Write $\mathscr{C}_1 = U_0^{(1)} \cup U_1^{(1)}$ where

$$U_0^{(1)} = \operatorname{Spec}(k[x, y, t]/(t^2)), \quad y^2 = F(x);$$
$$U_1^{(1)} = \operatorname{Spec}(k[u, v, t]/(t^2)),$$

with $u = \Phi_\eta(x^{-1})$, $v = \Phi_\eta(x^{-g-1}y)$, so

$$u = x^{-1} - ty\sum_{i=1}^{g-2}(b_i x^{i-2})$$

and

$$v = x^{-g-1}\left(y - ty^2\sum_{i=1}^{g-2}(g+1)b_i x^{-i-1} + \tfrac{1}{2}tF'(x)\sum_{i=1}^{g-2}(b_i x^{-i})\right).$$

In this way one obtains, that

$$\tilde{\omega}_k = -\left(1 + tv\sum_{j=1}^{g-k}(g-k-j)b_{g-j-1}u^{-j-1}\right) u^{g-k}\, du/v$$

so it is indeed regular on the whole of $\mathscr{C}_1$. Moreover π_1 is smooth and the forms

$$\omega_k = \tilde{\omega}_k \bmod t$$

form a basis for $H^0(C_0, \omega_{C_0})$, according to (2.12), so the lemma follows by Nakayama's lemma.

Choose any formal deformation $\pi_\infty : \mathscr{C}_\infty \to S$ restricting to $\mathscr{C}_2$ over S_2.

(3.10) LEMMA: *The class* $\kappa_{\pi_\infty}(\partial/\partial t) \bmod t^2$ *in* $H^1(\mathscr{C}_1, \mathscr{T}_{\mathscr{C}_1/S_1})$ *is represented by the derivation*

$$\tilde{D} : A_{01}[[t]]/(t^2) \longrightarrow A_{01}[[t]]/(t^2)$$

with

$$\tilde{D}(f_0 + f_1 t) = D(f_0) + tD(f_1).$$

PROOF: By construction, a representative for $\kappa_{\pi_\infty}(\partial/\partial t) \bmod t^2$ is given by the derivation

$$a \longmapsto \Phi_\eta^{-1} \frac{\mathrm{d}}{\mathrm{d}t} \Phi_\eta a - \frac{\mathrm{d}a}{\mathrm{d}t} \bmod t^2$$

of $A_{01}[[t]]/(t^2)$, and if $a = a_0 + a_1 t$, $a_i \in A_{01}$, then this equals

$$\begin{aligned}
&\Phi_\eta^{-1}(a_1 + Da_0 + t(2Da_1 + D^2 a_0)) - a_1 \\
&\quad = a_1 + Da_0 + t(2Da_1 + D^2 a_0 - Da_1 - D^2 a_0) - a_1 \\
&\quad = Da_0 + tDa_1 = \tilde{D}(a_0 + ta_1).
\end{aligned}$$

REMARK: The reason that we introduce π_∞ rather than work with π_2, lies in the fact that there is a difference between the sheaves

$$\mathcal{T}_S \otimes_{\mathcal{O}_S} \mathcal{O}_{S_2} \quad \text{and} \quad \mathcal{T}_{S_2}.$$

(3.11) LEMMA: *Denote by $\tilde{z}_i \in H^1(\mathscr{C}_1, \mathcal{O}_{\mathscr{C}_1})$ the class represented by $x^{-i}y \in A_{01}[[t]]/(t^2)$. Then*

$$\tilde{z}_1, \ldots, \tilde{z}_g$$

form a basis for $H^1(\mathscr{C}_1, \mathcal{O}_{\mathscr{C}_1})$.

PROOF: This is an immediate consequence of Lemma (2.15).

(3.12): To finish step (d) of the proof, we compute a representative of the class

$$c(\kappa_{\pi_\infty}(\partial/\partial t)) \bmod t^2.$$

Remark that if $m \geq 0$, then x^m represents the zero class in $H^1(\mathscr{C}_1, \mathcal{O}_{\mathscr{C}_1})$, and that for $m < 0$ one has

$$x^m \sim -mtyx^{m-1} \sum_{i=1}^{g-2} (b_i x^{-i})$$

because then $u^{-m} \in \Gamma(U_1^{(1)}, \mathcal{O}_{\mathscr{C}_1})$ and $u^{-m} = x^m + tD(x^m)$.

Hence $\kappa_{\pi_\infty}(\partial/\partial t) \cup \tilde{\omega}_k$ is represented by

$$\langle \tilde{D}, \tilde{\omega}_k \rangle = \left(\sum_{i=1}^{g-2} (b_i x^{-i}) \right) \cdot \left(x^{k-1} - ty \sum_{j=1}^{k-2} (j - k + 1) b_j x^{k-j-2} \right)$$

$$\sim \left(\sum_{i=k}^{g-2} (b_i x^{k-i-1})\right) - ty \left(\sum_{i=1}^{g-2}\sum_{j=1}^{k-2} (j-k+1) b_i b_j x^{k-i-j-2}\right)$$

$$\sim \left(-ty \sum_{i=k}^{g-2}\sum_{j=1}^{g-2} (k-i-1) b_i b_j x^{k-i-j-2}\right) - ty \left(\sum_{i=1}^{g-2}\sum_{j=1}^{k-2} (j-k+1) b_i b_j x^{k-i-j-2}\right)$$

$$= ty \sum_{i,j=1}^{g-2} |k-i-1| b_i b_j x^{k-i-j-2}.$$

So

$$\lambda'(0) = \tfrac{1}{2} \sum_{k=1}^{g} \sum_{\ell=k}^{g} \sum_{i+j=k+\ell-2} |k-i-1| b_i b_j z_k z_\ell.$$

Put

$$\alpha_m = d\iota_y(A_m), \qquad 1 \le m \le 2g-1;$$

$$\beta_{rr} = d\iota_y(B_{rr}), \qquad 1 \le r \le g-2;$$

$$\beta_{rs} = d\iota_y(B_{rs} - B_{rr} - B_{ss}), \quad 1 \le r \le s \le g-2.$$

Then one obtains (with $z_i = 0$ if $i \notin \{1, \dots, g\}$):

$$\alpha_m = \sum_{i+j=m+1} z_i z_j \quad \text{(see (2.16))};$$

$$\beta_{rr} = \tfrac{1}{2} \sum_{k \ge 1} k z_{r+1-k} z_{r+1+k};$$

$$\beta_{rs} = \{(s-r)/2\} \cdot \sum_{k=0}^{(s-r)/2} z_{r+1+k} z_{s+1-k} + \sum_{k=1}^{r} \{(s-r+2k)/2\} z_{r+1-k} z_{s+1+k}.$$

For the final step (e) of the proof, define $W_m \subset T_{A,z}$ the subspace spanned by all $z_i z_j$ with $i \le j$ and $i+j = m+1$, for $m = 1, 2, \dots, 2g-1$. Then $\alpha_m \in W_m$ and $\beta_{rs} \in W_{r+s+1}$. In particular, to show that the tangent vectors

$$\{\alpha_m, \beta_{rs} \mid 1 \le m \le 2g-1, 1 \le r \le s \le g-2\}$$

are linearly independent, it is necessary and sufficient to show, that for every m, the vectors

$$\{\alpha_m, \beta_{rs} \mid 1 \le r \le s \le g-2 \text{ and } r+s = m-1\}$$

are linearly independent, or equivalently, that they form a basis of W_m. To do this, we distinguish four cases, namely

m even, $m \leq g$,

m even, $m > g$,

m odd, $m \leq g$,

m odd, $m > g$.

Case $m = 2n,\ 3 \leq m \leq g$

W_m has the basis

$$z_n z_{n+1},\ z_{n-1} z_{n+2},\ \ldots,\ z_1 z_{2n}.$$

On this basis we write as row vectors of an $(n \times n)$ matrix the elements

$$-\beta_{n-1,n} + \beta_{n-2,n+1},\ -\beta_{n-2,n+1} + \beta_{n-3,n+2},\ \ldots,\ -\beta_{2,2n-3} + \beta_{1,2n-2},$$
$$-\beta_{1,2n-2} + (2n+1)/2\alpha_{2n},\ \alpha_{2n}$$

to get the matrix

$$H_n = \begin{pmatrix} 1 & 0 & \cdots & \cdots & 0 \\ 1 & 1 & 0 & & \vdots \\ \vdots & & \ddots & \ddots & \vdots \\ \vdots & & & \ddots & 0 \\ 1 & \cdots & \cdots & \cdots & 1 \end{pmatrix}$$

Case $m = 2n,\ g < m \leq 2g - 3$

W_m has the basis

$$z_n z_{n+1},\ z_{n-1} z_{n+2},\ \ldots,\ z_{2n+1-g} z_g$$

and one obtains the matrix H_{g-n} using the vectors

$$-\beta_{n-1,n} + \beta_{n-2,n+1},\ \ldots,\ -\beta_{2n+2-g,g-3} + \beta_{2n+1-g,g-2},\ -\beta_{2n+1-g,g-2}$$
$$+ (2g - 2n - 1)/2\alpha_{2n},\ \alpha_{2n}.$$

Case $m = 2n + 1,\ 3 \leq m \leq g$

W_m has the basis

$$z_{n+1}^2,\ z_n z_{n+2},\ \ldots,\ z_1 z_{2n+1}$$

and one now obtains H_n from the elements

$$-2\beta_{nn}+\beta_{n-1,n+1}, -\beta_{n-1,n+1}+\beta_{n-2,n+2}, \ldots, -\beta_{2,2n-2}+\beta_{1,2n-1},$$
$$-\beta_{1,2n-1}+n\alpha_{2n+1}, \alpha_{2n+1}.$$

Case $m = 2n+1,\ g < m \le 2g-3$

W_m has the basis

$$z^2_{n+1}, z_n z_{n+2}, \ldots, z_{2n+2-g} z_g$$

and H_{g-n} is obtained from the elements

$$-2\beta_{nn}+\beta_{n-1,n+1}, \ldots, -\beta_{2n+3-g,g-3}+\beta_{2n+2-g,g-2}, -\beta_{2n+2-g,g-2}$$
$$+(g-n-1)\alpha_{2n+1}, \alpha_{2n+1}.$$

Remark that the remaining cases $m = 1, 2, 2g-2$ or $2g-1$ are trivial because then $W_m = k \cdot \alpha_m$.

§4. The local Torelli problem: case $p \neq 2$, second proof

In this section we present another proof of our main theorem. This second approach will be an "infinitesimalized version" of the beautiful proof by Andreotti of the Torelli theorem, cf. [1].

Let $x \in M^{(n)}_g$, let $\mathcal{O}$ be its local ring and

$$x \in M = \mathrm{Spec}(\mathcal{O}) \subset M^{(n)}_g$$

be the local scheme concentrated at $x \in M^{(n)}_g$. In the same way we define

$$q(x) = y \in V = \mathrm{Spec}(\mathcal{O}_{V^{(n)},y}) \subset V^{(n)}$$

and

$$\iota(y) = j^{(n)}(x) = z \in A = \mathrm{Spec}(\mathcal{O}_{\ldots}) \subset A^{(n)}_{g,1}.$$

Thus we arrive at a commutative diagram of local schemes

$$\begin{array}{ccc} M & \xrightarrow{j^{(n)}} & A. \\ & {\scriptstyle q}\searrow \quad \swarrow{\scriptstyle \iota} & \\ & V & \end{array}$$

We denote by $\mathcal{X} \to A$ the universal family and by $\mathcal{T}_0(\mathcal{X} \to A)$ the sheaf of germs of tangent vectors along the zero section of $\mathcal{X}$. We fix a basis for the free A-Module

$$(\mathcal{T}_0(\mathcal{X} \longrightarrow A)),$$

and we have a canonical isomorphism

$$\mathbf{P}(\Gamma(\mathcal{T}_0(\mathcal{X} \longrightarrow A)))^{\vee} = \mathbf{P} \cong \mathbf{P}_A^{g-1}$$

(**P**: the associated projective space, and $\ldots^{\vee}$ denotes the dual space). Let $\mathscr{C} \to M$ be the universal curve. We have a canonical isomorphism

$$H^0(\mathscr{C}, \Omega^1_{\mathscr{C}/M}) \cong (j^{(n)})^* \Gamma(\mathcal{T}_0(\mathcal{X} \longrightarrow A))$$

thus the previous choice provides a basis for the free M-Module $H^0(\mathscr{C}, \Omega^1_{\mathscr{C}/M})$ (note that if $J = \mathrm{Jac}(C)$, then

$$T_0 J \cong H^1(C, \mathcal{O}_c) \cong H^0(C, \Omega^1_c)^D.$$

The canonical morphism is

$$\phi_K : \mathscr{C} \longrightarrow \mathbf{P}_M^{g-1} \cong \mathbf{P}(H^0(\Omega^1_{\mathscr{C}/M})^D) = \mathbf{P}^{\vee}$$

(canonical: defined by the canonical divisor; D stands for linear dual).

The central idea in the proof by Andreotti, cf. [1], is the following. Let C be a curve of genus g. Then we construct a hypersurface

$$h \subset \mathbf{P} = \mathbf{P}(H^0(\Omega^1_C))$$

of degree $6g-6$ in two ways.

First construction. Let h be the "set of $\omega \in H^0(\Omega^1_c)$ which have a double zero on C", i.e. let

$$\Gamma \subset C \times \mathbf{P}, \quad \mathbf{P} = \mathbf{P}(H^0(\Omega^1_c))$$

be the set of zeros; define h to be the branch locus of

$$\mathrm{proj}: \Gamma \to \mathbf{P}$$

(this map is a covering of degree $2g-2$). Note that in case D is non-hyperelliptic, then $h = h_D$ is

$$h = (\phi_K D)^{\vee},$$

the dual variety of the canonical curve. If C is hyperelliptic then

$$h = (\phi_K C)^{\vee 2} \cup W_1 \cup \cdots \cup W_{2g+2}$$

(here W_i correspond to the hyperplanes ω which have a zero at a (hyperelliptic) Weierstrass point on C).

Second construction: Let (X, W) be a polarized abelian variety (thus $W \subset X$ is a divisor). Consider the Gauss mapping

$$\mathfrak{g}: W \dashrightarrow \mathrm{Grass}(g-1, T_0X) \cong (\mathbf{P}^{g-1})^{\vee} \cong \mathbf{P}^{g-1}$$

which associates to a smooth point $w \in W$ the tangent space to W at w, translated to $0 \in X$, and considered as a hyperplane in projective space (we denote by $\dashrightarrow$ a rational mapping). Define h as the branch locus of this rational mapping (this makes sense in our case, cf. [1], page 820).

Crucial fact. Suppose

$$(X, W) = (\mathrm{Jac}(C), \Theta);$$

then $T_0X \cong H^0(C, \Omega^1_C)^D$ (D means the linear dual),

$$\mathrm{Grass}(g-1, T_0X) \cong \mathbf{P}(H^0(\Omega^1_C)).$$

Both constructions produce some $h \subset \mathbf{P}$, and it turns out that the two results are the same! (Cf. [1], the last six lines of page 811). Thus we arrive at the following situation:

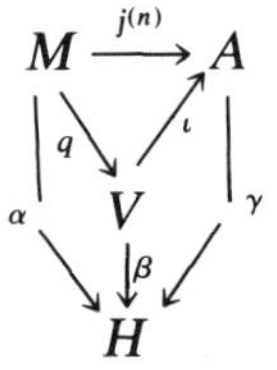

is a commutative diagram, where

$$\mathbf{P} = \mathbf{P}(H^0(\Omega^1_{\mathscr{C}/M})) \cong \mathbf{P}^{g-1}_M$$

($\cong$ by choice of basis), H is the Hilbert scheme of hypersurfaces of degree $6g-6$ in $\mathbf{P}^{g-1}$; the morphism α is defined as follows: consider

$$\Gamma \subset \mathscr{C} \times \mathbf{P}$$

the set of zeros, the branch locus of

$$\mathrm{proj}: \Gamma \longrightarrow \mathbf{P}$$

is a family of hypersurfaces of degree $6g-6$ parametrized by M, and this defines α; in an analogous way γ is defined (cf. above); note that Σ induces multiplication by -1 on $H^0(\Omega^1_{\mathscr{C}/M})$, hence q factors α; the commutativity of the diagram is true "for any curve" by the results of Andreotti, hence

$$\gamma \cdot j^{(n)} = \alpha,$$

and the commutativity involving β follows; finally we prove that the degree of $h \subset \mathbf{P}^{g-1}$ indeed equals $6g-6$; first let D be a non-hyperelliptic curve, then

$$\phi_K : D \longrightarrow \mathbf{P}^\vee, \quad \mathbf{P} = \mathbf{P}(H^0(D, \Omega^1_D))$$

is an immersion; we compute the degree of

$$(\phi_K(D))^\vee = h \subset \mathbf{P}$$

by choosing a line $L \subset \mathbf{P}$; the result is a covering

$$D \cong \phi_K(D) \longrightarrow L \cong \mathbf{P}^1$$

of degree $2g-2=\deg(K)$; by the Hurwitz formula we see that the number of branch points equals

$$2g-2-(0-2)(2g-2)=6g-6,$$

and this is the number of points in $L \cdot h$ (counted with multiplicity). Further let C be a hyperelliptic curve; then h decomposes

$$h = (\phi_K C)^{\vee 2} \cup W_1 \cup \cdots \cup W_{2g+2}$$

(cf. [1], page 821, Proposition 10(ii)), and in the same way as we did above we see that

$$\deg((\phi_K C)^\vee) = 2g-4;$$

hence

$$\deg(h) = 2(2g-4)+2g+2 = 6g-6.$$

(4.1) PROPOSITION: *The tangent mapping*

$$d\beta : T_{V,y} \longrightarrow T_{H,\beta(y)}$$

is injective.

Note that this proposition implies Theorem (3.1):

$$d\beta = d\gamma \cdot d\iota,$$

thus injectivity of $d\beta$ implies the same for $d\iota$.

The proof of (4.1) will be carried out by giving a more or less explicit computation. We follow the notations of the previous section, in particular we define the hyperelliptic curve C by the affine equations as in (2.11); we suppose that degree$(F) = 2g+2$, i.e. $\tilde{F}(0) \neq 0$, and we choose the coordinate X in such a way that $F(0) = 0$; we write $P = (0, 0)$, this is a Weierstrass point on $U_0 \subset C$, and C has no Weierstrass points outside U_0. We take a basis for $H^0(\Omega^1_C)$ by

$$2\omega_i = \frac{dx}{y} \cdot x^{i-1}, \quad 1 \leq i \leq g;$$

we write $\alpha_1, \ldots, \alpha_g$ for the coordinate functions in

$$\mathbf{P} = \mathbf{P}(H^0(\Omega^1_C)),$$

and we study the variation of h locally in the neighbourhood of the point

$$(0 : \ldots : 0 : 1) \in \mathbf{P};$$

We suppose that the integer n used in order to define $M_g^{(n)}$ is even (or we take a suitable étale covering of M, or we take the formal completion of M). A Weierstrass point on a hyperelliptic curve C is a point of order 2 on Jac(C) if $C \to \text{Jac}(C)$ sends a Weierstrass point on C to $0 \in \text{Jac}(C)$. Hence our assumption n to be even implies that all Weierstrass points on the hyperelliptic fibres in $\mathscr{C} \to M_g^{(n)}$ form regular sections. Thus we can choose the basis for $H^0(\Omega^1_{\mathscr{C}/M})$ in such a way that $P = (0, 0)$ is a Weierstrass point for all hyperelliptic deformations of C. The hyperplane

$$\sum_{i=1}^{g} \alpha_i \omega_i = 0$$

has a zero (and even a double zero) at P if $\alpha_1 = 0$. Thus we have achieved that equation h is divisible by α_1 for all hyperelliptic deformations (we use h both for the hypersurface $h \subset \mathbf{P}$ and for its equation). We summarize this as follows:

(4.2): Let

$$\theta_i = [y^2 x^{-i} \partial_x] \in H^1(C, \mathscr{T}_c), \quad 1 \leq i \leq 2g-1,$$

and

$$\begin{array}{ccc} S_1 = \mathrm{Spec}(k[t]/t^2) & \xrightarrow{\theta_i} & M \\ & \searrow^{A_i} & \downarrow q \\ & & V; \end{array}$$

then A_i defines

$$h_i \in k[\alpha_1, \ldots, \alpha_g, t], \quad t^2 = 0,$$

and

$$\alpha_1 \text{ divides } h_i, \quad 1 \leq i \leq 2g-1.$$

Now we choose

$$v_1 = \sum_{s=1}^{g-2} b_s \eta_s,$$

$$\eta_s = [yx^{-s}\partial_x] \in H^1(C, \mathcal{T}_c), \quad 1 \leq s \leq g-2;$$

the $b_1, \ldots, b_{g-2}$ will be considered as indeterminates for the time being. Any choice $b_s \in k$, $s = 1, 2, \ldots, g-2$, leads to a family

$$\mathscr{C}_1 \longrightarrow \mathrm{Spec}(k[t]/t^2);$$

this can be extended to $S_2 = \mathrm{Spec}(k[t]/t^3)$, and we obtain

$$\begin{array}{ccc} S_2 = \mathrm{Spec}(k[t]/t^3) & \xrightarrow{\bar{v}} & M \\ \downarrow & & \downarrow q \\ S_1 = \mathrm{Spec}(k[t^2]/t^3) & \xrightarrow[\rho(v)]{} & V. \end{array}$$

We are going to compute h_v, we shall see that it depends modulo α_1 only on $b_1, \ldots, b_{g-2}$ and not on the choice of the extension v to S_2.

Note that we can choose a basis $\{\omega_1, \ldots, \omega_g\}$ for $H^0(\Omega^1_{\mathscr{C}_2/S_2})$ in the form

$$2\omega_1 = \frac{\mathrm{d}x}{y}\{1 + t^2 G_1\},$$

$$2\omega_2 = \frac{\mathrm{d}x}{y}\{x + t^2 G_2\},$$

$$2\omega_3 = \frac{dx}{y}\{x^2 - b_1ty + t^2G_3\},$$
$$\vdots$$
$$2\omega_i = \frac{dx}{y}\phi_i = \frac{dx}{y}\{x^{i-1} - ty((i-2)b_1x^{i-3} + \cdots + b_{i-2}) + t^2G_i\},$$
$$\vdots$$
$$2\omega_g = \frac{dx}{y}\{x^{g-1} - ty((g-2)b_1x^{g-3} + \cdots + b_{g-2}) + t^2G_g\}$$

(cf. Lemma (3.9)); note that we can choose $G_2, \ldots, G_g$ to be polynomials in x and y with constant term equal to zero (if necessary change these with a multiple of ω_1). We define (from now on we write $\alpha_g = 1$):

$$R = k[[b_1, \ldots, b_{g-2}, \alpha_1, \ldots, \alpha_{g-1}]][t], \quad t^3 = 0,$$

and

$$T = R[[X, Y]]\Big/\left(Y^2 - F, \sum_{i=1}^{g} \alpha_i\phi_i\right).$$

We want to compute the discriminant h_v; we are interested only in this in a neighborhood of the point $\alpha_1 = 0$, $\alpha_2 = 0, \ldots, \alpha_{g-1} = 0$, $\alpha_g = 1$. Note that Γ above this point $(0:0:\ldots:0:1) \in \mathbf{P}$ is given by the equations

$$y^2 - F = 0 \quad \text{and} \quad \sum_{i=1}^{g} \alpha_i\phi_i = 0$$

thus h_v generates the discriminant ideal of the ring extension $R \subset T$.

Although not necessary, let us note that this again gives a computation for the discriminant h_0 in case of a hyperelliptic curve: take $b_1 = 0 = b_2 = \cdots = b_{g-2}$, $t = 0$,

$$R_0 \subset U = R_0[[X]]\Big/\left(\sum_{i=1}^{g} \alpha_i\phi_i\right) \subset T_0,$$
$$T_0 = U[[Y]]/(Y^2 - F);$$

these extensions are defined by monic equations, and

$$\Delta_{T_0/R_0} = \pm(\Delta_{U/R_0})^2 \cdot N_{U/R_0}(\Delta(T_0/U));$$

thus

$$h_0 = Q_0^2 \times W_1 \times \cdots \times W_{2g+2}.$$

We return to the previous situation, $t^3 = 0$. Note that T is free of rank $2g - 2$ over R. Let f be the characteristic polynomial of multiplication by y

on T;

$$f = \det(\lambda \cdot I - (\cdot\, y)).$$

Of course $f(y) = 0$ (Cayley–Hamilton); moreover

$$T \cong R[[Y]]/(\ldots)$$

because

$$x \in k[[y^2]] \text{ in } k[[X, Y]]/(Y^2 - F).$$

Thus f is the minimum polynomial of y over R, and

$$T \cong R[[Y]]/(f).$$

Hence

$$\Delta_{T/R} = (\Delta(f))$$

(here $\Delta(f)$ stands for the discriminant of the polynomial f).

(4.3) LEMMA: *α_1 divides $f(0)$.*

PROOF: Note that

$$T \cdot y \subset Rx + \cdots + Rx^{g-2} + Rxy + \cdots + Rx^{g-2}y + \alpha_1 \cdot T,$$

hence

$$-f(0) = \det(\cdot\, y) \equiv 0 \pmod{\alpha_1}.$$ Q.E.D.

We write

$$L = b_1\alpha_3 + b_2\alpha_4 + \cdots + b_{g-3}\alpha_{g-1} + b_{g-2}\alpha_g,$$

and we prove:

(4.4) LEMMA: *Let*

$$f = \lambda^{2g-2} + \cdots + d_1\lambda + d_0;$$

there exists $\xi \in R^*$ such that

$$d_1 \equiv t\xi L \,(\mathrm{mod}(\alpha_1, t^2)).$$

PROOF: We can write

$$x = x(y) \in k[[X, Y]]/(y^2 - F)$$

(which is an even power series). This we substitute in

$$2\left(\sum_{i=1}^{g} \alpha_i \phi_i\right) = 0,$$

and we obtain

$$\begin{aligned}&\alpha_1(1+t^2G_1) + \alpha_2(x(y) + t^2G_2) + \alpha_3((x(y))^2 - b_1ty + t^2G_3) + \cdots + \\ &\quad + \cdots + (x(y))^{g-1} - ty(\cdots + b_{g-2}) + t^2G_g = \alpha_1(1+t^2G_1) + \\ &\quad + y(-Lt + t^2(\ldots)) + y^2(\alpha_2 + \cdots) + \cdots;\end{aligned}$$

this power series in y we call ψ. Note that $\psi(y)$ is zero in T, hence f divides ψ, thus there exists

$$W \in R[[y]]$$

such that

$$\psi = W \cdot f(y).$$

We write

$$W = e_0 + e_1 y + \cdots,$$

and we see

$$\alpha_1(1+t^2G_1) = e_0 d_0,$$

and

$$-Lt \equiv e_0 d_1 + e_1 d_0 \pmod{t^2}.$$

By (4.3) we know that α_1 divides d_0, hence e_0 is a unit in R and

$$d_1 \equiv (-e_0)^{-1} Lt \ (\mathrm{mod}(t^2, \alpha_1)). \qquad \text{Q.E.D.}$$

Computation of $\Delta(f)$:

(4.5): $\Delta(f) \equiv \Delta(f - d_0) \pmod{d_0}$.

PROOF: The discriminant is a polynomial in the coefficients, hence the result follows.

We write

$$f - d_0 = \lambda \cdot g.$$

(4.6): $\Delta(f - d_0) = d_1^2 \cdot \Delta(g)$.

PROOF: This is a "universal formula" on discriminants. Suppose

$$H = \prod_{i=1}^{N} (\lambda - \lambda_i),$$

then

$$\Delta H = \prod_{i<j} (\lambda_i - \lambda_j)^2,$$

and

$$\Delta(\lambda \cdot H) = q^2 \cdot \prod_{i<j} (\lambda_i - \lambda_j)^2, \quad \text{with} \quad q = \prod_{i=1}^{N} \lambda_i = \pm H(0):$$

this will apply in case $H = g$.

Combination of (4.3), (4.4), (4.5) and (4.6) yields:

(4.7): $\Delta(f) \equiv (t\xi L)^2 \cdot \Delta(g) \pmod{\alpha_1}$.

Note that $t^3 = 0$, thus

(4.8): $\Delta(f) = (t\xi L)^2 \cdot \Delta(g_0) \pmod{\alpha_1}$

where

$$f_0 = f(t = 0) = b_0(t = 0) + \lambda \cdot q_0.$$

We notice that f_0 is an even polynomial, hence

$$g_0 = \lambda \cdot ((f_0 - d_0)/\lambda^2) = \lambda \cdot g_0.$$

(4.9): $d_2 \equiv \eta\alpha_2 \pmod{(t, \alpha_1)}$

for some unit η in R.

PROOF:

$$W_0 f_0 = \alpha_1 + \alpha_2(x(y)) + \alpha_3(x(y))^2 + \cdots$$

and W_0 and f_0 are even; moreover e_0 is a unit in R and α_1 divides d_0.

Q.E.D.

Using the reasoning as in (4.5) and (4.6) we conclude

(4.10): $\Delta(f) = (t\xi\eta L\alpha_2)^2\Delta(q_0)\,(\text{mod}\ \alpha_1)$.

(4.11) LEMMA: $\Delta(q_0) \not\equiv 0\,(\text{mod}\ \alpha_1)$.

PROOF: Consider the hyperplane $W \subset \mathbf{P} \cong \mathbf{P}^{g-1}$ defined by $\alpha_1 = 0$, and define

$$\Gamma' \subset C_0 \times W$$

in the following way: the equation $Y^2 - F = 0$ and

$$0.1 + \alpha_2 \cdot X + \cdots + \alpha_{g-1}X^{g-2} + X^{g-1} = 0$$

have a double zero along $x = 0 = y$, and Γ' is the remaining set of zeros. It is a finite covering

$$\Gamma' \longrightarrow W \cong \mathbf{P}^{g-2} \subset \mathbf{P}$$

of degree $2g-4$. Its discriminant modulo α_1 is given by $\Delta(q_0)$ mod α_1. This discriminant is not zero modulo α_1: the canonical image $\phi_K(C_0) \subset \mathbf{P}^{g-1}$ has $2g+2$ points which are images of the Weierstrass points; choose a hyperplane

$$0 \cdot \omega_1 + a_2 \cdot \omega_1 + \cdots + a_{g-2}\omega_{g-1} + \omega_g$$

which passes through the image of the point $(x = 0,\ y = 0)$, and which has further $g-2$ different intersection points which are not images of Weierstrass points; then

$$\Delta(q_0)(0, a_2, \ldots, a_{g-2}, 1) \neq 0$$

because Γ' has $2g-4$ mutually different points above $(0{:}a_2{:}\ldots{:}a_{g-2}{:}1)$.

Q.E.D.

PROOF OF PROPOSITION (4.1): The tangent space $T_{V,y}$ has a basis

$$\{A_k,\ 1 \leq k \leq 2g-1;\ B_{s,t},\ 1 \leq s \leq t \leq g-2\}$$

(cf. (3.4)). Any non-zero linear combination of $A_1, \ldots, A_{2g-1}$ gives a hyperelliptic, non-trivial deformation of C_0, hence h is not constant, and thus

$$(\mathrm{d}\beta)(A_1), \ldots, (\mathrm{d}\beta)(A_{2g-1})$$

are linearly independent. (One could also use directly the M. Noether lemma, conclude that $(d\iota)(A_1), \ldots, (d\iota)A_{2g-1}$ are linearly independent, and prove (4.1) in this way). Further, by (4.2), we have

$$(d\beta)(A_k) \equiv 0 \pmod{\alpha_1}.$$

Thus it suffices to show that

$$(d\beta)(B_{st}), \quad 1 \leq s \leq t \leq g-2$$

are linearly independent modulo α_1. For a given $v_1 = \Sigma\, b_s\eta_s$, and an arbitrary extension $\mathscr{C}_2 \to S_2$ of this first order deformation, and for $B = \rho(v) \in T_{V,y}$ we have shown that

$$(d\beta)(B) \equiv t^2 \cdot L^2 \cdot P \pmod{\alpha_1},$$

$$P = (\xi\eta\alpha_2)^2 \Delta(q_0) \not\equiv 0 \pmod{\alpha_1}.$$

Thus we obtain

$$(d\beta)(B_{ss}) \equiv t^2 \cdot \alpha_{s+2}^2 \cdot P \pmod{\alpha_1},$$

and

$$(d\beta)(B_{su}) \equiv t^2(\alpha_{s+2} + \alpha_{u+2})^2 \cdot P \pmod{\alpha_1}, \quad s < u$$

(with $\alpha_g = 1$); clearly

$$\{(d\beta)(B_{ss});\ 1 \leq s \leq g-2;\ (d\beta)(B_{su}),\ 1 \leq s \leq u \leq g-2\}$$

is a linearly independent set modulo α_1. This ends the proof of Proposition (4.1), and hence the second proof of Theorem (3.1) is concluded.

§5. Some remarks, and the case char$(k) = 2$

The problem studied in this paper could have been stated from the moment M_g was constructed. Construction of this moduli space at least over **C**, was already known around 1960. From the lines 4 ~ 7 of page 43, of [5] we thought it was classically known that M_g embeds into $A_{g,1}$. Also from [20], page 143, line 8–7 from below we concluded the same. However, Baily and Mumford kindly pointed out to us that an answer to the local Torelli problem for curves seemed unknown in the classical literature.

Let us denote by J_g the image of $j: M_g \to A_{g,1}$. An equivalent form of the local Torelli problem is the question whether J_g (over a field) is a *normal* variety. We have not been able to prove this directly, but over **C**, it follows once we have the result of this paper. Another approach of the local Torelli problem is the following. Consider M_g and $A_{g,1}$ over Spec(**Z**). Take for J_g the image of j over Spec(**Z**). If one knows that $j \otimes \mathbf{Q}$ is an immersion (which it is) and if one knows that $J_g \otimes k$ is a reduced scheme, then it would follow that $j \otimes k$ is an immersion; also this approach has failed up to now, and it certainly does not hold if k is a field of characteristic $p = 2$, and $q \geq 5$:

$$\mathrm{Im}(j: M_g \longrightarrow A_{g,1}) \otimes \mathbf{F}_2$$

is not equal to

$$\mathrm{Im}(j \otimes \mathbf{F}_2 : M_g \otimes \mathbf{F}_2 \longrightarrow A_{g,1} \otimes \mathbf{F}_2).$$

We have not settled the case $\mathrm{char}(k) = p > 2$ completely. At points $x \in M_g$ corresponding to curves which have an automorphism group of order divisible by p, we do not know whether the Torelli mapping is an immersion. The difficulty is the following ($p > 2$): we know that

$$\iota : V^{(n)} \longrightarrow A_{g,1}^{(n)}$$

is an immersion (cf. Theorem (3.1)). There is a group G acting on this such that

$$V^{(n)}/G = M_g \xrightarrow{j} A_{g,1} = A_{g,1}^{(n)}/G.$$

However, it is easy to construct an immersion

$$W_1 \longrightarrow W_2$$

of schemes (in characteristic p), and a finite group G (with p dividing the order of G) such that

$$W_1/G \longrightarrow W_2/G$$

is not an immersion. We have not been able to decide whether such a situation can occur for the case $\iota/G = j$.

We have constructed an example of a non-hyperelliptic curve C (genus $= 6$ with $\mathrm{char}(k) = 5$, its affine equation is of the form $y^5 - y = P/x^2(x^2 - 1)$ with $\deg(P) = 3$), such that there exists a tangent vector $L \neq 0$,

$$L \in T_{M_g^{(n)}}$$

such that L is invariant under the isotropy group, and such that L is going to zero under

$$M_g^{(n)} \longrightarrow A_{g,1}^{(n)} \longrightarrow A_{g,1};$$

however we do not know whether its image under

$$M_g^{(n)} \longrightarrow M_g$$

is non-zero, and we have been unable (even in this case) to settle the local Torelli problem.

For the rest of this section we restrict to $\operatorname{char}(k) = 2$.

(5.1) THEOREM: *Let* $\operatorname{char}(k) = 2$, *and* $g \geq 5$. *Then for any hyperelliptic C with moduli point* $x \in M_g^{(n)}$, $qx = y \in V^{(n)}$

$$\dim T_{V^{(n)},y} > \tfrac{1}{2}g(g+1) = \dim(A_{g,1}).$$

(5.2) COROLLARY: *If* $g \geq 5$ (*and* $\operatorname{char}(k) = 2$, *and n an odd large integer, and C hyperelliptic*) *then*

$$(d\iota)_y : T_{V^{(n)},y} \longrightarrow T_{A^n_{g,1},z}$$

is not injective.

(5.3) COROLLARY: *If* $g \geq 5$, $\operatorname{char}(k) = 2$, *the* local Torelli problem *has a negative answer for all hyperelliptic curves C with* $4 \nmid$ *order*$(\operatorname{Aut}(C))$.

We now give a proof for (5.1). Note that any hyperelliptic curve in $\operatorname{char}(k) = 2$ can be given in the form

$$Y^2 + BY = A,$$
$$A, B \in k[X],$$
$$\deg(B) = g+1,$$
$$\deg(A) \leq 2g+2,$$

this is the affine piece U_0, and U_1 is given by

$$V^2 + \tilde{B}V = \tilde{A}$$

while the glueing is given by

$$X = U^{-1}, \quad \tilde{B}(U) = U^{g+1}B(U^{-1}),$$
$$\tilde{A}(U) = U^{2g+2}A(U^{-1}),$$
$$Y = U^{-g-1}V$$

(cf. (2.10); cf. [12]; cf. [16]).

REMARK: The normal form used above is useful if we want to study all hyperelliptic curves of genus g in characteristic $p = 2$. If we focus on the curve we can "shift the branch points to the $(y = 0)$-axis", then we have a normal form

$$Y^2 + B \cdot Y = A$$

with the extra property that for any x with $B(x) = 0$, the polynomial A has a simple zero at x. This is the normal form used by Igusa (up to a transformation putting one of the branch points in ∞; cf. [13], page 615). This last form is convenient to work with. However irreducibility of the hyperelliptic locus in $\mathrm{char}(k) = 2$ can be seen from the first form (cf. [16]).

We take the normal form above and we define a derivation of

$$k[X, Y]/(Y^2 + BY + A)$$

by

$$(B\partial_x)(x) = B$$

$$(B\partial_x)(y) = yB' + A'$$

(here A' and B' stand for the derivatives of the polynomials A and B). Clearly this is a derivation. Claim:

$$\theta_j = yB\partial_x/x^j, \quad 1 \leq j \leq 2g - 1,$$

$$\eta_i = B\partial_x/x^i, \quad 1 \leq i \leq g - 2$$

provide a k-basis for $H^1(C, \mathcal{T}_C)$ (the proof is straightforward, and we omit it). Further we notice that the hyperelliptic involution σ on C is given by

$$\sigma^* x = x$$

$$\sigma^* y = y + B.$$

We see:

$$\sigma^* \eta_i = \eta_i$$

(note the difference here with characteristic zero); proof:

$$\sigma^*(\eta_i) = \sigma^* \cdot \eta_i \cdot \sigma^*,$$

thus

$$\left(\sigma^*(\eta_i) : x \longmapsto x \longmapsto \frac{B}{x^i} \longmapsto \frac{B}{x^i}\right) = \eta_i,$$

etc. Next we describe $\sigma^*(\theta_j)$. We write

$$B = \beta_0 + \beta_1 x + \cdots + \beta_{g+1}x^{g+1};$$

note that genus$(C) = g$ implies $\beta_{g+1} \neq 0$. We claim

$$\sigma^*(\theta_j) = \theta_j + \sum \beta_{j-1}\eta_i$$

(summation over $1 \leq i \leq g-2$ and $0 \leq j-i \leq g+1$), i.e.

$$\begin{aligned}
\sigma^*\theta_1 &= \theta_1 + \beta_0\eta_1,\\
\sigma^*\theta_2 &= \theta_2 + \beta_1\eta_1 + \beta_0\eta_2,\\
&\vdots\\
\sigma^*\theta_{g-2} &= \theta_{g-2} + \beta_{g-3}\eta_1 + \cdots + \beta_0\eta_{g-2},\\
&\vdots\\
\sigma^*\theta_{g+2} &= \theta_{g+2} + \beta_{g+1}\eta_1 + \cdots + \beta_4\eta_{g-2},\\
\sigma^*\theta_{g+3} &= \theta_{g+3} + 0\cdot\eta_1 + \beta_{g+1}\eta_2 + \cdots + \beta_5\eta_{g-2},\\
&\vdots\\
\sigma^*\theta_{2g-1} &= \theta_{2g-1} + 0 + \cdots + 0 + \beta_{g+1}\eta_{g-2}.
\end{aligned}$$

For the proof, we note that

$$0 = [B\partial_x/x^\alpha] \in H^1(C, \mathcal{T}_C)$$

if $\alpha < 1$, or $\alpha > g-2$. Thus

$$\sigma^*\theta_j = \theta_j + \left(\sum_0^{g+1} \beta_k x^k\right)\cdot B\partial_x/x^j, \text{ etc.}$$

Consider the $(g-2)\times(2g-1)$ matrix consisting of the coefficients of the η_i in the $\sigma^*(\theta_j) - \theta_j$. We note that this matrix has maximal rank: its left-hand lower triangle has zeros and $\beta_{g+1} \neq 0$ on all the places of the hypothenusa. Hence:

Claim. There exists a k-basis

$$\{a_1, \ldots, a_{g-2}, a_{g-1}, \ldots, a_{2g-1}, b_1, \ldots, b_{g-2}\}$$

for $H^1(C, \mathcal{T}_C)$ with

$$\begin{aligned}
\sigma^*a_i &= a_i, & 1 \leq i \leq 2g-1,\\
\sigma^*b_j &= b_j + a_j, & 1 \leq j \leq g-2.
\end{aligned}$$

PROOF:

$$a_i = \eta_i, \quad 1 \le i \le g-2,$$
$$a_{g-1} = \theta_1 + (\beta_0/\beta_{g+1})\theta_{g+2}, \text{ etc.},$$
$$b_1 = (\beta_0/\beta_{g+1})\theta_{g+2}, \text{ etc.}$$

In other words: we have shown that the action of Σ on the local ring of x on $M_g^{(n)}$ can be linearized. We change coordinates:

$$w_j = a_{j-(g-2)}, \quad 1 \le j \le g+1,$$
$$u_i = b_i, \quad 1 \le i \le g-2,$$
$$v_i = b_i + u_i, \quad 1 \le i \le g-2$$

(with w_j invariant, and σ interchanges u_i and v_i).

(5.4) LEMMA: *Let $\mathcal{O}$ be the local ring*

$$\mathcal{O} = k[[u_1, \dots, u_n, v_1, \dots, v_n]],$$
$$\operatorname{char}(k) = 2,$$
$$\sigma u_i = v_i, \quad \sigma v_i = u_i,$$

and $\mathcal{O}^\sigma$ the ring of invariants, with maximal ideal M. Then

$$\dim M/M^2 > 2n + \tfrac{1}{2}n(n-1) \quad \textit{if } n > 2.$$

PROOF: Clearly

$$u_i + v_i,\ u_iv_i,\ u_iv_j + u_jv_i \in M$$

($1 \le i \le g-2$ and $i < j$). Every form in $\mathcal{O}$ which has degree at most 2, and which is invariant under σ is a linear combination of these. Hence these elements are linearly independent modulo M^2. Now suppose $n > 2$. It is easy to see that the element

$$u_1u_2u_3 + v_1v_2v_3$$

is linearly independent of the previous forms modulo M^2. Q.E.D.

From Lemma (5.4) and from the description

$$\mathcal{O}_{M_g^{(n)},x} \cong k[[\dots w_j \dots, \dots u_i \dots, \dots v_i \dots]]$$

the result (5.1) immediately follows.

REMARK: It seems plausible that

$$(d\iota)_y$$

is surjective at all hyperelliptic points (and hence bijective if $g \leq 4$), but we have been unable to prove this in $\mathrm{char}(k) = 2$. We could not imitate the proof we used in $\mathrm{char}(k) \neq 2$, because we could not lift tangent vectors to $V^{(n)}$ at y to higher order jets to $M_g^{(n)}$ (like we did for B_{su}, lifting them to certain $S_2 \to M_g^{(n)}$). Is there a proof of the surjectivity of $(d\iota)_y$ at hyperelliptic points in characteristic $p > 0$ using the same result in characteristic zero? Also this we were unable to settle.

REFERENCES

[1] A. ANDREOTTI: On a theorem of Torelli. Amer. J. of Math. *80* (1958) 801–828.

[2] W.L. BAILY, JR.: On the moduli of Jacobian varieties and curves. Internat. Colloq. Tata Inst. F.R., Bombay, 1960: pp. 51–62.

[3] W.L. BAILY, JR.: On the theory of θ-functions, the moduli of abelian varieties, and the moduli of curves. Ann. Math. *75* (1962) 342–381.

[4] A. BEAUVILLE: Variétés de Prym et Jacobiennes intermédiaires. Ann. Sc. Ec. Norm. Sup. *10* (1977) 309–391.

[5] L. BERS: Uniformization and moduli. Internat. Colloq. Tata Inst. F.R. Bombay 1960, pp. 41–49.

[6] J. DIEUDONNÉ and A. GROTHENDIECK: Eléments de géométrie algébrique, III[1]. Publ. Math. IHES *11* (1961).

[7] R. GODEMENT: Topologie algébrique et théorie des faisceaux. Hermann, Paris 1958.

[8] PH. A. GRIFFITHS: Periods of integrals on algebraic manifolds II. Amer. J. Math. *90* (1968) 805–863.

[9] A. GROTHENDIECK: Géométrie formelle et géométrie algébrique. Sém. Bourbaki, 11e année 1958/1959, Exposé 182. Secrétariat Math. Paris 1959.

[10] A. GROTHENDIECK: Revêtements étales et groupe fondamental, SGA 1. Lecture Notes in Math. 224, Springer-Verlag 1971.

[11] W.J. HABOUSH: Reductive groups are geometrically reductive: a proof of the Mumford conjecture. Ann. Math. *102* (1975) 67–83.

[12] H. HASSE: Theorie der relativ-zyklischen algebraischen Funktionen-Körper, insbesondere bei endlichem Konstantenkörper. Journ. reine angew. Math. (Crelle) *172* (1935) 37–54.

[13] J.-I. IGUSA: Arithmetic variety of moduli for genus two. Ann. Math. *72* (1960) 621–649.

[14] F. KNUDSEN and D. MUMFORD: Projectivity of the moduli space of stable curves, I. Math. Scand. *39* (1976) 19–55.

[15] O.A. LAUDAL and K. LØNSTED: Deformations of curves I. Moduli for hyperelliptic curves. Algebraic Geometry, Proceedings Tromsø, Norway 1977. Lecture Notes in Math. 687, Springer-Verlag 1978.

[16] K. LØNSTED: The hyperelliptic locus with special reference to characteristic two. Math. Ann. *222* (1976) 55–61.

[17] K. LØNSTED and S.L. KLEIMAN: Basics on families of hyperelliptic curves. Compos. Math. *38* (1979) 83–111.

[18] H.H. MARTENS: A new proof of Torelli's theorem. Ann. Math. *78* (1963) 107–111.

[19] T. MATSUSAKA: On a theorem of Torelli. Amer. J. of Math. *80* (1958) 784–800.

[20] D. MUMFORD: Geometric invariant theory,. Springer-Verlag 1965.

[21] D. MUMFORD: Abelian varieties. Tata Inst. F.R. Studies in Math. 5. Oxford Univ. Press, 1974.

[22] D. MUMFORD: Stability of projective varieties. L'enseignement mathématique IIe série, *23* (1977) 39–110.

[23] F. OORT: Commutative group schemes. Lecture Notes in Math. 15, Springer-Verlag 1966.

[24] F. OORT: Finite group schemes, local moduli for abelian varieties and lifting problems. In: Algebraic geometry, Oslo 1970; Wolters-Noordhoff 1972. Also: Compos. Math. *23* (1972) 265–296.

[25] F. OORT and K. UENO: Principally polarized abelian varieties of dimension two or three are Jacobian varieties. Journ. Fac. Sc. Univ. Tokyo, *20* (1973) 377–381.

[26] H. POPP: Moduli theory and classification theory of algebraic varieties. Lecture Notes in Math. 620, Springer-Verlag 1977.

[27] C.S. SESHADRI: Geometric reductivity over arbitrary base. Advances in Math. *26* (1977) 225–274.

[28] B. SAINT-DONAT: On Petri's analysis of the linear system of quadrics through a canonical curve. Math. Ann. *206* (1973) 157–175.

[29] J.-P. SERRE: Groupes algébriques et corps de classes. Act. Sc. Ind. 1264, Hermann, Paris 1959.

[30] R. TORELLI: Sulla varietà de Jacobi. Rend. Acc. Lincei, *22* (1913) 98–103.

[31] A. WEIL: Zum Beweis des Torellischen Satzes. Nachr. Akad. Wissensch. Göttingen 1957, 33–53.

Joseph Steenbrink
Mathematisch Instituut
Wassenaarseweg 80
2333 AL Leiden
The Netherlands

Frans Oort
Mathematisch Instituut
Budapestlaan 6
3508 TA Utrecht
The Netherlands

Surfaces

SUR LE NOMBRE MAXIMUM DE POINTS DOUBLES D'UNE SURFACE DANS $\mathbf{P}^3$ ($\mu(5) = 31$)

Arnaud Beauville

1. Historique du problème

On note classiquement $\mu(n)$ le nombre maximum de points doubles possible pour une surface de degré n dans $\mathbf{P}^3_{\mathbf{C}}$, n'ayant que des points doubles ordinaires. On a évidemment $\mu(1) = 0$, $\mu(2) = 1$.

On obtient $\mu(3)$ et $\mu(4)$ en considérant le degré c de la surface duale; pour une surface de degré n dans $\mathbf{P}^3$, avec comme seules singularités μ points doubles ordinaires, il est donné par la formule $c = n(n-1)^2 - 2\mu$. En exprimant que c est supérieur à 3, on trouve $\mu \leqslant 4$ pour une cubique, $\mu \leqslant 16$ pour une quartique. On a en fait $\mu(3) = 4$: toute cubique à 4 points doubles ordinaires est projectivement isomorphe à la cubique de Cayley $\Sigma_{i=0}^{3} 1/X_i = 0$. On a $\mu(4) = 16$ (Kummer, 1864), l'égalité étant atteinte pour la surface de Kummer.

Le problème devient très difficile dès le degré 5; de fait le résultat principal de cet exposé ($\mu(5) = 31$) est à ma connaissance nouveau. La question a fait cependant l'objet de nombreux travaux, qu'on va essayer de résumer ici.

En utilisant comme précédemment le degré de la surface duale, on obtient facilement la majoration $\mu(n) \leqslant \frac{1}{2}n^2(n-2)$. Une meilleure inégalité—en fait, la meilleure connue jusqu'à présent pour n grand—a été obtenue en 1906 par A. Basset [1]. En projetant la surface depuis un point générique de $\mathbf{P}^3$, il obtient un morphisme fini de la surface sur $\mathbf{P}^2$, dont le lieu de ramification Δ a le même nombre de points doubles et la même classe que la surface. On peut alors appliquer les formules de Plücker à la courbe Δ; en exprimant que le nombre de bitangentes à Δ est positif, on trouve pour $n \geqslant 5$ l'inégalité

$$\mu(n) \leqslant \tfrac{1}{2}[n(n-1)^2 - 5 - \sqrt{n(n-1)(3n-14)+25}]$$

par exemple: $\mu(5) \leqslant 34$, $\mu(6) \leqslant 66$, $\mu(7) \leqslant 114 \ldots$

En 1946, Severi propose la majoration $\mu(n) \leqslant \binom{n+2}{3} - 4$ [8], bien meilleure que celle de Basset: la borne de Severi est asymptotiquement en $n^3/6$, alors que celle de Basset est en $n^3/2$. Mais Severi base sa démonstration sur un postulat qu'il considère comme intuitivement évident (tout en reconnaissant qu'il n'en a pas de démonstration rigoureuse): l'acquisition de μ points doubles ordinaires diminue d'au moins μ le nombre de modules de la surface. On sait aujourd'hui qu'il faut être prudent avec ce genre d'arguments; un an plus tard, B. Segre donne des contre-exemples à la majoration de Severi [7]. L'erreur de Severi a été analysée récemment par Burns et Wahl: les surfaces qui ne vérifient pas la majoration de Severi donnent des exemples très simples de surfaces "obstruées" au sens de la théorie des déformations [3].

Indiquons l'exemple le plus simple de Segre, pour n pair. Soient P une forme de degré $n/2$ et $L_1, \dots, L_n$ des formes linéaires en $X_0, \dots, X_3$; considérons la surface d'équation $P^2 + \Pi_{i=1}^n L_i = 0$. Il est clair qu'elle est singulière aux $\frac{1}{4}n^2(n-1)$ points d'équations $P = L_i = L_j = 0$ $(1 \leqslant i < j \leqslant n)$; un argument du type Bertini montre qu'elle n'a pas d'autres singularités pour un choix générique de P et des L_i, et que ces points sont des points doubles ordinaires. On a donc $\mu(n) \geqslant \frac{1}{4}n^2(n-1)$ pour n pair.

De meilleures minorations ont été obtenues pour des petites valeurs de n : $\mu(5) \geqslant 31$, $\mu(6) \geqslant 63$[1], ... Nous allons expliquer ici la construction, due à Togliatti [10], d'une quintique avec 31 points doubles.

Partons d'une hypersurface cubique V lisse dans $\mathbf{P}^5$ et d'une droite l contenue dans V. Soit V_l la variété obtenue par éclatement de l dans V; la projection de sommet l définit un morphisme $f: V_l \to \mathbf{P}^3$. La fibre générique de f est une conique lisse; le lieu des points t de $\mathbf{P}^3$ tels que la fibre $f^{-1}(t)$ soit singulière est une quintique $\Sigma \subset \mathbf{P}^3$. Cette surface admet 16 points doubles ordinaires, qui sont les points où la fibre est une droite double.

Aux points lisses de Σ, la fibre de f est réunion de deux droites distinctes; remarquons incidemment que l'ensemble de ces droites (i.e. l'ensemble des droites contenues dans V et incidentes à l) est une surface lisse X, munie d'un revêtement double $X \to \Sigma$ ramifié uniquement aux 16 points doubles de Σ.

Supposons maintenant que la cubique V acquière des points doubles ordinaires; on vérifie sans peine que pour un choix générique de la droite l, ces points doubles correspondent par projection à de nouveaux points doubles de Σ. Or Togliatti a montré que le nombre maximum de points doubles possible pour une hypersurface cubique dans $\mathbf{P}^5$ est 15 [9]. En appliquant la construction précédente à une cubique dans $\mathbf{P}^5$ avec 15 points doubles ordinaires, on obtient la surface de Togliatti: c'est une surface de degré 5 ayant comme seules singularités 31 points doubles ordinaires.

On a donc $\mu(5) \geqslant 31$; le reste de l'exposé est consacré à la démonstration de l'égalité $\mu(5) = 31$.

---

[1]E. Catanese et G. Ceresa ont construit récemment une sextique avec 64 points doubles (Constructing surfaces with a given number of nodes, à paraître au Journal of pure and applied Algebra).

2. $\mu(5)=31$

Soit Σ une surface de degré 5 dans $\mathbf{P}^3$, ayant pour seules singularités μ points doubles ordinaires $s_1,\dots,s_\mu$. Notons S la surface obtenue en éclatant $s_1,\dots,s_\mu$; pour $1\leq i\leq\mu$, soit E_i la courbe exceptionnelle correspondant à s_i. On a

$$E_i^2=-2 \qquad E_i.E_j=0 \text{ pour } i\neq j.$$

La surface S est difféomorphe à une quintique lisse dans $\mathbf{P}^3$, et vérifie en particulier

$$K_S^2=5 \qquad \chi(\mathcal{O}_S)=5 \qquad b_2(S)=53.$$

L'idée de la démonstration consiste à regarder les classes des E_i dans $H^2(S,\mathbf{Z}/2)$; elles engendrent un sous-espace totalement isotrope pour la forme d'intersection, donc de dimension ≤ 26. Toutefois les classes des E_i ne sont pas nécessairement linéairement indépendantes dans $H^2(S,\mathbf{Z}/2)$; il peut exister entre elles des relations, c'est-à-dire des ensembles $I\subset[1,\mu]$ tels que $\Sigma_{i\in I}[E_i]=0$ dans $H^2(S,\mathbf{Z}/2)$.

DÉFINITION: *On dit qu'un ensemble $\{s_i\}_{i\in I}$ de points doubles de Σ est pair si la sommes des droites exceptionnelles correspondantes est nulle dans $H^2(S,\mathbf{Z}/2)$.*

Il est immédiat qu'on peut aussi définir les ensembles pairs par l'une des deux conditions suivantes (cf. [2], 2.2):

(a) Il existe un élément δ de Pic(S) tel que $\Sigma_{i\in I}E_i\equiv 2\delta$ dans Pic(S).

(b) Il existe un revêtement double $\pi:X\to S$ dont le lieu de ramification (dans S) est $\cup_{i\in I}E_i$.

La clé de la démonstration réside dans la proposition suivante.

PROPOSITION: *Soit I un ensemble pair non vide de points doubles de Σ. Alors I a* 16 *ou* 20 *éléments.*

Nous avons déjà rencontré au n° 1 des ensembles pairs de 16 points doubles, en partant d'une cubique lisse dans $\mathbf{P}^5$. Un ensemble pair de 20 points doubles est construit dans [2] à partir d'un système linéaire de quadriques dans $\mathbf{P}^4$. Tous les ensembles pairs de 16 ou 20 éléments sont obtenus par ces deux constructions [4].

La proposition est démontrée au n° 3; indiquons maintenant comment on peut en déduire le résultat que nous avons en vue. Il nous faut pour cela rappeler quelques définitions de théorie des codes.

Un *code* est un sous-espace vectoriel de $\mathbf{F}_2^n$. Si x est un vecteur de $\mathbf{F}_2^n$, le nombre de coordonnées non nulles de x s'appelle le *poids* de x, et se note $w(x)$; les poids des vecteurs non nuls d'un code sont appelés les poids du code. Deux codes dans $\mathbf{F}_2^n$ sont dits isomorphes s'ils se déduisent l'un de l'autre par une permutation des coordonnées; ils ont alors les mêmes poids.

Voici deux exemples de codes ayant très peu de poids:

(i) Soient V un espace vectoriel de dimension p sur $\mathbf{F}_2$, Ω l'ensemble des formes linéaires non nulles sur V. L'homomorphisme $x \mapsto (f(x))_{f\in\Omega}$ de V dans $\mathbf{F}_2^\Omega$ identifie V à un code $\mathscr{C}_p \subset \mathbf{F}_2^n$, avec $\dim \mathscr{C}_p = p$, $n = 2^p - 1$. Le code $\mathscr{C}_p$ a pour seul poids $2^{p-1} = (n+1)/2$.

(ii) Posons $n = 2^{p-1}$, et notons simplement $\mathbf{F}_2^{n-1}$ le sous-espace de $\mathbf{F}_2^n$ engendré par les $(n-1)$ premiers vecteurs de base. Dans $\mathbf{F}_2^n$, considérons le code $\mathscr{D}_p$ engendré par $\mathscr{C}_{p-1} \subset \mathbf{F}_2^{n-1}$ et par le vecteur de poids n. C'est un code de dimension p dans $\mathbf{F}_2^n$, qui a pour poids $n/2$ et n.

Le seul résultat de théorie des codes que nous utiliserons est le lemme suivant, qui m'a été indiqué par M. Teissier-Daguenet:

LEMME 1: *Soit $V \subset \mathbf{F}_2^n$ un code de dimension p.*

(i) *Supposons que les poids de V soient strictement supérieurs à $n/2$. On a alors $n \geq 2^p - 1$; s'il y a égalité, le code V est isomorphe à $\mathscr{C}_p$.*

(ii) *Supposons les poids de V supérieurs à $n/2$; on a alors $n \geq 2^{p-1}$. En cas d'égalité, V est isomorphe au code $\mathscr{D}_p$.*

Le lemme 1 sera démontré au n° 4; montrons tout de suite comment le théorème résulte du lemme 1 et de la proposition.

THÉORÈME: *On a $\mu(5) = 31$.*

DÉMONSTRATION: Supposons que la quintique $\Sigma \subset \mathbf{P}^3$ admette (au moins) 32 points doubles. On associe aux droites $E_1, \ldots, E_{32}$ correspondantes un homomorphisme $\varphi: \mathbf{F}_2^{32} \to H^2(S, \mathbf{Z}/2)$.

Considérons le code $K = \operatorname{Ker}(\varphi)$. Puisque $H^2(S, \mathbf{Z}/2)$ est un $\mathbf{F}_2$-espace vectoriel de dimension 53 et que $\operatorname{Im}(\varphi)$ en est un sous-espace totalement isotrope, on a $\dim \operatorname{Im}(\varphi) \leq 26$, d'où $\dim(K) \geq 6$. D'autre part la proposition signifie que le code $K \subset \mathbf{F}_2^{32}$ a pour seuls poids 16 et 20. L'assertion (ii) du lemme 1 conduit alors à une contradiction.

REMARQUES: (1) On obtient de plus une description précise des relations entre les E_i pour une quintique avec 31 points doubles: il résulte en effet du lemme 1 (i) que le noyau de l'homomorphisme $\mathbf{F}_2^{31} \to H^2(S, \mathbf{Z}/2)$ est isomorphe au code $\mathscr{C}_5$. Une telle quintique contient donc 31 ensembles pairs de points doubles, ayant chacun 16 éléments; en particulier, compte tenu du résultat de Catanese cité après l'énoncé de la proposition, une quintique à 31 points doubles est toujours une surface de Togliatti.

(2) La même méthode s'applique aux surfaces K3, et permet de retrouver les résultats classiques sur les "ensembles syzygétiques" de points doubles: si Σ est une surface $K3$ avec 16 points doubles ordinaires et S sa

désingularisée, le noyau de l'homomorphisme $\mathbf{F}_2^{16} \to H^2(S, \mathbf{Z}/2)$ est isomorphe au code $\mathcal{D}_5$. Cette idée avait déjà été utilisée par Nikulin [16], pour prouver notamment que Σ est alors une surface de Kummer.

(3) La méthode devrait a priori s'appliquer également aux surfaces de degré n dans $\mathbf{P}^3$. Mais aussi bien la partie géométrique (caractérisation des ensembles pairs de points doubles) que la partie combinatoire (majoration de la dimension d'un code de poids donnés) semblent nettement plus difficiles en degré $\geqslant 6$.

3. Démonstration de la proposition

LEMME 2: *Soient X, S deux surfaces lisses, $(C_i)_{i\in I}$ une famille de courbes lisses disjointes sur S, $\pi: X \to S$ un revêtement double ramifié le long des C_i. Soient $\varphi: \mathbf{F}_2^I \to \mathrm{Pic}(S) \otimes_{\mathbf{Z}} \mathbf{F}_2$ l'homomorphisme associé aux C_i, et e l'élément de $\mathbf{F}_2^I$ somme des vecteurs de la base canonique. Supposons que* Pic(S) *n'ait pas de 2-torsion; alors le groupe de 2-torsion de* Pic(X) *est isomorphe à* $\mathrm{Ker}(\varphi)/\mathbf{F}_2 e$.

Notons K_X (resp. K_S) le corps des fonctions rationnelles sur X (resp. S), et G le groupe de Galois (cyclique d'ordre 2) du revêtement $X \to S$. Considérons la suite exacte de G-modules:

$$0 \to K_X^*/\mathbf{C}^* \to \mathrm{Div}(X) \to \mathrm{Pic}(X) \to 0.$$

Comme les groupes $H^2(G, \mathbf{C}^*)$ et $H^1(G, K_X^*)$ sont nuls (théorème 90), on a $H^1(G, K_X^*/\mathbf{C}^*) = 0$, d'où un diagramme de suites exactes:

$$\begin{array}{ccccccccc} 0 & \longrightarrow & K_S^*/\mathbf{C}^* & \longrightarrow & \mathrm{Div}(S) & \longrightarrow & \mathrm{Pic}(S) & \longrightarrow & 0 \\ & & \downarrow{\scriptstyle\alpha} & & \downarrow{\scriptstyle\beta} & & \downarrow{\scriptstyle\gamma} & & \\ 0 & \to & (K_X^*/\mathbf{C}^*)^G & \to & \mathrm{Div}(X)^G & \to & \mathrm{Pic}(X)^G & \to & 0 \end{array}$$

où α, β, γ sont les homomorphismes d'image réciproque par π.

Le conoyau de α est isomorphe à $H^1(G, \mathbf{C}^*) \cong \mathbf{Z}/2$; on vérifie aisément que le conoyau de β est isomorphe à $\mathbf{F}_2^I$ et que l'homomorphisme naturel $\mathrm{Coker}(\alpha) \to \mathrm{Coker}(\beta)$ a pour image $\mathbf{F}_2 e$. On a donc une suite exacte

$$0 \to \mathrm{Pic}(S) \to \mathrm{Pic}(X)^G \to \mathbf{F}_2^I/\mathbf{F}_2 e \to 0.$$

En appliquant le lemme du serpent à la multiplication par 2 dans cette suite, on obtient un isomorphisme de $\mathrm{Ker}(\varphi)/\mathbf{F}_2 e$ sur le sous-groupe de Pic(X) formé des éléments d'ordre 2 et invariants par G. Il reste à montrer que tout faisceau inversible L sur X d'ordre 2 est invariant par l'involution σ de X qui échange les deux feuillets du revêtement $X \to S$; on a en effet $\mathrm{Nm}(L) \cong \mathcal{O}_S$ puisque Pic(S) n'a pas de 2-torsion, d'où $\pi^* \mathrm{Nm}(L) = \sigma^* L \otimes L = \mathcal{O}_X$ et $\sigma^* L = L$.

DÉMONSTRATION DE LA PROPOSITION: Soit $(s_i)_{i\in I}$ un ensemble pair de points doubles, de cardinal p, et soit $\pi:\hat{X}\to S$ le revêtement double ramifié le long de $\cup_{i\in I} E_i$. Notons $\tilde{E}_i$ la courbe sur $\hat{X}$ telle que $\pi(\tilde{E}_i)=E_i$; les $\tilde{E}_i$ sont des courbes rationnelles de carré (-1), donc il existe une surface lisse X et un morphisme birationnel $\epsilon:\hat{X}\to X$ qui contracte chaque $\tilde{E}_i$ sur un point.

Rappelons ([2], 2.2) qu'on a

$$\pi_*\mathcal{O}_{\hat{X}}=\mathcal{O}_S\oplus\mathcal{O}_S(-\delta)\qquad K_{\hat{X}}\equiv\pi^*(K_S+\delta)$$

où δ est l'élément de Pic(S) tel que $2\delta\equiv\Sigma_{i\in I}E_i$; comme $\delta^2=\frac{1}{4}(\Sigma E_i)^2=-p/2$ et $\delta.K=0$, on trouve: $\chi(\mathcal{O}_X)=\chi(\mathcal{O}_{\hat{X}})=2\chi(\mathcal{O}_S)+\frac{1}{2}(\delta^2+\delta K)$ par Riemann-Roch,

$$\text{soit } \chi(\mathcal{O}_X)=10-\frac{p}{4}\tag{1}$$

$$(K_{\hat{X}})^2=2(K_S+\delta)^2=2K_S^2-p$$

$$\text{d'où } K_X^2=2K_S^2=10\tag{2}$$

Notons qu'il résulte de (1) que p est divisible par 4.

(1) Le cas $p<16$

Supposons $p<16$; on a alors $\chi(\mathcal{O}_X)\geqslant 7$, et en particulier $p_g(X)>p_g(S)$. Puisque l'application canonique de S est un morphisme birationnel de S sur son image dans $\mathbf{P}^3$, il en résulte aussitôt que l'application canonique de X est birationnelle. Ceci entraîne ([2], th. 5.5) que la surface X satisfait à l'inégalité de Castelnuovo $K_X^2\geqslant 3p_g(X)-7\geqslant 3\chi(\mathcal{O}_X)-10$. En reportant dans cette inégalité les formules (1) et (2), on obtient $p\geqslant 16$, d'où une contradiction.

(2) Le cas $p>20$

Supposons $p>20$; on trouve alors $\chi(\mathcal{O}_X)\leqslant 4$, d'où en particulier $q(X)\geqslant 1$. Le groupe Pic(X) contient alors des éléments de 2-torsion; d'après le lemme 2, ceci entraîne qu'il existe un sous-ensemble J de I, non vide et distinct de I, qui est lui-même pair. Mais alors l'ensemble I–J est également pair; ceci élimine les cas $p=24$ et $p=28$, car l'un des deux ensembles J ou I–J devrait être de cardinal <16.

Supposons $p=32$. On a alors $q(X)\geqslant 3$, d'où $\dim_{\mathbf{F}_2}(\mathrm{Pic}(X)_2)\geqslant 6$. Considérons l'homomorphisme $\varphi:\mathbf{F}_2^{32}\to\mathrm{Pic}(X)\otimes_{\mathbf{Z}}\mathbf{F}_2$ associé aux E_i $(i\in I)$ et le code $K=\mathrm{Ker}(\varphi)$ dans $\mathbf{F}_2^{32}$. Le lemme 2 entraîne $\dim(K)\geqslant 7$, et d'après ce qui précède les poids de K sont supérieurs à 16: ceci contredit le lemme 1.

Enfin si $p \geqslant 36$ on obtient $K_X^2 = 10$ et $\chi(\mathcal{O}_X) \leqslant 1$, ce qui contredit l'inégalité de Miyaoka $K_X^2 \leqslant 9\chi(\mathcal{O}_X)$ [5].

4. Démonstration du lemme 1

(a) Soit V un code de dimension p dans $\mathbf{F}_2^n$, de poids strictement supérieurs à $n/2$; démontrons l'inégalité $n \geqslant 2^p - 1$ par récurrence sur p (l'inégalité étant triviale pour $p = 1$). Soit a un élément de V de poids minimal; si $a = \Sigma_{i \in A} e_i$, notons E_1 (resp. E_0) le sous-espace de $\mathbf{F}_2^n$ engendré par les e_i pour $i \in A$ (resp. $i \notin A$), de sorte que $\mathbf{F}_2^n = E_0 \oplus E_1$. Soit p_0 la projection de $\mathbf{F}_2^n$ sur E_0 parallèlement à E_1. On a $\mathrm{Ker}(p_{0|_V}) = E_1 \cap V = \mathbf{F}_2 a$ puisque a est de poids minimal. Ainsi $p_0(V)$ est un code de dimension $p - 1$ dans E_0 (muni de la base $(e_i)_{i \notin A}$).

Montrons que $p_0(V)$ satisfait l'hypothèse de récurrence. Soit x un élément de V distinct de a, qui se décompose en $x = x_0 + x_1$, avec $x_0 \in E_0$, $x_1 \in E_1$. On a

$$w(a + x) = w(x_0) + w(a + x_1) = w(x_0) + w(a) - w(x_1).$$

La minimalité de $w(a)$ entraîne $w(x_0) \geqslant w(x_1)$, d'où

$$w(x_0) \geqslant \tfrac{1}{2} w(x) > \frac{n}{4} \tag{3}$$

D'autre part on a

$$\dim(E_1) = w(a) > n/2, \text{ d'où } n > 2 \dim(E_0) \tag{4}$$

Il résulte de (3) et (4) qu'on a $w(x_0) > \frac{1}{2} \dim(E_0)$, c'est-à-dire que le code $p_0(V) \subset E_0$ satisfait à l'hypothèse de récurrence. On a donc

$$\dim(E_0) \geqslant 2^{p-1} - 1$$

et d'après (4): $n > 2(2^{p-1} - 1)$, autrement dit $n \geqslant 2^p - 1$.

(b) L'inégalité $n \geqslant 2^{p-1}$ dans le cas (ii) se démontre de manière identique, les inégalités strictes dans (3) et (4) étant remplacées par des inégalités larges.

(c) Nous allons démontrer maintenant les assertions suivantes:

(i_p): *Soit V un code de dimension p dans $\mathbf{F}_2^n$, avec $n = 2^p - 1$, de poids strictement supérieurs à $n/2$. Alors V est isomorphe à $\mathscr{C}_p$.*

(ii_p): *Soit V un code de dimension p dans $\mathbf{F}_2^n$, avec $n = 2^{p-1}$, de poids supérieurs à $n/2$. Alors V est isomorphe à $\mathscr{D}_p$.*

Notons que les deux assertions sont triviales pour $p = 1$.

d) Remarquons d'abord que (i_{p-1}) entraîne (ii_p).

Soit en effet $W \subset \mathbf{F}_2^n$ $(n = 2^{p-1})$ un code de dimension p et de poids $\geqslant n/2$. Pour tout i $(1 \leqslant i \leqslant n)$, le code $W \cap \operatorname{Ker}(e_i^*) \subset \mathbf{F}_2^{n-1}$ est de dimension $(p-1)$ et de poids strictement supérieurs à $(n-1)/2$; d'après (i_{p-1}), il est isomorphe à $\mathscr{C}_{p-1}$, et en particulier tous ses vecteurs non nuls sont de poids $n/2$. Il en résulte que tous les vecteurs non nuls de W sont de poids $n/2$, sauf peut-être le vecteur $e = \Sigma_{i=1}^n e_i$. Mais si e n'appartenait pas à W, le code $W + \mathbf{F}_2 e$ serait de dimension $p+1$ dans $\mathbf{F}_2^n$ $(n = 2^{p-1})$ et de poids $\geqslant n/2$, ce qui contredirait b). Donc W contient e, et $W = \mathscr{C}_{p-1} + \mathbf{F}_2 e$ est isomorphe à $\mathscr{D}_p$.

(e) Montrons maintenant que (i_{p-1}) entraine (i_p).

Reprenons la construction de a), sous l'hypothèse supplémentaire $n = 2^p - 1$. On a ici d'après (4) $\dim(E_0) = 2^{p-1} - 1$ et $\dim(E_1) = 2^{p-1}$. D'après (i_{p-1}), le code $p_0(V) \subset E_0$ est isomorphe à $\mathscr{C}_{p-1}$; nous allons considérer aussi la projection $p_1 : \mathbf{F}_2^n \to E_1$ de noyau E_0, et montrer que le code $p_1(V) \subset E_1$ est isomorphe à $\mathscr{D}_p$. Soit $x = x_0 + x_1$ un élément de V, avec $x_i \in E_i$ $(i = 0, 1)$. Si x_0 est non nul, on a $w(x_0) = 2^{p-2}$, d'où aussi $w(x_1) = 2^{p-2}$ par (3); si x_0 est nul, x est égal à 0 ou a. Ainsi les poids du code $p_1(V) \subset E_1$ sont supérieurs à $\frac{1}{2}\dim E_1$. Puisque les poids de V sont strictement supérieurs à $\dim E_0$, on a $E_0 \cap V = (0)$, de sorte que $p_1(V)$ est de dimension p. On déduit alors de (i_{p-1}) et de d) que le code $p_1(V) \subset E_1$ est isomorphe à $\mathscr{D}_p$.

Ceci étant, pour montrer que le code V est isomorphe à $\mathscr{C}_p$, il suffit de prouver que pour tout couple d'entiers distincts (i, j) de $[1, n]$, les formes linéaires e_i^* et e_j^*, restreintes à V, sont différentes. Si $i, j \in A$ (resp. $i, j \notin A$), cela résulte du fait que les restrictions de e_i^* et e_j^* à $p_1(V)$ (resp. à $p_0(V)$) sont distinctes, par construction des codes $\mathscr{C}_{p-1}$ et $\mathscr{D}_p$. Si $i \in A$ et $j \notin A$, on a $\langle e_i^*, a\rangle = 1$ et $\langle e_j^*, a\rangle = 0$, d'où le résultat.

Ceci prouve l'implication $(i_{p-1}) \Rightarrow (i_p)$ et donc, par récurrence sur p, l'assertion (i_p) pour tout p. L'assertion (ii_p) résulte alors de d); ceci achève la démonstration du lemme.

BIBLIOGRAPHIE

[1] A.B. Basset: The maximum number of double points on a surface. Nature, *73* (1906) 246.

[2] A. Beauville: L'application canonique pour les surfaces de type général. Inventiones math. *55* (1979) 121–140.

[3] D. Burns–J. Wahl: Local contributions to global deformations of surfaces. Inventiones math. *26* (1974) 67–88.

[4] F. Catanese: Babbage's conjecture, contact of surfaces, symmetric determinantal varieties and applications (à paraître).

[5] Y. Miyaoka: On the Chern numbers of surfaces of general type. Inventiones math. *42* (1977) 225–237.

[6] V.V. Nikulin: On Kummer surfaces. Math. USSR Izvestija *9*, (1975) 261–275.

[7] B. Segre: Sul massimo numero di nodi delle superficie di dato ordine. Bull. U.M.I. *2* (1947) 204–212.

[8] F. SEVERI: Sul massimo numero di node di una superficie di dato ordine dello spazio ordinario o di una forma di un iperspazio. Annali di Mat. *25* (1946) 1–41.

[9] E. TOGLIATTI: Sulle forme cubiche dello spazio a cinque dimensioni aventi il massimo numero finito di punti doppi. Scritti offerti a L. Berzolari p. 577–593, Pavia (1936).

[10] E. TOGLIATTI: Una notevole superficie di 5° ordine con soli punti doppi isolati. Festchrift R. Fueter Zürich (1940), 127–132.

Faculté des Sciences
Boulevard Lavoisier
49045 Angers Cédex (France)

ON AN ARGUMENT OF MUMFORD IN THE THEORY OF ALGEBRAIC CYCLES*

S. Bloch**

Let X be a smooth projective surface over an algebraically closed field k. Let $CH_0(X)$ denote the Chow group of zero cycles modulo rational equivalence on X, and let $A_0(X) \subset CH_0(X)$ be the subgroup of cycles of degree 0. I will say that $A_0(X)$ is finite-dimensional if there exists a complete smooth (but possibly disconnected) curve C mapping to X such that the map $J(C) = \text{jacobian}(C) \to A_0(X)$ is surjective. Some years ago, Mumford proved, in the case $k = \mathbf{C}$, that $p_g(X) > 0$ implies $A_0(X)$ is not finite-dimensional. The purpose of the present note is to prove an analogue of this result applicable in all characteristics. The role of the geometric genus, which is not a good invariant in characteristic p, is played by the "transcendental part" of $H^2_{\acute{e}t}(X, \mathbf{Q}_\ell)$. The present proof also reveals the influence of the hypothesis of the finite-dimensionality of the Chow group on the structure of the "motive" of X.

The idea that one could deduce interesting information about the Chow group by considering the generic 0-cycle was suggested by Colliot-Thélène. I am indebted to him for letting me steal it.

LEMMA 1: *Let X be a smooth variety over an algebraically closed field k, Y any k-variety. Let $n \geq 0$ be an integer. Then, writing $K = k(Y)$,*

$$CH^n(X_K) \cong \varinjlim_{U \subset Y \text{ open}} CH^n(X \times_k U).$$

(CH^n = cod. n cycles mod. rational equivalence).

PROOF: For any variety W and any integer m, let W^m = set of points of codimension m on W. One has

$$(X_K)^m = \varprojlim_U (X \times_k U)^m.$$

*My talk at the conference on Chow groups of rational surfaces was based on a paper previously submitted to the Annals of the École Normale Supérieure. The present note grew out of conversations with Colliot-Thélène at the conference, so it seems reasonable to include it in the proceedings.

**The author gratefully acknowledges support from the NSF and from the conference.

Since

$$CH^n(X_K) = \operatorname{Coker}\Big(\coprod_{x\in(X_K)^{n-1}} K(x)^* \longrightarrow \coprod_{x\in(X_K)^n} \mathbb{Z}\Big)$$

$$CH^n(X\times U) = \operatorname{Coker}\Big(\coprod_{y\in(X\times U)^{n-1}} k(y)^* \longrightarrow \coprod_{y\in(X\times U)^n} \mathbb{Z}\Big),$$

the desired result is immediate. Q.E.D.

PROPOSITION 2: *Let X be a smooth projective surface over k and let $\Omega \supset k$ be a universal domain in the sense of Weil. Assume $A_0(X_\Omega)$ is finite-dimensional. Then there exist one-dimensional subschemes $C', C'' \subset X$ and a 2-cycle Γ supported on $(C'\times X)\cup(X\times C'')$ such that some non-zero multiple of the diagonal Δ on $X\times_k X$ is rationally equivalent to Γ.*

PROOF: Let $C \to X$ be such that the map $J(C_\Omega)\to A_0(C_\Omega)$ is onto, and let $C'\subset X$ be the image of C. Enlarging k, we may assume C' defined over k.

LEMMA 3: *Let $k\subset K\subset K'$ be extensions of fields. Then the kernel of $CH^2(X_K)\to CH^2(X_{K'})$ is torsion.*

PROOF: If $[K':K]<\infty$ this follows from the existence of a norm $CH^2(X_{K'})\to CH^2(X_K)$. The case K' algebraic over K follows by a limit argument. Enlarging K and K', we may thus assume K algebraically closed. In this case, $CH^2(X_{K'})$ is a limit of Chow groups $CH^2(X\times_K U)$, where U is a K-variety of finite type. A K-point of U gives a section of $CH^2(X)\to CH^2(X\times_K U)$ so the lemma follows.

PROOF OF PROPOSITION 2: Let K be the function field of X over k, and fix an embedding $K\hookrightarrow\Omega$. Let $P\in X(\Omega)$ be the corresponding point. Our hypotheses imply $CH^2((X-C')_\Omega)=(0)$, so by Lemma 3, there exists $N\geq 1$ such that $N(P)=0$ in $CH^2((X-C')_K)$. (We abuse notation by writing P also for the generic point of X. In other words, $P=$ image in $CH^2(X_K)$ of the diagonal Δ in $X\times X$.) By Lemma 1, there exists $U\subset X$ open $\neq\emptyset$ such that $N\cdot\Delta$ is rationally equivalent to zero on $(X-C')\times_k U$. Let $C''\subset X$ be a subscheme of codimension 1 containing $X-U$. The proposition now follows from the exact sequence

$$\begin{Bmatrix}\text{cycles supported on}\\ (C'\times X)\cup(X\times C'')\end{Bmatrix}$$

$$\longrightarrow CH^2(X\times_k X)\longrightarrow CH^2((X-C')\times_k(X-C''))\longrightarrow 0.$$ Q.E.D.

Exercise 4. Generalize the definition of finite dimensionality for $A_0(X)$ to varieties X of dimension >2 and prove the analogue of Proposition 2.

Fix now an ℓ prime to char p, and define

$$H^2(X)_{trans} = H^2_{ét}(X, \mathbf{Q}_\ell)/\mathrm{Image}(NS(X) \otimes \mathbf{Q}_\ell \longrightarrow H^2_{ét}(X, \mathbf{Q}_\ell)),$$

where $NS(X)$ = Neron–Severi group. We will systematically ignore twisting by roots of 1. An alternate description of $H^2(X)_{trans}$ is given by

$$H^2(X)_{trans} = \mathrm{Image}(H^2_{ét}(X, \mathbf{Q}_\ell) \longrightarrow H^2_{gal}(\bar{K}/K, \mathbf{Q}_\ell)).$$

Note $H^2_{ét}(X)$, $NS(X)$, and hence also $H^2(X)_{trans}$ are functorial for correspondences, which we will take to be elements in $CH^2(X \times_k X)$. In particular, the diagonal induces the identity on $H^2(X)_{trans}$.

LEMMA 5: *With notation as above, let Γ be a codimension 2 cycle on $X \times X$ supported on $(C' \times X) \cup (X \times C'')$. Then the correspondence $\Gamma_* : H^2(X)_{trans} \to H^2(X)_{trans}$ is zero.*

PROOF: It is convenient to work with étale homology, which is defined for any k-scheme Y which can be embedded in a smooth k-variety. Write $\Gamma = \Gamma' + \Gamma''$ with $\mathrm{Supp}\,\Gamma' \subset C' \times X$ and $\mathrm{supp}\,\Gamma'' \subset X \times C''$. Let i': $C' \times X \hookrightarrow X \times X$, $i'': X \times C'' \hookrightarrow X \times X$, and write $[\Gamma'] \in H_4(C' \times X, \mathbf{Q}_\ell)$, $[\Gamma''] \in H_4(X \times C'', \mathbf{Q}_\ell)$. For $\alpha \in H^2_{ét}(X, \mathbf{Q}_\ell)$, we have

$$\begin{aligned}\Gamma'_*(\alpha) &= pr_{2*}(pr_1^*(\alpha) \cdot i'_*[\Gamma']) = pr_{2*}i'_*(i'^*pr_1^*(\alpha) \cdot \Gamma') \\ &= pr_{2*}i'_*(pr_1^* i'^*(\alpha) \cdot \Gamma)\end{aligned}$$

with morphisms labeled as indicated:

$$\begin{array}{ccc} C' \times X & \overset{i'}{\hookrightarrow} & X \times X \\ \downarrow{\scriptstyle pr_1} & & \downarrow{\scriptstyle pr_1} \\ C' & \overset{i'}{\hookrightarrow} & X \end{array}$$

Note that $NS(X) \otimes \mathbf{Q}_\ell$ is self-dual under the cup product pairing on $H^2_{ét}(X, \mathbf{Q}_\ell)$. Without changing the image of α in $H^2(X)_{trans}$ we may assume, therefore, that α is perpendicular to $NS(X)$. Since $H_2(C', \mathbf{Q}_\ell)$ is generated by the classes of components of C', we find $i'^*(\alpha) = 0$ in $H^2(C', \mathbf{Q}_\ell)$, so $\Gamma'_*(\alpha) = 0$.

It remains to show $\Gamma''_*(\alpha) = 0$ where Γ'' is supported on $X \times C''$. We have a diagram

$$\begin{array}{ccc} X \times C'' & \xrightarrow{i''} & X \times X \\ {\scriptstyle pr_2}\downarrow & & \downarrow{\scriptstyle pr_2} \\ C'' & \xrightarrow{i''} & X \end{array}$$

and the projection formula gives

$$\begin{aligned} \Gamma''_*\alpha &= pr_{2*}(i''_*\Gamma'' \cdot pr_1^*\alpha) \\ &= i''_* pr_{2*}(\Gamma'' \cdot i''^* pr_1^*\alpha) \\ &\qquad \in \operatorname{Image}(H_2(C'', \mathbf{Q}_\ell) \longrightarrow H^2(X, \mathbf{Q}_\ell)) \subset NS(X) \otimes \mathbf{Q}_\ell \end{aligned}$$

so $\Gamma''_*\alpha = 0$ in $H^2(X)_{trans}$. Q.E.D.

We have proven:

THEOREM 6: *Let X be a smooth projective surface over an algebraically closed field k, and let $k \subset \Omega$ be a universal domain. If $H^2_{ét}(X, \mathbf{Q}_\ell) \neq NS(X) \otimes \mathbf{Q}_\ell$ ($\ell \neq$ char k) then $A_0(X_\Omega)$ is not finite-dimensional.*

Exercise 7. Formulate and prove an analogous result for varieties of dimension >2.

Question 8. If E is a supersingular elliptic curve over a field k of characteristic $\neq 0, \ell$; then $H^2_{ét}(E \times E, \mathbf{Q}_\ell) = NS(E \times E) \otimes \mathbf{Q}_\ell$. Is $A_0(E \times E)$ finite-dimensional? Shioda has shown (using his proof of unirationality for the corresponding Kummer surface) that the answer is yes. The analogous question for a general supersingular surface remains open.

When $k = \mathbf{C}$, $H^2_{ét}(X) \neq NS(X) \otimes \mathbf{Q}_\ell$ if and only if $p_g(X) > 0$, so we recover Mumford's result:

COROLLARY 9: *When $k = \mathbf{C}$, $p_g(X) > 0$ implies $A_0(X_\mathbf{C})$ not finite-dimensional.*

REFERENCES

D. MUMFORD: Rational equivalence of zero cycles on surfaces. J. Math., Kyoto Univ., *9* (1968) 195–204.

A.A. ROITMAN: Rational equivalence of zero-cycles, Math. USSR Sb. *18* No. 4, (1972) 571–588.

Department of Math.
University of Chicago
5734 University Avenue
Chicago, Illinois 60637 (USA)

LA DESCENTE SUR LES VARIETES RATIONNELLES

J.-L. Colliot-Thélène et J.-J. Sansuc

Table des matières

I. Introduction 223
II. La méthode de la descente 224
III. L'exemple des tores 228
IV. L'exemple des surfaces de Châtelet 230
V. Questions sur les surfaces rationnelles 232
VI. Quelques résultats récents 233

I. Introduction

On note k un corps, $\bar{k}$ une clôture séparable et $\mathfrak{g} = \mathrm{Gal}(\bar{k}/k)$. Une k-variété algébrique géométriquement intègre X est dite k-rationnelle si son corps des fonctions $k(X)$ est transcendant pur sur k, et elle est dite *rationnelle* si $\bar{X} = X \times_k \bar{k}$ est $\bar{k}$-rationnelle. Les surfaces cubiques lisses dans $\mathbf{P}^3$, les surfaces fibrées en coniques sur la droite projective $\mathbf{P}^1$, les groupes algébriques linéaires sont, au moins pour k parfait, des exemples familiers de variétés rationnelles.

Dans la description de l'ensemble $X(k)$ des points k-rationnels d'une k-variété rationnelle lisse X, on rencontre plusieurs types de problèmes, qu'on peut grossièrement classer comme suit:

(i) problème de l'existence d'un point k-rationnel; par exemple, si k est un corps de nombres, problème de la validité du principe de Hasse (si X a un point dans chaque complété de k, a-t-elle un point dans k?)

(ii) la k-variété X est-elle k-rationnelle? k-unirationnelle? si k est un corps de nombres, X satisfait-elle l'approximation faible (pour tout ensemble fini Σ de places de k, a-t-on $X(k)$ dense dans $\prod_{v\in\Sigma} X(k_v)$?)?

(iii) quand X n'est pas k-rationnelle avec cependant $X(k) \neq \emptyset$, trouver une description raisonnable de $X(k)$, ou, du moins, définir, et étudier, des relations d'équivalence sur $X(k)$ permettant d'approcher une telle description (R-équivalence, [16], chap. II, §4, engendrée par définition par la relation: $A \sim B$ s'il existe un ouvert U de $\mathbf{P}^1_k$ et un k-morphisme $U \xrightarrow{\varphi} X$ tels que A et B appartiennent à $\varphi[U(k)]$; équivalence de Brauer, [16], chap. VI, §3) et de donner une "mesure" de la non-k-rationalité de X; de ce point de vue, il est également intéressant de considérer, outre $X(k)/R$, le quotient $A_0(X)$ du groupe $Z_0(X)$ des 0-cycles de X par l'équivalence rationnelle.

Pour X une k-variété rationnelle, propre et lisse, le module galoisien $\operatorname{Pic}\bar{X}$ est $\mathbf{Z}$-libre de type fini. Ce $\mathfrak{g}$-module, ainsi que le groupe $H^1(\mathfrak{g}, \operatorname{Pic}\bar{X})$ qui est intimement lié au groupe de Brauer de X, ont déjà joué un grand rôle dans les travaux de Manin [16] et Voskresenskiĭ [20]. La première observation à cet égard est la suivante: si X est k-rationnelle, le $\mathfrak{g}$-module $\operatorname{Pic}\bar{X}$ est stablement de permutation, i.e. il existe M_1 et M_2 deux $\mathfrak{g}$-modules $\mathbf{Z}$-libres de rang fini admettant chacun une $\mathbf{Z}$-base permutée par $\mathfrak{g}$, tels que $\operatorname{Pic}\bar{X} \oplus M_1 \approx M_2$, ce qui implique $H^1(\mathfrak{g}, \operatorname{Pic}\bar{X}) = 0$. On obtient ainsi des obstructions à la k-rationalité. Par ailleurs, si k est un corps de nombres, Manin ([15], [16] chap. VI) a obtenu grâce au groupe $H^1(\mathfrak{g}, \operatorname{Pic}\bar{X})$ une analyse générale de divers contre-exemples connus au principe de Hasse.

Les Notes [5, 6, 7] exploitent également le module galoisien $\operatorname{Pic}\bar{X}$, mais par l'intermédiaire du k-tore dual S_0. On y introduit les torseurs universels, qui sont des torseurs sur X sous S_0 et qui donnent sur $X(k)$ une information plus précise que la simple considération du groupe $H^1(\mathfrak{g}, \operatorname{Pic}\bar{X})$.

Les §§II–IV apportent des compléments et illustrations à la méthode de la descente résumée dans les Notes. Cette méthode ne suffit pas en général au traitement des problèmes (i)–(iii) pour les variétés rationnelles, mais il est possible qu'elle suffise pour les surfaces rationnelles. Ceci aurait de nombreuses conséquences qui sont inventoriées au §V et discutées au §VI à la lumière de quelques résultats récents.

II. La méthode de la descente

A. *L'idée de la méthode*

Soient S un k-tore et X un k-schéma. Etant donné un point P de X, on note $k(P)$ son corps résiduel. Si $\mathcal{T}$ est un torseur (= espace principal homogène) sur X sous S, on note $p_{\mathcal{T}}: \mathcal{T} \to X$ la projection canonique et $[\mathcal{T}]$ la classe de $\mathcal{T}$ dans le groupe de cohomologie étale $H^1(X, S)$, qui classe précisément ces torseurs. L'accouplement

$$\begin{aligned} X(k) \times H^1(X, S) &\longrightarrow H^1(k, S) \\ (P, [\mathcal{T}]) &\longmapsto [\mathcal{T}_P] \end{aligned}$$

(où $\mathcal{T}_P = \mathcal{T} \times_X k(P)$) a la propriété:

$$[\mathcal{T}_P] = 0 \Longleftrightarrow P \text{ appartient à } p_{\mathcal{T}}(\mathcal{T}(k)).$$

Soit $\mathcal{T}$ un torseur sur X sous S. Il définit l'application $\theta_{\mathcal{T}}: X(k) \to H^1(k, S)$ par $\theta_{\mathcal{T}}(P) = [\mathcal{T}_P]$. Pour $\alpha \in H^1(k, S)$, notons $p_\alpha: \mathcal{T}_\alpha \to X$ un torseur de classe $[\mathcal{T}] - \alpha$ dans $H^1(X, S)$. On a la relation

$$X(k) = \bigcup_{\alpha \in \operatorname{im}\theta_{\mathcal{T}}} p_\alpha(\mathcal{T}_\alpha(k)) = \bigcup_{\substack{\alpha \in H^1(k, S) \\ \mathcal{T}_\alpha(k) \neq \emptyset}} p_\alpha(\mathcal{T}_\alpha(k)).$$

Lorsque X est propre sur k, l'application $\theta_{\mathcal{T}}$ a de bonnes propriétés (cf. B). Pour X une k-variété rationnelle, propre et lisse, la méthode de la descente consiste à bien choisir S et $\mathcal{T}$ de telle sorte que les k-variétés rationnelles $\mathcal{T}_\alpha$, bien que de dimension plus grande, aient néanmoins une k-géométrie plus simple que X.

Dans le suite on identifie la plupart du temps, implicitement, un torseur et sa classe, ce qui, pour les problèmes considérés ici, ne présente pas d'inconvénient; ainsi, l'expression "torseur unique" signifie "torseur unique à isomorphisme, non unique, près".

B. *Propriétés générales*

L'application $\theta_{\mathcal{T}}$ définie ci-dessus se prolonge en un homomorphisme

$$\theta_{\mathcal{T}}: Z_0(X) \longrightarrow H^1(k, S)$$

tel que $\theta_{\mathcal{T}}(P) = \mathrm{cor}_{k(P)/k}([\mathcal{T}_P])$ pour tout point fermé P de X.

Proposition 1 ([8] prop. 12, p. 198): *Pour X une k-variété propre, l'homomorphisme $\theta_{\mathcal{T}}$ passe au quotient par l'équivalence rationnelle des 0-cycles.*

A fortiori, pour X propre sur k, l'application $\theta_{\mathcal{T}}: X(k) \to H^1(k, S)$ passe au quotient par la R-équivalence.

Proposition 2: *Pour k un corps de type fini sur le corps premier et X une k-variété propre, l'image de l'homomorphisme $\theta_{\mathcal{T}}$ est finie.*

A fortiori, l'application $\theta_{\mathcal{T}}: X(k) \to H^1(k, S)$ a une image finie ([5], proposition 1).

Démonstration: On sait (d'après Nagata, cf. [12] EGA IV 6.12.6, et d'après EGA IV 8.8.3) qu'il existe un anneau régulier A, contenu dans k et de type fini comme $\mathbf{Z}$-algèbre, un A-tore $\tilde{S}$, un A-schéma propre $\tilde{X}$ et un torseur $\tilde{\mathcal{T}}$ sur $\tilde{X}$ sous $\tilde{S}$, tels que $(\tilde{S}, \tilde{X}, \tilde{\mathcal{T}}) \times_A k = (S, X, \mathcal{T})$. Soient P un point fermé de X, puis $K = k(P)$ et B la clôture intégrale de A dans K, qui est finie sur A (d'après Nagata, $\mathbf{Z}$ est excellent, EGA IV 7.8.3). Le k-morphisme $\mathrm{Spec}\, K \to X$ défini par P se prolonge en un A-morphisme $W \to \tilde{X}$, où W est un ouvert régulier de $\mathrm{Spec}\, B$ contenant tous les points de codimension 1 (on utilise la propreté de $\tilde{X}$ sur A et la fermeture du lieu singulier de $\mathrm{Spec}\, B$). Soient F le fermé de $\mathrm{Spec}\, A$, image par le morphisme fini $q: \mathrm{Spec}\, B \to \mathrm{Spec}\, A$ du fermé complémentaire de W dans $\mathrm{Spec}\, B$, puis U l'ouvert complémentaire de F dans $\mathrm{Spec}\, A$ et $q_U: V \to U$ la restriction de q à $V = q^{-1}(U)$. Ainsi, l'ouvert U contient tous les points de codimension 1 de $\mathrm{Spec}\, A$. L'ouvert V, contenu dans W, est régulier. On en déduit ([12] EGA IV 15.4.2) que le morphisme fini q_U est plat. Les propriétés fonctorielles de la trace d'un torseur par un morphisme fini et plat ([1] SGA 4, XVII 6.3.26) montrent alors que $\theta_{\mathcal{T}}(P) = \mathrm{cor}_{K/k}([\mathcal{T}_P])$ est l'image, par l'application $H^1(U, \tilde{S}) \to H^1(k, S)$, de $\mathrm{cor}_{V/U}([\tilde{\mathcal{T}} \times_{\tilde{X}} V])$. Comme A est régulier et que U contient tous les points de codimension 1 de Spec A, un théorème de pureté facile (par descente on se ramène aux résultats de pureté pour $H^0(\ ,\mathbf{G}_m)$ et $H^1(\ ,\mathbf{G}_m)$) montre l'égalité $H^1(U, \tilde{S}) = H^1(A, \tilde{S})$. On a ainsi montré que $\theta_{\mathcal{T}}$ envoie $Z_0(X)$ dans l'image de $H^1(A, \tilde{S})$ dans $H^1(k, S)$. Comme $H^1(k, S)$ est de torsion, il suffit, pour établir la proposition, de prouver que $H^1(A, \tilde{S})$ est un groupe de type fini. Or ceci se ramène, par descente galoisienne, à la même assertion pour $H^0(\ ,\mathbf{G}_m)$ et $H^1(\ ,\mathbf{G}_m)$ sur un anneau régulier de type fini sur $\mathbf{Z}$ (Roquette, via le théorème des unités de Dirichlet et celui de Mordell–Weil–Néron).

En résumé, pour X une k-variété propre, les applications $\theta_{\mathcal{T}}$ induisent des applications $X(k)/R \to H^1(k, S)$ et $A_0(X) \to H^1(k, S)$, d'images finies pour k de type fini sur le corps premier.

Notons que, pour k un corps de nombres, un certain nombre d'arguments de la démonstration ci-dessus se simplifient; de plus, si l'on considère l'application $X(k) \to \mathrm{Br}\, k$ définie par un élément de $H^2(X, \mathbf{G}_m)$, on peut alors utiliser des arguments analogues, en particulier ([1] SGA 4, VII 5.9), pour montrer que son image est finie (comparer avec Manin, [16] chap. VI §4).

C. *Torseurs universels*

On note $S \mapsto \hat{S}$ la dualité qui à un k-tore S associe son $\mathfrak{g}$-module $\hat{S}$ des caractères sur $\bar{k}$, qui est un module galoisien $\mathbf{Z}$-libre de type fini. Etant donné X une k-variété rationnelle, propre et lisse, et S_0 le k-tore dual du $\mathfrak{g}$-module Pic $\bar{X}$, on a défini dans les Notes [5, 6] les *torseurs universels* sur X comme les torseurs sur X sous S_0 dont l'image par l'application χ: $H^1(X, S_0) \to \mathrm{Hom}_{\mathfrak{g}}(\hat{S}_0, \mathrm{Pic}\,\bar{X})$ est l'application identité de $\hat{S}_0 = \mathrm{Pic}\,\bar{X}$ (l'application χ associe à un torseur $\mathcal{T}$ sur X sous S_0 et à un caractère λ: $\bar{S}_0 \to \bar{\mathbf{G}}_m$ le torseur sur $\bar{X}$ sous $\bar{\mathbf{G}}_m$ déduit de $\bar{\mathcal{T}}$ par λ) et deux tels torseurs universels diffèrent par un unique élément de $H^1(k, S_0)$.

THÉORÈME 1 ([6]): *Etant donné une k-variété X rationnelle, propre et lisse, et $\mathcal{T}^c$ une k-compactification lisse d'un torseur universel sur X, la k-variété $\mathcal{T}^c$ est une variété rationnelle telle que le $\mathfrak{g}$-module* Pic $\bar{\mathcal{T}}^c$ *soit un module de permutation.*

Ainsi, le passage à une k-compactification lisse d'un torseur universel améliore complètement la situation du point de vue du groupe de Picard. On dispose par ailleurs de *descriptions locales* des torseurs universels, qui permettent d'en donner des "équations". Soient X une k-variété rationnelle, propre et lisse, U un ouvert de X tel que Pic $\bar{U} = 0$ (de tels ouverts forment une base d'ouverts de X), et F le fermé complémentaire. Notons respectivement T et M les k-tores duaux des $\mathfrak{g}$-modules $\mathbf{Z}$-libres de type fini $\bar{k}[U]^*/\bar{k}^*$ (où $\bar{k}[U] = \Gamma(\bar{U}, \mathcal{O}_{\bar{X}})$) et $\mathrm{Div}_{\bar{F}}\bar{X}$ (qui est le $\mathfrak{g}$-module de permutation des diviseurs de $\bar{X}$ à support dans $\bar{F}$). De la suite exacte de $\mathfrak{g}$-modules

$$0 \longrightarrow \bar{k}[U]^*/\bar{k}^* \longrightarrow \mathrm{Div}_{\bar{F}}\bar{X} \longrightarrow \mathrm{Pic}\,\bar{X} \longrightarrow 0 \tag{1}$$

on tire par dualité la suite exacte de k-tores

$$1 \longrightarrow S_0 \longrightarrow M \longrightarrow T \longrightarrow 1 \tag{2}$$

qui fait de M un torseur sur T sous S_0. Soit σ une $\mathfrak{g}$-section de l'homomorphisme $\bar{k}[U]^* \twoheadrightarrow \bar{k}[U]^*/\bar{k}^* = \hat{T}$, par exemple, pour $P \in U(k)$, la section σ_P donnée par $\sigma_P(\mathrm{cl}\, f) = f/f(P)$. Une telle section σ définit un $\mathfrak{g}$-homomorphisme de $\bar{k}$-algèbres $\bar{k}[T] = \bar{k}[\hat{T}] \to \bar{k}[U]$, donc un k-morphisme φ_σ: $U \to T$ qui, pour $\sigma = \sigma_P$, envoie P sur l'élément neutre de T.

ASSERTION 1: *Etant donné U et σ comme ci-dessus, le torseur sur U sous S_0, déduit via φ_σ: $U \to T$ du torseur M sur T défini par* (2), *est la restriction à U d'un unique torseur universel $\mathcal{T}^\sigma$ sur X. Inversement, tout torseur universel sur X admet une telle description locale sur tout ouvert U tel que* $\mathrm{Pic}\,\bar{U} = 0$. *Pour $\sigma = \sigma_P$, le torseur $\mathcal{T}^\sigma$ n'est autre que le torseur universel $\mathcal{T}^P$ de fibre triviale en P.*

D. *La descente*

Le processus de la descente sur une k-variété rationnelle X, propre et lisse, est décrit en détail dans les Notes [6, 7]. Il est essentiellement résumé dans la décomposition

$$X(k) = \bigcup_{\alpha \in I} p_\alpha(\mathcal{T}_\alpha(k))$$

qui constitue une partition de $X(k)$ définie par les torseurs universels (cf. A): $\mathcal{T}$ désigne un torseur universel et I la partie de $H^1(k, S_0)$ formée des α pour lesquels le torseur universel $\mathcal{T}_\alpha$ a un point k-rationnel. L'existence d'un torseur universel équivaut (cf. [6]) à l'existence d'une $\mathfrak{g}$-rétraction pour l'injection $\bar{k}^* \hookrightarrow \bar{k}(X)^*$. L'existence d'un torseur universel ayant un point k-rationnel équivaut à $X(k) \neq \emptyset$. Lorsqu'il en est ainsi, la décomposition ci-dessus, qui est finie lorsque k est de type fini sur le corps premier (proposition 2), est plus grossière que celle donnée par la R-équivalence et découpe $X(k)$ en classes paramétrées sur k par des k-variétés à k-géométrie plus simple que celle de X (théorème 1). On peut considérer que cette méthode réussit parfaitement lorsque la question suivante a une réponse affirmative:

(Q$_1$) *Les torseurs universels sur X qui ont un point k-rationnel sont-ils k-rationnels?*

Dans le cas d'un corps de nombres, la descente se précise ainsi: on peut se limiter au cas où X a un point dans chaque complété de k (sinon $X(k)=\emptyset$); si, pour chaque torseur universel $\mathcal{T}$ sur X, il existe une place v de k avec $\mathcal{T}(k_v)=\emptyset$, la décomposition ci-dessus montre que $X(k)$ est vide; cette obstruction au principe de Hasse, dont le calcul ne requiert qu'un nombre fini de vérifications, équivaut en fait ([7], théorème) à l'obstruction associée par Manin au groupe $H^1(\mathfrak{g}, \operatorname{Pic} \bar{X})$. Si cette obstruction est vide, il existe donc un nombre fini non nul de torseurs universels $\mathcal{T}_\alpha$ ayant un point dans chaque complété de k. La question qui se pose alors est de savoir si l'un de ces $\mathcal{T}_\alpha$ au moins a un point k-rationnel. D'après le théorème 1, l'obstruction de Manin au principe de Hasse sur les k-compactifications lisses des $\mathcal{T}_\alpha$ est vide. On est ainsi amené à poser la question suivante:

($\mathbf{Q}_2$) *Les k-compactifications lisses des torseurs universels sur X satisfont-elles le principe de Hasse?*

De même, l'analyse des contre-exemples à l'approximation faible donnée dans [7] soulève la question suivante, plus faible que ($\mathbf{Q}_1$):

($\mathbf{Q}_3$) *Les torseurs universels sur X qui ont un point k-rationnel satisfont-ils l'approximation faible?*

La question ($\mathbf{Q}_1$) n'a en général pas de réponse affirmative pour $\dim X \geq 3$ (cf. [6]): l'intersection lisse X de deux quadriques dans $\mathbf{P}^5_{\mathbf{R}}$ avec coordonnées (x, y, z, t, u, v) définie par

$$x^2+y^2+z^2=uv$$

$$x^2+2y^2+t^2=(u-v)(u-2v)$$

est $\mathbf{C}$-rationnelle, $\operatorname{Pic} X_{\mathbf{C}}$ est le $\mathfrak{g}$-module trivial $\mathbf{Z}$ et pourtant $X(\mathbf{R})$ a deux composantes connexes réelles; cette dernière propriété vaut donc pour toute $\mathbf{R}$-compactification lisse d'un torseur universel sur X et celle-ci ne peut être $\mathbf{R}$-rationnelle.

III. L'exemple des tores

Nous décrivons ici, de façon plus détaillée que dans [5], comment la méthode de la descente donne les résultats qualitatifs sur la R-équivalence sur les tores, par une méthode différente de celle utilisée dans [8].

Soient T un k-tore et X une k-compactification lisse de T: d'après Brylinski [3], il en existe toujours une (et même équivariante). Soit F le fermé complémentaire de T dans X. Comme $\operatorname{Pic} \bar{T}=0$ et $\hat{T}=\bar{k}[T]^*/\bar{k}^*$, on a la suite exacte de Voskresenskiĭ (cf. (1))

$$0 \longrightarrow \hat{T} \longrightarrow \operatorname{Div}_{\bar{F}} \bar{X} \longrightarrow \operatorname{Pic} \bar{X} \longrightarrow 0 \tag{3}$$

et la suite duale de k-tores

$$1 \longrightarrow S_0 \longrightarrow M \longrightarrow T \longrightarrow 1. \tag{4}$$

Comme le $\mathfrak{g}$-module $\operatorname{Div}_{\bar F} \bar X$ est de permutation, M est un k-tore quasi-trivial; ainsi, M est k-rationnelle, $M(k)/R = \{0\}$ et $H^1(k, M) = 0$. La suite (4) donne la suite exacte de groupes

$$M(k) \longrightarrow T(k) \xrightarrow{\partial} H^1(k, S_0) \longrightarrow 0. \tag{5}$$

La description locale des torseurs universels donnée en II.C s'applique ici en prenant $U = T$. Pour $P \in T(k)$, il existe donc un k-morphisme $\varphi_P : T \to T$ tel que le torseur sur T sous S_0 déduit par le changement de base φ_P du torseur M sur T défini par (4) soit la restriction du torseur universel $\mathcal{T}^P$ de fibre triviale en P. Comme la section $\sigma_P : \hat T = \bar k[T]^*/\bar k^* \to \bar k[T]^*$ associe au caractère f la fonction $x \mapsto f(x - P)$ et que $\bar k[T] = \bar k[\hat T]$, le morphisme φ_P n'est autre que la translation par $-P$. Pour $P = O$ l'élément neutre de $T(k)$, on voit ainsi que le diagramme

$$\begin{array}{ccc} T(k) & \xrightarrow{\ \partial\ } & \\ \big\downarrow i & & H^1(k, S_0) \\ X(k) & \xrightarrow{\ \theta_{\mathcal{T}^O}\ } & \end{array}$$

commute. Comme X est propre sur k, la proposition 1 montre que $\theta_{\mathcal{T}^O}$ passe au quotient par la R-équivalence; il en est donc de même de ∂. Par ailleurs, deux éléments de $T(k)$ de même image par ∂ diffèrent, d'après (5), d'un élément de $M(k)$, et sont donc R-équivalents. Ainsi, $T(k)/R \xrightarrow{\approx} H^1(k, S_0)$ et on a le triangle commutatif

$$\begin{array}{ccc} T(k)/R & \xrightarrow[\ \partial\ ]{\approx} & \\ \big\downarrow & & H^1(k, S_0). \\ X(k)/R & \xrightarrow{\ \theta\ } & \end{array}$$

De plus, pour k de type fini sur le corps premier, la proposition 2 montre que l'image de $\theta = \theta_{\mathcal{T}^O}$ est finie: ainsi, $T(k)/R$ et $H^1(k, S_0)$ sont alors finis. Pour k quelconque, l'assertion 1, la surjectivité de ∂ et le calcul ci-dessus de φ_P montrent que tout torseur universel $\mathcal{T}$ sur X est une variété k-rationnelle: $\mathcal{T}$ étant donné, il existe $P \in T(k)$ tel que $\mathcal{T}_T \cong M \times_T T$ avec pour flèche structurale $T \to T$ la translation par $-P$; ainsi $\mathcal{T}_T$ est une k-variété k-isomorphe à M, qui est k-rationnelle. En caractéristique 0, on peut établir ([8], proposition 13, p. 203) l'égalité $T(k)/R = X(k)/R$; la décomposition de $X(k)$ donnée en II.D décrit alors exactement la R-équivalence.

En résumé, le torseur M sur T sous S_0 défini par (4) calcule la R-équivalence sur $T(k)$ et celle-ci est finie pour k de type fini sur le corps premier. La méthode que nous venons d'utiliser pour obtenir ce résultat est fondée sur l'existence d'un prolongement de ce torseur M en un torseur sur la k-variété *propre* et lisse X et ce prolongement nous est ici donné comme un torseur universel; mais, au moins en caractéristique 0, les propositions 8 et 9 de [8] assurent en fait a priori l'existence d'un tel prolongement.

On peut enfin dire que dans le cas des k-compactifications lisses de tores, la méthode de la descente réussit parfaitement: la question ($\mathbf{Q}_1$) a une réponse affirmative, la question ($\mathbf{Q}_2$) aussi, mais de façon triviale, car tout torseur universel a alors un point k-rationnel.

IV. L'exemple des surfaces de Châtelet

L'objet de ce paragraphe est de décrire explicitement les torseurs universels sur certaines surfaces rationnelles parmi lesquelles les surfaces de Châtelet [4] et de donner ainsi une interprétation des résultats de Châtelet du point de vue de la descente (à comparer avec l'interprétation de Manin [16], chap. VI §5, qui utilise le groupe de Brauer).

Soit k un corps de caractéristique différente de 2. On considère $a \in k^*$ non carré, $K = k(\sqrt{a})$ et $G = \mathrm{Gal}(K/k) = \mathbf{Z}/2$. On note par $\sim$ la conjugaison de K/k. On désigne par $R\mathbf{G}_m$ le k-tore dual du $\mathfrak{g}$-module $\mathbf{Z}[G]$ et par $R^1\mathbf{G}_m$ le k-tore dual du $\mathfrak{g}$-module $\mathbf{Z}[G]/\mathbf{Z}$: c'est aussi le noyau de la norme $N: R\mathbf{G}_m \to \mathbf{G}_m$.

Soient n un entier ≥ 2 et $\{e_1, \ldots, e_{2n-1}\}$ des éléments de k distincts deux à deux. Soient X_1 la k-surface définie dans $\mathbf{P}^2 \times \mathbf{A}^1$, avec coordonnées $(y, z, t; \lambda)$, par l'équation

$$y^2 - az^2 = (\lambda - e_1) \ldots (\lambda - e_{2n-1})t^2$$

et X_2 la k-surface définie dans $\mathbf{P}^2 \times \mathbf{A}^1$, avec coordonnées $(Y, Z, T; \mu)$, par l'équation

$$Y^2 - aZ^2 = \mu(1 - e_1\mu) \ldots (1 - e_{2n-1}\mu)T^2.$$

Soit alors X la k-surface propre et lisse obtenue par recollement de X_1 et X_2 via

$$\begin{aligned} X_1 - \{\lambda \neq 0\} &\longrightarrow X_2 - \{\mu \neq 0\} \\ (y, z, t; \lambda) &\longmapsto (\lambda^{-n}y, \lambda^{-n}z, t; \lambda^{-1}). \end{aligned}$$

Les projections sur $\mathbf{A}^1$ se recollent en un k-morphisme $X \to \mathbf{P}^1$ qui fait de X

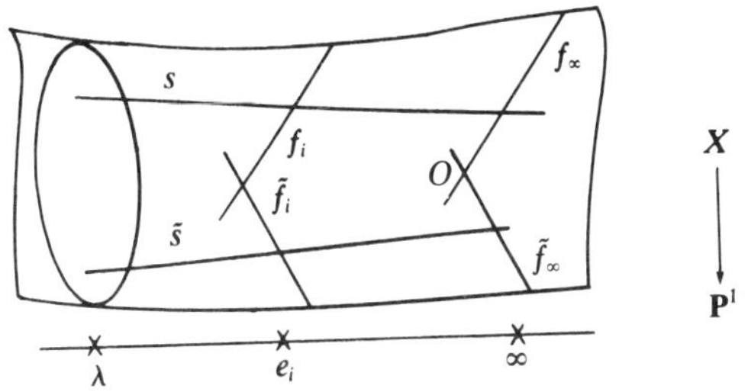

une k-surface fibrée en coniques au-dessus de $\mathbf{P}^1$. Cette fibration possède sur K des sections s et $\tilde{s}$ définies par $(y, z, t; \lambda) = (\pm\sqrt{a}, 1, 0; \lambda)$. Elle a $2n$ fibres dégénérées $f_i + \tilde{f}_i$ pour $i = 1, \ldots, 2n-1, \infty$ en les points $\lambda = e_1, \ldots, e_{2n-1}, \infty$.

Soit U l'ouvert complémentaire du k-fermé F union de s, $\tilde{s}$, et des f_i, $\tilde{f}_i$, pour $i \neq 2n-1$. On a Pic $\bar{U} = 0$. La suite exacte (1) s'écrit ici

$$0 \longrightarrow \bigoplus_{i=1}^{2n-2} \mathbf{Z}\alpha_i \xrightarrow{\rho} \bigoplus_{i=1}^{2n-2} [\mathbf{Z}(f_i - f_\infty) \oplus \mathbf{Z}(\tilde{f}_i - \tilde{f}_\infty)] \oplus \mathbf{Z}f_\infty \oplus \mathbf{Z}\tilde{f}_\infty \oplus \mathbf{Z}s \oplus \mathbf{Z}\tilde{s}$$
$$\longrightarrow \operatorname{Pic} \bar{X} \longrightarrow 0$$

où α_i est la classe de la fonction $\lambda - e_i$, qui a pour diviseur $f_i + \tilde{f}_i - f_\infty - \tilde{f}_\infty$. L'homomorphisme de k-tores dual de ρ s'identifie au k-morphisme

$$(R\mathbf{G}_m)^{2n-2} \times (R\mathbf{G}_m)^2 \longrightarrow \mathbf{G}_m^{2n-2}$$
$$(x, y) \longmapsto N(x).$$

Pour chaque $(c_i) \in (k^*)^{2n-2}$, la famille $\{c_i(\lambda - e_i)\}$, pour $i = 1, \ldots, 2n-2$, définit une $\mathfrak{g}$-section de la projection $\bar{k}[U]^* \to \bar{k}[U]^*/\bar{k}^* \approx \mathbf{Z}^{2n-2}$, et toute $\mathfrak{g}$-section est de ce type. La description locale des torseurs universels (II.C) montre alors que la restriction de tout torseur universel à U admet des équations du type suivant:

$$0 \neq \lambda - e_i = c_i(u_i^2 - av_i^2) \qquad i = 1, \ldots, 2n-2$$
$$y^2 - az^2 = (\lambda - e_1) \ldots (\lambda - e_{2n-1})$$
$$0 \neq x_j^2 - aw_j^2 \qquad j = 1, 2$$

dans l'espace affine de coordonnées $(\lambda, y, z, u_i, v_i, x_j, w_j)$. Si l'on se restreint de plus à l'ouvert $\lambda \neq e_{2n-1}$, on obtient le produit de $(R\mathbf{G}_m)^2$ par la variété $X((c_i))$ d'équations

$$0 \neq \lambda - e_i = c_i(u_i^2 - av_i^2) \qquad i = 1, \ldots, 2n-1 \tag{6}$$

dans l'espace affine de coordonnées (λ, u_i, v_i) avec $c_{2n-1} = \prod_{i=1}^{2n-2} c_i$.

La question $(\mathbf{Q}_1)$ est ainsi très proche ici de la question de savoir si les variétés $X((c_i))$ qui ont un point k-rationnel sont k-rationnelles. Pour $n=2$, Châtelet a montré [4] que $X(1, 1)$ est k-rationnelle, ce qui implique alors, au moins en caractéristique 0, que les $X((c_i))$ qui ont un point k-rationnel sont k-stablement k-rationnelles; dans ce cas, la descente réussit encore très bien: les classes pour la R-équivalence sur $X(k)$ coïncident avec les fibres de l'application $\theta_{\mathcal{T}^O}$ associée au torseur universel trivial au point double O de la fibre à l'infini; cette application est définie sur $U(k)$ par

$$\begin{aligned} U(k) &\longrightarrow H^1(k, S_0) = (k^*/N(K^*))^2 \\ (y, z, t; \lambda) &\longmapsto (\lambda - e_1, \lambda - e_2) \end{aligned}$$

et son image est finie quand k est de type fini sur le corps premier.

La question $(\mathbf{Q}_2)$ pour les surfaces X considérées dans ce paragraphe se ramène à la question de la validité du principe de Hasse pour les variétés $X((c_i))$: pour $n=2$, on retrouve une question de Manin ([16] chap. VI, fin du §5). La question $(\mathbf{Q}_3)$ pour ces surfaces X se ramène à la question de la validité de l'approximation faible pour les variétés $X((c_i))$ qui ont un point dans k.

V. Questions sur les surfaces rationnelles

Nous dressons ici une liste de questions naturelles sur les surfaces rationnelles qui se trouvent dériver pour la plupart de l'une des questions $(\mathbf{Q}_1)$, $(\mathbf{Q}_2)$, $(\mathbf{Q}_3)$.

On considère une k-surface rationnelle X, propre et lisse. On suppose en outre $X(k) \neq \emptyset$ dans les questions **(a)–(j)**. On note $\tilde{A}_0(X)$ le sous-groupe de $A_0(X)$ formé des classes des 0-cycles de degré 0 et $\varphi : \tilde{A}_0(X) \to H^1(k, S_0)$ la restriction, indépendante de $\mathcal{T}$, de l'homomorphisme $\theta_{\mathcal{T}}$ défini par un torseur universel $\mathcal{T}$.

$(\mathbf{a}_1)$ X est-elle k-unirationnelle?

$(\mathbf{a}_2)$ Pour k infini, $X(k)$ est-il Zariski-dense dans X?

(b) Si $\operatorname{Pic} \bar{X}$ est un $\mathfrak{g}$-module stablement de permutation, existe-t-il des entiers m et n tels que $X \times_k \mathbf{A}_k^m$ soit k-birationnelle à $\mathbf{A}_k^n$?

$(\mathbf{c}_1)$ Si $\operatorname{Pic} \bar{X}$ est un facteur direct d'un $\mathfrak{g}$-module de permutation, existe-t-il une k-variété rationnelle Y telle que $X \times_k Y$ soit k-rationnelle?

$(\mathbf{c}_2)$ Sous la même hypothèse sur $\operatorname{Pic} \bar{X}$, a-t-on $X(k)/R$ trivial?

$(\mathbf{c}_3)$ Sous la même hypothèse sur $\operatorname{Pic} \bar{X}$, a-t-on $\tilde{A}_0(X)$ trivial?

Pour les trois questions $(\mathbf{d}_i)$ suivantes, on considère en outre un point $O \in X(k)$ et le triangle commutatif

$$
\begin{array}{ccc}
X(k)/R & \xrightarrow{\theta} & H^1(k, S_0) \\
\downarrow{\scriptstyle i} & \nearrow{\scriptstyle \varphi} & \\
\tilde{A}_0(X) & &
\end{array}
$$

où θ est induite par $\theta_{\mathcal{T}^0}$ et où i associe à $P \in X(k)$ la classe du 0-cycle $P - O$.

($\mathbf{d_1}$) θ est-elle injective?

($\mathbf{d_2}$) i est-elle injective?

($\mathbf{d_3}$) φ est-elle injective?

($\mathbf{e_1}$) A-t-on, pour k de dimension cohomologique ≤ 1, $X(k)/R = \{O\}$?

($\mathbf{e_2}$) Sous la même hypothèse sur k, a-t-on $\tilde{A}_0(X) = 0$?

(f) Chaque classe pour la R-équivalence sur $X(k)$ est-elle paramétrée par les points à valeurs dans k d'une k-variété k-rationnelle?

($\mathbf{g_1}$) Pour k de type fini sur le corps premier, $X(k)/R$ est-il fini?

($\mathbf{g_2}$) Sous la même hypothèse sur k, a-t-on $\tilde{A}_0(X)$ fini?

(h) Pour k local, les classes pour la R-équivalence sur $X(k)$ sont-elles ouvertes pour la topologie définie par celle de k? En particulier, pour $k = \mathbf{R}$, ces classes coïncident-elles avec les composantes connexes réelles?

Dans les questions suivantes, k est un corps de nombres et Ω désigne l'ensemble des places de k.

(i) L'adhérence de $X(k)$ dans le produit topologique $\prod_{v \in \Omega} X(k_v)$ est-elle ouverte?

($\mathbf{j_1}$) L'obstruction à l'approximation faible définie dans [7] est-elle la seule?

($\mathbf{j_2}$) Si $H^1(\mathfrak{g}, \operatorname{Pic} \bar{X}) = 0$, l'approximation faible vaut-elle pour X?

($\mathbf{k_1}$) L'obstruction de Manin au principe de Hasse ([15], [16] chap. VI §1) est-elle la seule?

($\mathbf{k_2}$) Si $H^1(\mathfrak{g}, \operatorname{Pic} \bar{X}) = 0$, le principe de Hasse vaut-il pour X?

ASSERTION 2: *Au moins pour k de caractéristique* 0, *une réponse affirmative à* (**$\mathbf{Q_1}$**) *implique une réponse affirmative aux questions* **(a)–(i)** *à l'exclusion près de* (**$\mathbf{d_3}$**) *et* (**$\mathbf{g_2}$**); *pour* **(i)**, *il suffit d'une réponse affirmative à* (**$\mathbf{Q_3}$**); *une réponse affirmative à* (**$\mathbf{Q_2}$**) (*resp. à* (**$\mathbf{Q_2}$**) *et* (**$\mathbf{Q_3}$**)) *implique une réponse affirmative à* **(k)** (*resp.* **(j)**).

Ainsi, la méthode de la descente donne-t-elle une présentation unifiée de questions sur les surfaces rationnelles, déjà posées pour la plupart par divers auteurs (Manin, Iskovskikh, Swinnerton-Dyer).

VI. Quelques résultats récents

Nous passons en revue les questions précédentes à la lumière d'un certain nombre de résultats récents. Au moins en caractéristique 0, toutes les questions **(a)–(k)** sont invariantes k-birationnellement. On peut donc espérer obtenir, pour certaines de ces questions, une réponse par une analyse cas par cas, grâce au résultat suivant d'Iskovskikh, qui parachève une classification due à Enriques et Manin:

THÉORÈME 2 (Iskovskikh [14]): *Soient k un corps quelconque, X une k-surface rationnelle, propre et lisse. Alors, X est k-birationnelle à une k-surface de l'un des types suivants:*

(1) *une k-surface de Del Pezzo de degré n avec* $1 \leq n \leq 9$;

(2) *une k-surface fibrée en coniques.*

On sait par ailleurs que les surfaces de Del Pezzo de degrés 5 et 7 sont toujours k-rationnelles, du moins pour k infini. De plus, si X est une surface de Del Pezzo de degré 6, 8 ou 9, Pic $\bar{X}$ est toujours un $\mathfrak{g}$-module stablement de permutation, X est k-rationnelle dès que $X(k) \neq \emptyset$ et, si k est un corps de nombres, elle vérifie le principe de Hasse. Les surfaces de Del Pezzo de degré 4 sont les intersections lisses de deux quadriques dans $\mathbf{P}^4$. Celles de degré 3 sont les surfaces cubiques lisses dans $\mathbf{P}^3$. Parmi les surfaces fibrées en coniques pour k de caractéristique différente de 2, des exemples intéressants sont donnés par les équations affines

$$y^2 - az^2 = P(\lambda)$$

où $a \in k^*$ et où P est un polynôme sans facteur multiple.

($\mathbf{a}_1$) Pour les surfaces de Del Pezzo de degré ≥ 2 ayant un point dans k, on sait presque toujours montrer ([16] Chap. IV, §7) que ce sont des variétés k-unirationnelles. Pour les surfaces fibrées en coniques, la question reste largement ouverte, de même ($\mathbf{a}_2$), sauf pour $k = \mathbf{R}$ (Iskovskikh [13]).

(**b**) Soit X une k-surface rationnelle, propre et lisse, avec $X(k) \neq \emptyset$, telle que Pic $\bar{X}$ soit le $\mathfrak{g}$-module trivial $\mathbf{Z}^n$: on peut montrer, au moins pour k parfait, que X est alors k-rationnelle (ceci ne vaut plus pour dim $X \geq 3$, cf. fin du §II).

($\mathbf{c}_3$), ($\mathbf{d}_3$), ($\mathbf{e}_2$), ($\mathbf{g}_2$) Pour toute k-surface rationnelle X, propre et lisse, Bloch a défini et étudié [2], par des méthodes de K-théorie, un homomorphisme

$$\Phi: \tilde{A}_0(X) \longrightarrow H^1(k, S_0)$$

dont on peut montrer qu'il coïncide avec la restriction φ de l'homomorphisme $\theta_{\mathcal{T}}$ défini par un torseur universel $\mathcal{T}$ quelconque (II.B). Les résultats de Bloch [2] impliquent d'après [9] l'énoncé suivant:

THÉORÈME 3: *Soient k de caractéristique* $\neq 2$ *et X une k-surface fibrée en coniques telle que* $X(k) \neq \emptyset$. *L'homomorphisme* Φ *est alors injectif.*

On en déduit, en utilisant entre autres la proposition 2, une réponse affirmative aux quatre questions ci-dessus pour une telle surface. La finitude de $\tilde{A}_0(X)$ est déjà dans [2] pour k corps de nombres.

($\mathbf{d}_1$), ($\mathbf{d}_2$), ($\mathbf{e}_1$), ($\mathbf{g}_1$) Soit X une surface de Del Pezzo de degré 4 sur k de caractéristique 0 avec $X(k) \neq \emptyset$. Une telle surface est k-birationnelle à une surface fibrée en coniques. Une analyse précise des courbes tracées sur X a permis à Swinnerton-Dyer d'obtenir le résultat suivant:

THÉORÈME 4 (Swinnerton-Dyer [19]): *Pour une telle surface de Del Pezzo de degré* 4, *l'application naturelle* $X(k)/R \to A_0(X)$ *est injective et son image est l'ensemble des éléments de degré* 1.

D'où une réponse affirmative dans ce cas aux quatre questions ci-dessus.

($\mathbf{e}_1$) On a le résultat suivant:

THÉORÈME 5 (Swinnerton-Dyer [18]): *Soit* X *une surface cubique lisse dans* $\mathbf{P}_k^3$ *définie sur le corps fini* k. *Alors* $X(k)/R = \{0\}$.

(**h**) La réponse est affirmative pour une surface cubique (Manin [16], chap. II, §6). Pour $k = \mathbf{R}$, des résultats d'Iskovskikh [13] impliquent une réponse affirmative pour les **R**-surfaces fibrées en coniques.

(**i**) Soit X une surface cubique lisse sur un corps de nombres k. Un argument de Swinnerton-Dyer montre que l'adhérence de $X(k)$ dans tout produit fini de $X(k_v)$ est ouverte.

($\mathbf{k}_1$) Soit X une k-compactification lisse de la k-surface fibrée en coniques d'équation affine

$$y^2 - az^2 = \prod_{i=1}^{r} P_i(\lambda) \tag{7}$$

où $a \in k^*$ et où les P_i sont des polynômes irréductibles de degré pair, étrangers deux à deux. La méthode de la descente conduit à la question suivante: les k-compactifications lisses des k-variétés (lisses) définies par les équations affines

$$0 \neq P_i(\lambda) = c_i(u_i^2 - av_i^2) \quad i = 1, \ldots, r \tag{8}$$

où les c_i appartiennent à k^* et où les variables sont (λ, u_i, v_i) satisfont-elles le principe de Hasse? Si la réponse à cette question est affirmative, la réponse à ($\mathbf{k}_1$) pour X l'est aussi. C'est par cette démarche qu'on obtient (Coray et les auteurs) le résultat particulier suivant:

THÉORÈME 6 ([11]): *L'obstruction de Manin est la seule obstruction au principe de Hasse pour un k-modèle propre et lisse de la surface d'équation*

$$y^2 - az^2 = P_1(\lambda)P_2(\lambda)$$

où P_1 et P_2 sont deux polynômes du second degré irréductibles premiers entre eux.

(j_1) Pour X comme en (7), la question (j_1) aurait une réponse affirmative si les variétés du type (8) satisfaisaient, outre le principe de Hasse, l'approximation faible. On retrouve ainsi des questions déjà rencontrées à propos des équations (6), ce qui invite à poser la question générale suivante: soient k un corps de nombres, puis, pour $i = 1, \ldots, r$, des éléments a_i de k^* et des polynômes irréductibles $P_i \in k[\lambda]$; *les k-variétés d'équations affines*

$$0 \neq P_i(\lambda) = u_i^2 - a_i v_i^2 \qquad i = 1, \ldots, r$$

satisfont-elles le principe de Hasse et l'approximation faible?

Pour $k = \mathbf{Q}$, on peut montrer [10] que l'hypothèse (H) de Schinzel [17] (extrapolation hardie du théorème de la progression arithmétique) donne une réponse affirmative à cette dernière question.

BIBLIOGRAPHIE

[1] [SGA 4] M. Artin, A. Grothendieck et J.-L. Verdier: Théorie des Topos et Cohomologie étale des schémas. Lecture Notes in Math. n^{os} *270, 305*, Berlin–Heidelberg–New York: Springer 1972–73.

[2] S. Bloch: On the Chow group of certain rational surfaces (preprint).

[3] J.-L. Brylinski: Décomposition simpliciale d'un réseau, invariante par un groupe fini d'automorphismes. C.R. Acad. Sci. Paris 288, série A (1979) 137–139.

[4] F. Châtelet: Points rationnels sur certaines courbes et surfaces cubiques. Enseignement Math. *5* (1959) 153–170.

[5] J.-L. Colliot-Thélène et J.-J. Sansuc: Torseurs sous des groupes de type multiplicatif. C.R. Acad. Sci. Paris 282, série A, (1976) 1113–1116.

[6] J.-L. Colliot-Thélène et J.-J. Sansuc: Variétés de première descente attachées aux variétés rationnelles. C.R. Acad. Sci. Paris 284, série A, (1977) 967–970.

[7] J.-L. Colliot-Thélène et J.-J. Sansuc: La descente sur une variété rationnelle définie sur un corps de nombres. C.R. Acad. Sci. Paris 284, série A, (1977) 1215–1218.

[8] J.-L. Colliot-Thélène et J.-J. Sansuc: La R-équivalence sur les tores. Ann. scient. Ec. Norm. Sup., 4^{è} série, *10* (1977) 175–229.

[9] J.-L. Colliot-Thélène et J.-J. Sansuc: Remarques sur "On the Chow group of certain rational surfaces" par S. Bloch (en préparation).

[10] J.-L. Colliot-Thélène et J.-J. Sansuc: Sur le principe de Hasse et l'approximation faible, et sur une hypothèse de Schinzel, à paraître dans Acta Arithm. *41.*

[11] J.-L. Colliot-Thélène, D. Coray et J.-J. Sansuc: Descente et principe de Hasse pour certaines variétés rationnelles (prépublication).

[12] [EGA IV] A. Grothendieck et J. Dieudonné: Eléments de Géométrie Algébrique. Publications mathématiques I.H.E.S. *24, 28*, Paris 1965–66.

[13] V.A. Iskovskikh: Surfaces rationnelles munies d'un pinceau de courbes rationnelles.

Mat. Sbornik, *74* (116), (1967) 608–638 (trad. ang. Math. USSR-Sbornik, *3*, (1967) 563–587).
[14] V.A. ISKOVSKIKH: Modèles minimaux des surfaces rationnelles sur un corps quelconque, Izv. Akad. Nauk SSSR Ser. Math. *43*, (1979) 19–43.
[15] Y.I. MANIN: Le groupe de Brauer-Grothendieck en géométrie diophantienne, Actes du Congrès intern. math. Nice, *1*, (1970) 401–411.
[16] Y.I. MANIN: Formes cubiques, Nauka, Moscou 1972 (trad. ang., North-Holland, Amsterdam 1974).
[17] A. SCHINZEL et W. SIERPIŃSKI: Sur certaines hypothèses concernant les nombres premiers. Acta Arithm. *4*, (1958) 185–208. Corrig. ibid. *5*, (1959) 259.
[18] H.P.F. SWINNERTON-DYER: Universal equivalences for cubic surfaces over finite and local fields (preprint Rome 1979).
[19] H. P. F. SWINNERTON-DYER: Rational equivalence and R-equivalence on cubic surfaces (preprint).
[20] V.E. VOSKRESENSKIĬ: Tores algébriques, Nauka, Moscou 1977.

Jean-Louis Colliot-Thélène
C.N.R.S. Mathématiques
Bâtiment 425, Université de Paris-Sud
F-91405 Orsay, France

Jean-Jacques Sansuc
E.N.S., 45 rue d'Ulm
F-75230 Paris Cedex 05, France

ON THE MUMFORD–RAMANUJAM VANISHING THEOREM ON A SURFACE

Yoichi Miyaoka*

1. Introduction

In the theory of surfaces of general type, two vanishing theorems due to Mumford and Ramanujam play an important role. In spite of their striking similarity, however, they have been proven in quite different contexts. In the present paper, we generalize the following vanishing theorem of Mumford so that it implies Ramanujam vanishing theorem as well.

THEOREM (1.1) (Mumford [2]): *Let X be a surface and D be a divisor with the following properties*

(a) $D^2 > 0$;

(b) $DC \geqq 0$ *for any curve C on X.*

Then the first cohomology group $H^1(X, -D)$ vanishes.

In §2, we give some sufficient condition on D for $H^1(X, -D)$ to vanish. But, in general, $H^1(X, -D)$ is not trivial even if D is effective. Here it gives rise to the notion of "connectedness" of an effective divisor. We discuss how to estimate the dimension of the first cohomology group $H^1(X, -D)$ using this notion in §3.

Originally, the author proved his main result (2.7) under the additional condition $D^2 > 0$, by using Mumford's idea to prove Theorem 1.1 with the help of Bogomolov's criteria for unstability of vector bundles. The condition $D^2 > 0$ was essential for this method (for Mumford's idea, see Reid [4]).

In this paper, all varieties are assumed to be complete, smooth and defined over the complex number field **C** unless otherwise mentioned. We freely use standard terms and symbols from algebraic geometry. Further the following notation is used:

– If X is a variety and F is a coherent sheaf on X, then

$$h^i(X, F) = \dim_{\mathbf{C}} H^i(X, F).$$

– If D is a divisor on X, we often write $H^i(X, D)$ and $h^i(X, D)$ instead of $H^i(X, \mathcal{O}_X(D))$ and $h^i(X, \mathcal{O}_X(D))$, respectively.

*Partially supported by SFB 40 "Theoretische Mathematik".

– $NS(X)$ denotes the Néron–Severi group of X, and $NS_{\mathbf{Q}}(X) = NS(X) \otimes_{\mathbf{Z}} \mathbf{Q}$ is the $\mathbf{Q}$-vector space of the rational divisors or divisorial cycles on X (up to rational homology equivalence). For a divisor D, the corresponding element of $NS(X)$ or $NS_{\mathbf{Q}}(X)$ will be denoted also by D, by abuse of notation.

– By $\kappa(X, D)$ we denote the D-dimension. That is

$$\kappa(X, D) = \left\{ \begin{array}{l} \text{transcendence degree of the graded ring} \\ \bigoplus_{k \geq 0} H^0(X, kD). \end{array} \right\} - 1.$$

2. The vanishing theorems of Mumford and Ramanujam

We start with two well-known results:

THEOREM (2.1) (Bloch–Gieseker [1]): *Let X be a non-singular surface and $C_1, \ldots, C_r$ a finite number of smooth curves which intersect each other transversally. For any positive integer m, there exists a composition of finite Galois coverings $f: Y \to X$ such that*

(a) *Y is smooth;*
(b) *f^*C_i is a smooth irreducible curve;*
(c) *$f^*C_1, \ldots, f^*C_r$ intersect transversally;*
(d) *there exists a divisor $F_i \in \mathrm{Pic}(Y)$ with $f^*C_i = mF_i$.*

THEOREM (2.2) (Zariski [5]): *Let D be a divisor on a surface X with $\kappa(X, D) = 2$. Then there exists a unique decomposition*

$$D = D_+ + \sum m_i C_i$$

in $NS_{\mathbf{Q}}(X)$ which satisfies the following conditions:

(a) *$D_+C \geqq 0$ for any curve C;*
(b) *$D_+^2 > 0$;*
(c) *C_i is an irreducible curve with $D_+C_i = 0$;*
(d) *m_i is a positive rational number.*

We call this decomposition the "Zariski decomposition" of D.

PROPOSITION (2.3): *Let D be a divisor on a surface with $\kappa(X, D) = 2$ and $D = D_+ + \Sigma\, m_i C_i$ its Zariski decomposition. If each m_i is smaller than 1, then $H^1(X, -D)$ vanishes.*

Our proof proceeds step by step.

CLAIM (2.4): *Proposition 2.3 holds if there exists an integral divisor F_i numerically equivalent to m_iC_i for each i and the union of the curves C_i forms a normal crossing of smooth curves.*

PROOF: From our assumption, we can construct a series of finite cyclic coverings $f_i: X_i \to X_{i-1}$ of degree m_i, the branch locus of f_i being the smooth curve $(f_1 \ldots f_{i-1})^{-1}C_i$. Then $X_1, \ldots, X_r$ are smooth. We set $f'_i = f_1 \ldots f_i$ and $\bar{C}_i = f'^{-1}_r(C_i)$. We have

$$f'^*_r D = f'^*_r D_+ + \sum d_i\bar{C}_i$$

where d_i is an integer. From Theorem 1.1 we have $H^1(X_r, -f'^*_r D_+) = 0$. Now we observe the commutative diagram

$$\begin{array}{ccccccccc} 0 \longrightarrow & \mathcal{O}(-f'^*_r D_+ - k\bar{C}_r) & \xrightarrow{\iota} & \mathcal{O}(-f'^*_r D_+ - (k-1)\bar{C}_r) & \xrightarrow{\rho} & \mathcal{O}_{\bar{C}_r}(-f'^*_r D_+ - (k-1)\bar{C}_r) & \longrightarrow 0 \\ & \downarrow g & & \downarrow g & & \downarrow \text{identity} & \\ 0 \longrightarrow & \mathcal{O}(-f'^*_r D_+ - k\bar{C}_r) & \xrightarrow{\iota'} & \mathcal{O}(-f'^*_r D_+ - (k-1)\bar{C}_r) & \xrightarrow{\rho} & \mathcal{O}_{\bar{C}_r}(-f'^*_r D_+ - (k-1)\bar{C}_r) & \longrightarrow 0 \end{array}$$

where ι is the natural injection, $\iota'(*) = \omega\iota(*)$, ρ the restriction mapping and g the action of the generator of the Galois group over X_{r-1}. Here ω denotes a m_r-th root of unity. g acts naturally on the cohomology groups and they are decomposed into subspaces associated with the eigenvalues of g. In view of the commutative diagram above, we see that

$$\begin{aligned} &\ker\{H^1(-f'^*_r D_+ - k\bar{C}_r) \longrightarrow H^1(-f'^*_r D_+ - (k-1)\bar{C}_r)\} \\ &\quad \subset H^1(-f'^*_r D_+ - k\bar{C}_r)^{\omega^{-1}} \end{aligned}$$

where $H^1(-f'^*_r D_+ - k\bar{C}_r)^{\omega^{-1}}$ denotes the eigenspace with the eigenvalue ω^{-1} with respect to the action of g. Hence we have

$$\begin{aligned} &\ker\{H^1(-f'^*_r D_+ - k\bar{C}_r) \longrightarrow H^1(-f'^*_r D_+)\} \\ &\quad \subset \bigoplus_{1 \leq i \leq k} H^1(-f'^*_r D_+ - k\bar{C}_r)^{\omega^{-i}}. \end{aligned}$$

Since $f_r: X_r \to X_{r-1}$ is a finite cyclic covering, $H^1(X_{r-1}, -D_+ - F_r)$ is a subspace of $H^1(X_r, -f'^*_r D_+ - d_r\bar{C}_r)^1$. On the other hand, since $m_i < 1$, this subspace is mapped injectively into $H^1(X_r, -f'^*_r D_+) = 0$, so that $H^1(X_{r-1}, -f'^*_{r-1}(D_+ + F_r)) = 0$. Inductively, we get $H^1(X, -D) = 0$. Q.E.D.

CLAIM (2.5): *Proposition 2.3 holds if the union of the C_i forms a normal crossing of smooth curves.*

PROOF: By Theorem 2.1, we can construct a composition of Galois coverings $f: Y \to X$ such that (Y, f^*D) satisfies the condition of (2.4). Hence $H^1(Y, -f^*D) = 0$. On the other hand, $\mathcal{O}_X$ is a direct summand of $f_*\mathcal{O}_Y$ since f is a composition of finite Galois coverings. Hence
$H^1(X, -D) \subset H^1(X, f_*\mathcal{O}_Y \otimes \mathcal{O}_X(-D)) = H^1(Y, -f^*D) = 0.$ Q.E.D.

Now we prove Proposition 2.3. Let $p: \tilde{X} \to X$ be a composition of blowing-ups such that $p^{-1}(\Sigma\, C_i)$ forms a normal crossing of smooth curves. We denote by E_k $(k = 1, \ldots, s)$ the exceptional divisors on $\tilde{X}$. Then, for some non-negative integers $n_1, \ldots, n_s$, $p^*D - \Sigma\, n_iE_i$ satisfies the conditions of 2.3 and 2.5. Hence the assertion is reduced to the following

LEMMA (2.6): *Let $p: \bar{X} \to X$ be a blowing up at a point and E the exceptional divisor associated with p. For any divisor D on X, the canonical homomorphism $H^1(p^*D + (n-1)E) \to H^1(p^*D + nE)$ is injective $(n = 1, 2, \ldots)$.*

PROOF: Trivial.

As a consequence of Proposition 2.3, we get our main result.

THEOREM (2.7): *Let D be a divisor on a surface X with $\kappa(X, D) = 2$, and $D = D_+ + \Sigma\, m_iC_i$ the Zariski decomposition of D. If the first cohomology group $H^1(X, -D)$ is non-trivial, then there exists a tuple of integers $(\ldots, n_i, \ldots) \neq (\ldots, 0, \ldots)$ such that*

(a) $0 \leqq n_i \leqq m_i$;

(b) $(D - \Sigma\, n_iC_i)C_a \leqq 0$ *for any a with $n_a > 0$.*

PROOF: Let $(\ldots, n_i, \ldots)$ be a tuple of integers which satisfies the condition (a). If $(\ldots, n_i, \ldots)$ is minimal (in natural sense) so that $H^1(X, -D + \Sigma\, n_iC_i) = 0$, we have $H^1(X, -D + \Sigma\, n_iC_i - C_a) \neq 0$ for any a with $n_a > 0$. But this implies that $H^0(C_a, -D + \Sigma\, n_iC_i)$ is non-trivial and we get (b).

Q.E.D.

The condition (a) is characterized as follows:

PROPOSITION (2.8): *Let D and X be as in Theorem 2.7. Then an effective divisor N on X is of the form $\Sigma\, n_iC_i$, $0 \leqq n_i \leqq m_i$ if and only if the graded ring $\bigoplus_{k \geqq 0} H^0(X, k(D - N))$ is isomorphic to $\bigoplus_{k \geqq 0} H^0(X, kD)$ via the natural inclusion.*

PROOF: Easy.

In particular, if D is effective in Theorem 2.7, then $D - \Sigma\, n_iC_i$ is also effective. Thus we have the following

THEOREM (2.9) (Ramanujam's vanishing theorem, see [3]): *Let D be a divisor on a surface X. Assume that D is effective and 1-connected and that $\kappa(X, D) = 2$. Then $H^1(X, -D)$ vanishes.*

REMARK (2.10): In Theorem 2.9, D need not necessarily be effective. Assume that $\kappa(X, D) = 2$ and D is numerically equivalent to an effective 1-connected divisor. Then $H^1(X, -D) = 0$.

REMARK (2.11): If the ground field of X is an algebraically closed field of characteristic $p > 0$, the author does not know if Theorem 2.7 holds in general.

3. Estimation of the dimension of the first cohomology group

DEFINITION (3.1): *A series of irreducible curves $C_1, \ldots, C_r$ on a surface is called* connected composite series *with respect to D if $(D + C_1 + \cdots + C_{i-1})C_i > 0$ for $i = 1, \ldots, r$. An effective divisor D is called* s-connected *if there exists a decomposition $D = C_0 + C_1 + \cdots + C_r$ such that $C_1, \ldots, C_r$ is a connected composite series with respect to C_0.*

REMARK (3.2): If an effective divisor D is 1-connected, then D is s-connected. Assume that D is 1-connected. Take any irreducible component C_0 of D. Then, by definition, $C_0(D - C_0) > 0$. Hence, for some irreducible component C_1 of $D - C_0$, we have $C_0C_1 > 0$. Similarly, we can choose a series $(C_0, C_1, \ldots, C_r)$ so that D is s-connected. The converse does not hold, however, in general. Assume that three irreducible curves C_1, C_2, C_3 with self-intersection numbers $-1, -2, -3$, respectively, intersect each other as in the figure. Let $D = 2(C_1 + C_2 + C_3)$. Then D is s-connected but not 1-connected. In fact, $(C_1, C_2, C_3, C_1, C_2, C_3)$ is the desired sequence and $(C_1 + C_2 + C_3)^2 = 0$.

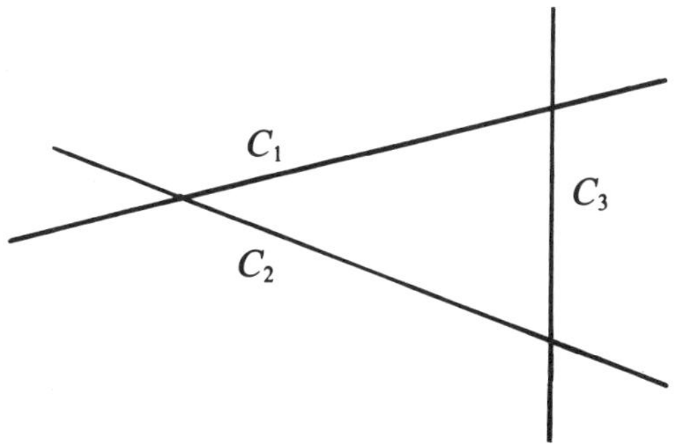

Proposition (3.3): *An effective divisor D is not s-connected if and only if there exists a non-trivial decomposition $D = D_1 + D_2$ into effective divisors such that $D_1 C_k \leqq 0$ for any irreducible component C_k of D_2.*

Proof: The "only if" part is trivial. Now assume that D is s-connected. There exists a series of curves $(C_0, \ldots, C_r)$ which satisfies the condition of 3.1. We take distinct irreducible curves C'_i so that $D = \Sigma\, c_i C'_i$, where c_i is a positive integer. Suppose that there exists a non-trivial decomposition as above. Put $D_1 = \Sigma\, d_i C'_i$ and $\Sigma\, e_i^{(k)} C'_i = C_0 + \cdots + C_k$. We take the maximal integer s such that $e_i^{(s)} \leqq d_i$ for any i. Then we have

$$0 < C_{s+1}(C_0 + \cdots + C_s) \leqq C_{s+1} D_1 \leqq 0,$$

which is a contradiction. Q.E.D.

Proposition (3.4): *Let $(C_1, \ldots, C_r)$ be a connected composite series with respect to a divisor D on a surface X. Then we have an estimation $h^1(X, -D - C_1 - \cdots - C_r) \leqq h^1(X, -D)$.*

Proof: Since $C_k(D + C_1 + \cdots + C_{k-1}) > 0$, we have $H^0(C_k, -D - C_1 - \cdots - C_{k-1}) = 0$, so that $H^1(X, -D - C_1 - \cdots - C_k) \to H^1(X, -D - C_1 - \cdots - C_{k-1})$ is injective. Hence $H^1(X, -D - C_1 - \cdots - C_r) \to H^1(X, -D)$ is injective. This gives our estimation. Q.E.D.

Corollary (3.5): *Let D be an effective s-connected divisor on a surface X. Then $h^1(X, -D) \leqq h^1(X, \mathcal{O})$.*

Proof: Let $(C_0, C_1, \ldots, C_r)$ be as in 3.1. Then we have $h^1(X, -D) \leqq h^1(X, -C_0)$. But we have the natural exact sequence

$$0 \longrightarrow H^0(X, \mathcal{O}) \xrightarrow{\approx} H^0(C_0, \mathcal{O}) \longrightarrow H^1(X, -C_0) \longrightarrow H^1(X, \mathcal{O}).$$

Hence we get an estimate $h^1(X, -C_0) \leqq h^1(X, \mathcal{O})$. Q.E.D.

Corollary (3.6): *Let D be an effective s-connected divisor on a surface. Then $h^0(D, \mathcal{O}_D) = 1$.*

Proof: Similar as the proof of 3.5.

Proposition (3.7): *Let D be an effective s-connected divisor on a surface X and L a divisor with $LC \geqq 0$ for any irreducible component C of D. Then we have*

$$\dim \ker(H^1(X, L - D) \longrightarrow H^1(X, L))$$

$$\leqq LD + 1 - h^0(X, L) + h^0(X, L - D).$$

PROOF: Take a series $(C_0, C_1, \dots, C_r)$ as in 3.1. Then we have a natural exact sequence

$$\begin{aligned} 0 \longrightarrow\ & H^0(X, L - C_0 - \cdots - C_k) \longrightarrow H^0(X, L - C_0 - \cdots - C_{k-1}) \\ \longrightarrow\ & H^0(C_k, L - C_0 - \cdots - C_{k-1}) \\ \longrightarrow\ & H^1(X, L - C_0 - \cdots - C_k) \longrightarrow H^1(X, L - C_0 - \cdots - C_{k-1}). \end{aligned}$$

Hence

$$\begin{aligned} &\dim \ker(H^1(X, L - C_0 - \cdots - C_k) \longrightarrow H^1(X, L - C_0 - \cdots - C_{k-1})) \\ &\quad = h^0(X, L - C_0 - \cdots - C_k) - h^0(X, L - C_0 - \cdots - C_{k-1}) \\ &\qquad + h^0(C_k, L - C_0 - \cdots - C_{k-1}). \end{aligned}$$

On the other hand, we have

$$h^0(C_k, L - C_0 - \cdots - C_{k-1}) \leqq \max(C_k(L - C_0 - \cdots - C_{k-1}) + 1, 0) \begin{cases} = C_0L + 1, & k = 0; \\ \leqq C_kL, & k > 0. \end{cases}$$

Thus we obtain

$$\begin{aligned} &\dim \ker(H^1(X, L - C_0 - \cdots - C_k) \longrightarrow H^1(X, L - C_0 - \cdots - C_{k-1})) \\ &\quad \leqq h^0(X, L - C_0 - \cdots - C_k) - h^0(X, L - C_0 - \cdots - C_{k-1}) + LC_k + e_k, \end{aligned}$$

where $e_0 = 1$, $e_1 = \cdots = e_r = 0$. Adding side by side, we get the desired inequality. Q.E.D.

COROLLARY (3.8): *Let D and L be as in Proposition* 3.7. *Then $h^0(D, L) \leqq LD + 1$.*

If D is an effective divisor on a surface, then D can be decomposed to s-connected components. Evidently, if D_1 and D_2 are distinct s-connected components of D, $D_1C \leqq 0$ for any irreducible component C of D_2.

COROLLARY (3.9): *Let D be an effective divisor on a surface X and $D_1, \dots, D_n$ its s-connected components. Then we have an estimate*

$$h^1(X, -D) \leqq h^1(X, \mathcal{O}) - \sum_{i<j} D_iD_j + n - 1.$$

PROOF: We prove the assertion by induction on n. We have proven for the case $n = 1$. Assume that the assertion holds for $D_1 + \cdots + D_{n-1}$. Then we have the exact sequence

$$H^0(D_n, -D_1 - \cdots - D_{n-1}) \longrightarrow H^1(X, -D) \longrightarrow H^1(X, -D_1 - \cdots - D_{n-1}).$$

Applying Corollary 3.8, we get our estimate. Q.E.D.

Similar argument shows

COROLLARY (3.10): *Let D be an effective divisor on a surface and $D_1, \ldots, D_n$ its s-connected components. Then the inequality*

$$h^0(D, L) \leqq LD - \sum_{i<j} D_iD_j + n$$

holds for a divisor L on the surface such that $LC \geqq 0$ for any irreducible component C of D.

REMARK (3.11): The condition on L in 3.7, 3.8 and 3.10 is essential. For example, let D be a fundamental cycle $C_1 + C_2$ associated with a rational double point of type A_2. Let L_0 be an ample divisor such that $\dim \operatorname{im}(H^0(X, L_0) \to H^0(C_2, L_0)) \geqq 2$. Then $h^0(D, L_0 + aC_1) \geqq 2$ for any positive integer a. But, since $C_1D = -1$, we can find a such that $(L_0 + aC_1)D = 0$.

REMARK (3.12): In $3.3, \ldots, 3.9$, we need no condition on the ground field.

We can reformulate Theorem 2.7 as follows:

THEOREM (3.13): *If an effective divisor D on a surface X is s-connected and satisfies $\kappa(X, D) = 2$, then $H^1(X, -D)$ vanishes.*

COROLLARY (3.14): *Let $D_1, \ldots, D_n$ be the s-connected components of an effective divisor D with $\kappa(X, D) = 2$. Then we have*

$$h^1(X, -D) \leqq -\sum_{i<j} D_iD_j + n - 1.$$

REMARK (3.15): It is easy to see that there is one and only one s-connected component D_i of D with $\kappa(X, D_i) = 2$, so that we have the corollary above.

REFERENCES

[1] S. BLOCH and D. GIESEKER: The positivity of the Chern classes of an ample vector bundle. Inv. Math. *12* (1971) 112–117.
[2] D. MUMFORD: Pathologies III, Amer. J. Math. *89* (1967) 94–104.
[3] C.P. RAMANUJAM: Remarks on the Kodaira vanishing theorem, Ind. J. of Math. N.S. *36* (1972) 41–51.
[4] M. REID: On Bogomolov's theorem $c_1^2 \leqq 4c_2$, Preprint.
[5] O. ZARISKI: The Theorem of Riemann–Roch for high multiples of an effective divisor on an algebraic surface, Ann. of Math. *76* (1962) 550–612.

Dept. of Math.
Tokyo Metropolitan University
Fukasawa, Tokyo 158
Japan

ON AUTOMORPHISMS OF COMPACT KÄHLER SURFACES

C.A.M. Peters

§1 Introduction

For any topological space X and homeomorphism g of X one can ask how much information on g survives on cohomology level. In the sequel I shall assume that X is a compact Kähler manifold and that g is a biholomorphic map of X into itself, for short: *automorphism.* S. Bochner and D. Montgomery ([4]) proved that for any compact complex manifold X the collection of its automorphisms form a finite dimensional Lie group $Aut(X)$, whose Lie algebra can be canonically identified with the Lie algebra of holomorphic vector fields on X. The Kähler condition, by [9], Prop. 2.2 implies that the group $Aut_*(X)$ consisting of automorphisms of X inducing the identity in cohomology (with complex coefficients) is of finite index in $Aut_0(X)$, the identity component of $Aut(X)$.

Now there is a natural representation of $Aut(X)/Aut_0(X)$ in the cohomology ring $H^*(X, \mathbf{C})$. A natural question in this context is whether it is faithful, or – in other words:

$$\text{Is } Aut_0(X) = Aut_*(X)? \tag{1.1}$$

Lieberman's result cited above in particular implies:

$$\text{If } Aut_0(X) = \{1\} \text{ and } g \in Aut_*(X), \text{ then } g \text{ has finite order,} \tag{1.2}$$

a property which will be useful in the sequel.

For curves $Aut_0(X) = Aut_*(X)$, a result going back to Hurwitz ([6]). In this note I want to extend the results of [12] to a much larger class of Kähler surfaces. The main result is stated in §2 together with a discussion of the exceptions and an outline of the proof. Also in this section there are some preliminaries about the classification of Kähler surfaces.

Notations

If C is any curve, $\pi(C)$ will stand for its genus.

If X is any compact Kähler surface I let:

$b_j(X) = \dim_{\mathbf{R}} H^j(X, \mathbf{R})$,

$e(X) = \sum_{j=0}^{4} (-1)^j b_j(X)$, the *Euler number* of X,
$c_1^2(X)$ and $c_2(X)$, the *Chern numbers* of $X(c_2(X) = e(X))$,
T_X the *holomorphic tangent bundle*,
$K_X = \Lambda^2 T_X^\vee$, the *canonical bundle*,
$\mathcal{O}_X$: the structural sheaf,
$q(X) = \dim_{\mathbf{C}} H^1(X, \mathcal{O}_X)$ $(=\frac{1}{2}b_1(X))$, the *irregularity* of X,
$p_g(X) = \dim_{\mathbf{C}} H^2(X, \mathcal{O}_X)$, the *geometric genus* of X,
$P_m(X) = \dim_{\mathbf{C}} H^0(X, K_K^{\otimes m})$, the m-th *plurigenus* of X,
(notice: $P_1(X) = p_g(X)$, by Serre's duality theorem)
$\mathrm{Pic}(X) = H^1(X, \mathcal{O}_X^*)$, the group of isomorphy classes of line bundles on X,
$\mathrm{Pic}^0(X)$ the subgroup corresponding to line bundles with zero Chern class.
The intersection-pairing on $\mathrm{Pic}(X)$ is denoted by (,); □ means "end of proof, "proof postponed" or "no proof".

Acknowledgements: I owe much to correspondence with David Lieberman, who in fact proposed to study the question in the form (1.1) rather than in the form as given in [12]. Also I want to thank Miles Reid for pointing out the relevance of [3] (cf. 3.7) and Kenji Ueno for help in the case of elliptic fibrations. Finally I want to thank Arnaud Beauville for his valuable suggestions, especially for the simplification of Proposition (5.3).

§2 The Main Theorem

For an understanding of the statement and proof of the main result, some preliminaries about the classification of Kähler surfaces are needed. An *exceptional curve* on a non-singular complex surface is a non-singular rational curve with self intersection -1. Blowing up a point means replacing that point by an exceptional curve such that the resulting surface is non-singular. Conversely, an exceptional curve can always be blown down, i.e. replaced by a point such that the result is a non-singular surface ([7], Theorem 6.1). The processes of blowing up and down preserve the set of Kählerian surfaces ([10], Proposition C). These remarks imply that for the purpose of classification it is sufficient to enumerate the Kählerian surfaces without exceptional curves, to be called *minimal*. The result is called Enriques classification ([8], Theorem 55):

(2.1) THEOREM: *Minimal compact Kähler surfaces fall in exactly one of the following classes*:
(1) *The projective plane* $\mathbf{P}^2$ *and the ruled surfaces*,
(2) *The 2-dimensional tori* ($c_1^2 = c_2 = 0$),
(3) *The Kählerian K3-surfaces* ($c_1^2 = 0$, $c_2 = 24$),
(4) *The minimal honestly elliptic Kählerian surfaces* ($c_1^2 = 0$, $c_2 \geq 0$),
(5) *The minimal surfaces of general type* ($c_1^2 > 0$, $c_2 > 0$). □

I shall now explain the terminology in this theorem: A *ruled surface* is the total space of a holomorphic $\mathbf{P}^1$-bundle over an algebraic curve. A surface X with K_X trivial and $b_1(X)=0$ is called a *K3-surface.*

Any surface admitting at least one elliptic fibration, i.e. a holomorphic map onto a smooth curve such that all but a finite number of fibres are smooth elliptic curves is called an *elliptic* surface. Some of the ruled surfaces, tori and $K3$-surfaces are also elliptic. Any elliptic surface not belonging to one of these classes is called "*honestly*" *elliptic.*

Finally a surface X is called "*of general type*" if for some natural number n the meromorphic map associated to the n-th canonical system $|nK_X|$ has a surface as its image.

Next I want to observe that a positive answer to (1.1) for a surface X implies a positive answer to (1.1) for any blow up $\sigma: X'\to X$ of X. Indeed let $g'\in Aut_*(X')$, let E be the exceptional curve on X' and let $p=\sigma(E)$. Since $g'(E)$ is an irreducible curve with $(g'(E), E)=(E, E)=-1$, necessarily $g'(E)=E$ and g' induces an automorphism g of X with $g(p)=p$. Now σ^* embeds $H^*(X, \mathbf{C})$ into $H^*(X', \mathbf{C})$, so $g\in Aut_*(X)$ and if (1.1) can be answered positively g belongs to $Aut_0(X)$ and moreover fixes p. This means that $g'\in Aut_0(X')$.

This discussion shows that in order to find positive answers to (1.1) it is sufficient to study minimal surfaces.

(2.2) MAIN THEOREM: *Let X be a minimal compact Kähler surface. Then $Aut_*(X)=Aut_0(X)$ except in the following cases*

(i) *X is an honestly elliptic surface with $c_2(X)=12$, $q(X)=p_g(X)=0$.*

(ii) *X is an honestly elliptic surface with $c_2(X)=0$ fibred over a curve of genus $\pi=q(X)-1$, with $\pi\leq 1$.*

(iii) *X is an honestly elliptic surface with $c_2(X)=0$ fibred over a curve of genus $\pi=q(X)$. (No restrictions on π.)*

(iv) *X is a surface of general type and either $|K_X|$ has non-empty base locus or its Chern numbers $c_1^2(X)$ and $c_2(X)$ satisfy $c_1^2(X)=2c_2(X)$ or $c_1^2(X)=3c_2(X)$.*

Moreover, in the cases (i), (ii) *and* (iii) *$Aut_0(X)$ coincides with the group $Aut_*(X, \mathbf{Z})$ of automorphisms inducing the identity in integral cohomology.* □

Some remarks concerning the exceptions mentioned in this theorem. For case (iv) I postpone this discussion to §3. Here I only give some examples illustrating why (i), (ii), (iii) really occur. The constructions are all very similar.

EXAMPLE 1: An elliptic surface $X\to C$, with $c_2(X)=0$, $q(X)=\pi(C)$ and $Aut_0(X)=\{1\}$, but $Aut_*(X)\neq\{1\}$.

Take any curve D with a fixed point free involution i such that $C=D/\langle 1, i\rangle$ is of genus $\pi\geq 2$. Let E be any elliptic curve, say $E=\mathbf{C}/(\mathbf{Z}+\mathbf{Z}\tau)$, and define $h\in Aut(E\times D)$ by $h(e, d)=(-e, i(d))$. The quotient

$X = (E \times D)/\{1, h\}$ is smooth and admits an elliptic fibration $F: X \to C$. Since h does not commute with translations, X does not have a holomorphic vector field parallel to the fibres, hence no vector field at all, hence $Aut_0(X) = \{1\}$. It is not difficult to show that $f^*: H^1(C, \mathbf{C}) \tilde{\to} H^1(X, \mathbf{C})$, hence $q(X) = \pi$. Define an automorphism $\bar{g}$ of $E \times D$ by $\bar{g}(e, d) = (e + \frac{1}{2}, d)$, where $\frac{1}{2}$ is the 2-torsion point on E coming from $\frac{1}{2} \in \mathbf{C}$. Clearly $\bar{g}$ and i commute, so $\bar{g}$ defines an automorphism g of X. It is also rather obvious that $\bar{g}$ induces the identity in $H^*(\bar{X}, \mathbf{C})$, hence g induces the identity in $H^*(X, \mathbf{C})$, which can be identified with the i-invariants of $H^*(\bar{X}, \mathbf{C})$.

EXAMPLE 2: An elliptic surface X with $e(X) = 0$, $\dim Aut_0(X) = 1$, and $Aut_*(X) \neq Aut_0(X)$.

Let C be the elliptic curve with periods 1 and i and let E be an elliptic curve with periods 1 and τ. Define an automorphism λ of $E \times C$ by $\lambda(e, c) = (e + \frac{1}{4}, ic)$. Define $X = E \times C/\{1, \lambda, \lambda^2, \lambda^3\}$. This is a so called *hyperelliptic surface.* It has two elliptic fibrations: $X \to E/\langle\lambda\rangle$ (with $E/\langle\lambda\rangle$ elliptic) *and* $X \to C/\langle\lambda\rangle = \mathbf{P}^1$. $Aut_0(X)$ consists of the automorphisms h_α defined by $h_\alpha(e, c) = (e + \alpha, c)$. Define the automorphism $\bar{g}$ of $E \times C$ by $\bar{g}(e, c) = (e, c + \frac{1}{2} + \frac{1}{2}i)$. It induces an automorphism g of X and as before one easily sees that $g \in Aut_*(X)$. Clearly $g \notin Aut_0(X)$ however.

EXAMPLE 3: (Due to Lieberman, cf. [12].) An elliptic surface X fibered over $\mathbf{P}^1$ with $c_2(X) = 12$, $q(X) = p_g(X) = 0$ and $Aut_0(X) = \{1\}$ whereas $Aut_*(X) \neq \{1\}$.

Let E be the elliptic curve with periods $\{1, i\}$ and let τ be the image in E of $\frac{1}{2}(1 + i)$. Let $X_1 = E \times E$ and X_2 the $K3$-surface obtained by resolving the 16 singularities of $X_1/\{\pm\mathrm{id}\}$. The automorphism $\lambda: (a, b) \mapsto (a + \tau, -b + \tau)$ induces a fixed point free involution on X_2 and its quotient $X = X_2/\{1, \lambda\}$ is a so called *Enriques surface.* The surface X has an elliptic fibration coming from the elliptic fibration $X_2 \to E/\{\pm\mathrm{id}\} = \mathbf{P}^1$. It is well known and easily checked in this example that $p_g(X) = q(X) = 0$ and $c_2(X) = 12$. The automorphism $\bar{g}$ of $E \times E$ defined as $\bar{g}(e, e') = (i \cdot e, i \cdot e')$ commutes with $-\mathrm{id}$ and with λ, so defines an automorphism g of X. The 10 distinct algebraic cycles $(E \times 0)/\{\pm\mathrm{id}\}$, $(0 \times E)/\{\pm\mathrm{id}\}$, and $C_j + C_{j+(\tau,\tau)}$ are λ-invariant cycles on X_2, where j is any of the 16 points of order 2 on $E \times E$ and C_j the nodal curve (i.e. $(C_j, C_j) = -2$ and C_j rational) corresponding to it. Since they are also g-invariant they define a 10-dimensional g-invariant subspace of $H^2(X, \mathbf{C})$. Since $b_2(X) = 10$, this shows that $g \in Aut_*(X)$.

Outline of the proof of (2.2):

The proof proceeds according to the classification of Kähler surfaces as given in (2.1).

Class (1). The case of projective plane is trivial and the case of ruled surfaces goes as follows:

$X = \mathbf{P}(E)$ with E an affine 2-bundle over a curve C of genus $\pi \geq 0$. In case $\pi = 0$, apart from the trivial case $X = \mathbf{P}^1 \times \mathbf{P}^1$, one has $Aut(X) = Aut_0(X)$, so there is nothing to prove. So one may assume that $\pi \geq 1$. Any automorphism of X then preserves the $\mathbf{P}^1$-bundle structure and I let h be the automorphism that $g \in Aut_*(X)$ induces on C. Since $H^1(X, \mathbf{C})$ and $H^1(C, \mathbf{C})$ are canonically isomorphic, $h \in Aut_*(C)$, so $h \in Aut_0(C)$ by Hurwitz result [6]. In case $\pi \geq 2$ this implies $h = \mathrm{id}$, hence g is in $\mathrm{PGL}(E)$, and $g \in Aut_0(X)$. In case $\pi = 1$, h is a translation. Since for any translation t of E there exists $\phi(t) \in Aut_0(X)$ inducing t one may assume that $h = \mathrm{id}$. But then $h \in \mathrm{PGL}(E)$ as before and $h \in Aut_0(X)$.

Class (2). The 2-dimensional tori.

Any 2-torus A is of the form $\mathbf{C}^2/\Gamma$, where Γ is a lattice. There is an isomorphism $\Gamma \simeq H_1(A, \mathbf{Z})$, unique once the up to a translation unique isomorphism $A \tilde{\to} \mathbf{C}^2/\Gamma$ is chosen. Hence since any $g \in Aut_*(X)$ induces the identity in $H_1(X, \mathbf{Z})$, g comes from a translation.

Class (3) and (5).

These two classes will be treated in §3 as consequences of the Lefschetz Fixed Point Formula.

Class (4).

These two classes will be treated in §4 (non-zero Euler number) and §5 (zero Euler number). The arrangement is in increasing order of difficulty. □

It will be clear from the nature of the question (1.1) that it is of importance to know which surfaces X have $Aut_0(X) \neq \{1\}$, or – equivalently – which admit vector fields. The result follows from [9], Theorem 4.9 and is stated here for later reference:

(2.3) PROPOSITION: *Any minimal compact Kähler surface admitting a non-trivial holomorphic vector field falls in one of the following classes:*

(A) *The classes* (1) *and* (2) *of Theorem* (2.1)

(B) *Some minimal honestly elliptic surfaces with zero Euler number. They are obtained as quotients of a holomorphic principal fibre bundle of elliptic curves with fibre F by a finite group G of fibre preserving automorphisms that commute with the F-action. They are called* (F–G) Seifert bundles. □

§3 Application of Lefschetz' Fixed Point Theorems

Assume X is a Kähler surface and g is a holomorphic automorphism of X of finite prime order p. Let g have μ isolated fixed points $p_1, \ldots, p_\mu$ and let $F_1, \ldots, F_\lambda$ be the irreducible fixed curves, i.e. $g(F_i) = F_i$ and $g \mid F_i = \mathrm{id}$. Recall Lefschetz fixed point formula, [13], Lemma 1.6:

$$\sum_{i=1}^{\lambda} e(F_i) + \mu = \sum_{k=0}^{4} (-1)^k \,[\mathrm{Trace}(g^* \mid H^k(X, \mathbf{Q}))]. \tag{3.1}$$

Since g has finite order the action of g can be linearized around any fixed point q ([5], p. 97) and hence diagonalized, say

$$(dg)_q = \begin{pmatrix} \rho^a & 0 \\ 0 & \rho^b \end{pmatrix},$$

where $\rho = \exp(2\pi i/p)$. Moreover not both a and b are divisible by p and if, say $a \equiv 0 \bmod p$ there is a fixed curve passing through q with tangent direction equal to the eigenvector with eigenvalue 1. In this situation we make the following useful observation:

(3.2) LEMMA: *Suppose that there exists a holomorphic 2-form on X which does not vanish at q. In case $g \in Aut_*(X)$, one has* $\det(dg)_q = 1$, *i.e.* $a + b \equiv 0 \bmod p$.

PROOF: By Hodge Theory $H^{2,0}(X) \cong H^0(X, K_X)$ is a direct factor of $H^2(X, \mathbf{C})$ hence g preserves the holomorphic 2-forms. In particular this means that $\det(dg)_q$, viewed as an action on the second exterior power $\Lambda^2 T^{\vee}_{X,q}$ of the cotangent space $T^{\vee}_{X,q}$ of X at q must be the identity. □

Since in the case at hand all fixed points are transversal and the induced action of g on the normal bundle of the fixed curves is non-trivial I may apply the Holomorphic Lefschetz Fixed Point Formula ([1], p. 567):

$$\sum_{i=1}^{\mu} [\det(1 - dg)_{p_i}]^{-1} + \sum_{j=1}^{\lambda} (1 - \pi(F_j))[1 - \rho_j]^{-1} - \sum_{j=1}^{\lambda} F_j^2[\rho_j(1 - \rho_j)^{-2}] \\ = \sum_{k=0}^{2} (-1)^k \,\mathrm{Trace}(g^* \mid H^{k,0}), \tag{3.3}$$

where ρ_j^{-1} is the eigenvalue of the induced action of g on the normal bundle of F_j.

I want to apply those formulas under the assumption that $1 \neq g \in Aut_*(X)$ is of finite prime order p. In order to make calculations, some further assumptions will be made. In view of the previous Lemma I call any isolated fixed point q with $\det(dg)_q = 1$ a *regular* fixed point.

The Lemma then may be restated as follows:

(3.4) *If g has any fixed curve or non-regular fixed points, they all must lie in the base locus of the canonical system $|K_X|$. In particular, if $|K_X|$ has no base points or fixed components g only has isolated regular fixed points.*

(3.5) PROPOSITION: *Let X be a surface and let $g \in Aut_*(X)$ have prime order p. Assume moreover that g has only fixed curves F that are elliptic* *with $F^2 = 0$ and fixed points that are regular. Then the following equality between the Chern numbers of X holds:*

$$c_1^2(X) = pc_2(X).$$

PROOF: The assumptions imply for (3.1) that $\mu = e(X) = c_2(X)$ and for (3.3):

$$(*) \quad \sum_{i=1}^{\mu} [\det(1 - dg)_{p_i}]^{-1} = 1 - q + p_g = \chi(\mathcal{O}_X) = \tfrac{1}{12}[c_1^2(X) + c_2(X)].$$

Now $\det(1 - dg)_{p_i} = (1 - \rho^{n_i})(1 - \rho^{-n_i})$, $\rho = \exp(2\pi i/p)$, and

$$\sum_{k=1}^{p-1} [\det(1 - (dg)^k)_{p_i}]^{-1} = \sum_{k=1}^{p-1} [(1 - \rho^k)(1 - \rho^{-k})]^{-1} = \tfrac{1}{12}[p^2 - 1].$$

Since one has for g^k an equality similar to $(*)$, upon summing them for $k = 1, \ldots, p - 1$ one finds:

$$\tfrac{1}{12}\mu[p^2 - 1] = [p - 1]\tfrac{1}{12}[c_1^2(X) + c_2(X)].$$

Using that $\mu = c_2(X)$ one finds: $(p + 1)c_2(X) = c_1^2(X) + c_2(X)$, i.e. $c_1^2(X) = pc_2(X)$.

I apply these considerations to some classes of surfaces:

(3.6) PROPOSITION: *We have $Aut_*(X) = \{1\}$ in the following cases:*
(i) *X is a Kähler K3-surface,*
(ii) *X is a minimal surface of general type, $|K_X|$ has no base locus and $c_1^2 \neq 2c_2$ or $3c_2$.*

PROOF: Since for a K3-surface K_X is trivial, $|K_X|$ has no base locus. Moreover, by 2.3. $Aut_0(X) = \{1\}$ in this case, so $Aut_*(X)$ is finite by (1.2), so (3.5) applies to any $g \in Aut_*(X)$, $g \neq 1$ having prime order p. Since this can always be assumed (one simply takes a suitable power of g) we find for our surface that $c_1^2(X) = pc_2(X)$, contrary to the values of the Chern numbers $c_1^2 = 0$, $c_2 = 24$.

Likewise this applies to minimal surfaces of general type with properties as stated. Since $c_1^2(X) \leq 3c_2(X)$ for these surfaces by Miyaoka [11] and since $c_1^2(X) > 0$ (cf. 2.1), p can only assume the value 2 if and only if $c_1^2(X) = 2c_2(X)$, and 3 if and only if $c_1^2(X) = 3c_2(X)$.

(3.7) REMARK: The proof shows that if any exceptions X of general type exist for $|K_X|$ without base locus they must satisfy:
(A) $c_1^2(X) = 2c_2(X)$ and $1 \neq g \in Aut_*(X)$ has order $2^\alpha, \alpha \in \mathbf{N}$

or

(B) $c_1^2(X) = 3c_2(X)$ and $1 \neq g \in Aut_*(X)$ has order 3^β, $\beta \in \mathbf{N}$.

Let me try to find examples of this situation. Assume that the canonical map Φ of X maps X onto a surface Σ. According to [3], Thm. 3.1. two cases occur:

(I) $p_g(\Sigma) = 0$,

(II) Σ is of general type.

In our situation the map Φ factors over $Y = X/\langle g\rangle$ and $p_g(Y) = p_g(X)$. There is the possibility that $Y = \Sigma$ and we are in case (II). Of this case however only one example is known ([3], Prop. 3.6) and for it deg $\Phi = 2$, so automatically $Y = \Sigma$. However, the involution g is easily seen not to act trivially in cohomology.

Let me now consider the possibility $Y \neq \Sigma$ and $p_g(\Sigma) = 0$. I know of only one possible candidate: take $X = C \times D$, the product of two hyperelliptic curves and g the involution arising from the hyperelliptic involution on both factors. Then for $Y = X/\langle g\rangle$, we have $p_g(Y) = p_g(X)$ and $\Sigma = \mathbf{P}^1 \times \mathbf{P}^1$. However g does not preserve the one forms.

So it seems difficult to find such examples; in fact, upon supplementing the methods of [3] it might very well be possible to show that they do not exist. For example, an application of [3], Prop. 4.1. gives:

In case (A) (I) $|g| = 2$ or 4 if $c_2 = 4A$ with $A \geq 31$,
(II) $|g| = 2$ if $c_2 = 4A$ with $A \geq 14$,
(B) one has $|g| = 3$ if $c_2 = 3B$ with $B \geq 31$ (Case I), resp. $B \geq 14$ (Case II).

(3.8) REMARK: In case X is of general type the results in [12] are stronger in that one already can conclude that $g = \text{id}$ if g acts trivially in the second cohomology – with the same exceptions. It seems however that the formulation as given here is more natural.

§4 Honestly elliptic surfaces with non-zero Euler number

I first study the local situation around a singular fibre. Let $D = \{z \in \mathbf{C};\ |z| < \epsilon\}$ for some $\epsilon > 0$ and let $\pi : U \to D$ be an elliptic fibration with fibres θ_t, smooth for $t \neq 0$ and singular for $t = 0$. Put $\theta = \theta_0$. Any fibre preserving automorphism of U will be called *fibering-automorphism*. It is called *rigid* if it induces the identity on D and it does not permute the components of θ. A rigid fibering-automorphism is called *direct*, resp. *indirect*, resp. *trigonal* if it has no, resp. 4, resp. 3 isolated fixed points on θ_t $(t \neq 0)$, i.e. it induces a translation, resp. a multiplication by -1, resp. a multiplication by a third root of unity on θ_t $(t \neq 0)$.

In [13], §3 there is a discussion of the possible rigid fibering-automorphisms, proceeding along Kodaira's list of types of singular fibres ([7], Theorem 6.2). Ueno's results can be summarized and extended to:

(4.1) PROPOSITION:

(i) *If g is a direct fibering-automorphism of π: $U \to D$, the singular fibre θ must be one of the types occurring in the following list*:

type	# fixed points	fixed curves
I_b $(b \geq 1)$	b	none
${}_mI_0$	0	none or θ
${}_mI_b$ $(b \geq 1)$	b	none
II	1	none

Moreover, the isolated fixed points q are regular, i.e. $\det(dg)_q = 1$.

(ii) *If g is indirect, there exists a fixed curve F intersecting $\theta_t (t \neq 0)$ transversally in* 4 *points and the isolated fixed points lie on θ. Moreover θ is one out of the following list*:

type	$\# \theta \cap F$	fixed curves
I_2	4	none
${}_2I_0$	2	none
${}_4I_0$	1	none
I^*_{2b}	2	$b+1$ nodal curves
II	2	none
II^*	2	4 nodal curves
III	3	none
III^*	3	3 nodal curves

Here, a nodal curve is a smooth irreducible rational curve E with $E^2 = -2$.

(iii) *In case g is trigonal the j-invariant of θ_t $(t \neq 0)$ is zero; θ is of type ${}_3I_0$ and there is one isolated fixed point on θ, the intersection of θ with the fixed curve of g.*

PROOF: Most of these statements are proved in [13], §3. It is for further applications necessary however to treat the case ${}_mI_b$ in detail. For sake of simplicity I only give the details for $m = 1$. The following description of π: $U \to D$ if θ is of type I_b stems from the theory of toroïdal embeddings and was pointed out to me by *K.* Ueno. Take for each $k \in \mathbf{Z}$ an open set $\mathcal{U}_k = \{(u_k, v_k) \in \mathbf{C}^2;\ |u_k v_k| < \epsilon\}$ and glue $\mathcal{U}_k$ and $\mathcal{U}_{k+1}$ via $u_{k+1} = v_k^{-1}$; $v_{k+1} = u_k v_k^2$. Let $\mathbf{Z}$ act on this manifold $\cup_k \mathcal{U}_k$ via $n.(u_k, v_k) = (u_{k+nb}, v_{k+nb})$. The quotient X is then isomorphic to U and π is given by $\pi(u_k, v_k) = u_k \cdot v_k$. The fibre θ looks as in fig. 1.

It is clear that any direct fibration automorphism g: $U \to U$ must be given by $g(u_k, v_k) = (\lambda_k u_k, \lambda_k^{-1} v_k)$ for some $\lambda_k \in \mathbf{C}^*$. It easily follows that $\lambda_k = \lambda$ and $\lambda^b = 1$. It means that there are exactly b isolated transversal fixed points.

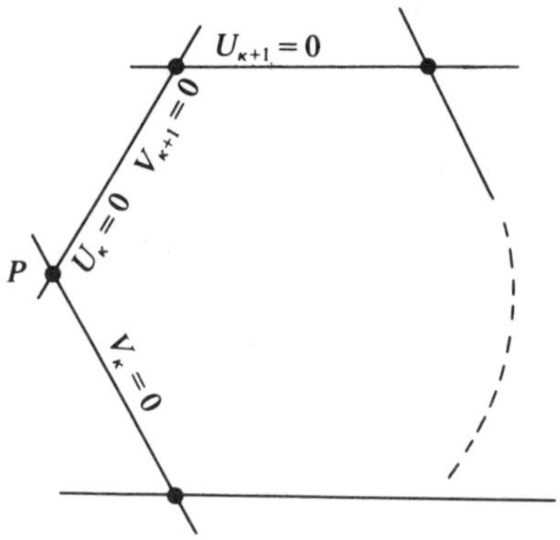

Fig. 1.

Finally I need to comment briefly on the trigonal case. In this case θ a priori also might be of type I_0^*, II, II*, IV or IV*. Using the description of their neighbourhoods in Kodaira [7], p. 585–593 it is not difficult to show that there are no trigonal fibering-automorphisms in these cases. □

(4.2) REMARK: In case θ is of type ${}_mI_b$ and g is a non-rigid fibre automorphism of prime order p with fixed set on θ, which maps each component of θ into itself the following possibilities occur:

(1) *Either* $p \mid b$ and exactly b/p nodal curves are fixed and there are $b-2b/p$ isolated fixed points. These fixed points form b/p sets of $p-2$ each and the eigenvalues (λ_i, λ_j) for $g, \ldots, g^{p-1}$ for each set assume all possibilities for p-th roots of unity λ_i and λ_j with $\lambda_i\lambda_j \neq 1$.

(2) *Or* $b=0$, and θ is fixed or – in case $\#g=2$ and $m=2$ there might be 2 transversal fixed points.

The assertion (1) can be checked easily for ${}_1I_b$, using the explicit construction given above, and from this the assertion (1) can be verified for ${}_mI_b$ using the explicit description of a neighbourhood of a singular fibre of type ${}_mI_b$ as given by Kodaira [9], p. 767–770. Likewise assertion (2) can be verified using this explicit description.

From now on I assume that X is a *minimal honestly elliptic Kähler surface with positive Euler number.* As follows from 2.3, X does not have vector fields, hence $Aut_0(X)=\{1\}$ and by (1.2) $Aut_*(X)$ is finite. *I assume that* $1 \neq g \in Aut_*(X)$ *has prime order p.* If necessary I will assume that $g \in Aut_*(X, \mathbf{Z})$.

(4.3) PROPOSITION: *The map g preserves the elliptic fibration* $f: X \to C$. *Moreover it induces the identity on C, except in case* $C=\mathbf{P}^1$, $p_g(X)=0$ *and* $g \notin Aut_*(X, \mathbf{Z})$.

PROOF: Since g maps a fibre F to an irreducible curve F' with $(F', F)=(F, F)=0$ it follows that F' is a fibre of f. Let $h: C \to C$ be the map induced on C. Since $f^*: H^1(C, \mathbf{C}) \to H^1(X, \mathbf{C})$ is an isomorphism and g acts trivially on $H^1(X, \mathbf{C})$, also h acts trivially in cohomology. Thus, since the answer to (1.1) is positive in the curve case, $h \in Aut_0(C)$ and $h = \mathrm{id}$ if $\pi(C) \geq 2$ and h is a translation if $\pi(C) = 1$. If in this case $h \neq \mathrm{id}$, g would have no fixed points, so – by the ordinary Lefschetz Fixed Point Formula (3.1) $c_2(X) = 0$, contrary to the assumptions. The case $C = \mathbf{P}^1$ is far more complicated. Assume $h \neq \mathrm{id}$.

CASE 1: $p_g(X) > 0$.
In this case $|K_X|$ has a base locus supported on the multiple fibres. Indeed, the canonical bundle formula ([8], Theorem 12) for elliptic fibrations f: $X \to C$ reads:

$$K_X = f_* \mathscr{L} \oplus \left[\sum_{j=1}^{k} (m_j - 1)\theta_j \right], \tag{4.3.1}$$

where $\mathscr{L}$ is a line bundle on C of degree $2\pi(C) - 2 + \chi(\mathcal{O}_X)$, and where $m_j\theta_j (j = 1, \ldots, k)$ are the multiple fibres of f. So $|K_X| = |f^*\mathscr{L}|$ and

$$\sum_{j=1}^{k} (m_j - 1)\theta_j$$

is the base locus in this case. And by (3.4) fixed curves of g must be components of multiple fibres and non-regular fixed points must lie on them. Since $h \neq \mathrm{id}$ and h is cyclic, h has two fixed points. Let θ_1 and θ_2 be the two fibres above these points. If both are non-multiple g has only regular fixed points and one may apply (3.5). For an elliptic surface $c_1^2(X) = 0$, so one finds that $c_2(X) = 0$, contrary to the assumptions. Suppose $\theta_\alpha (\alpha = 1, 2)$ is of type ${}_{m_\alpha}I_{b_\alpha}$ with $b_\alpha \neq 0$. I want to compute the contribution of θ_α to $\sum_{j=1}^{p-1} \sum_{i=0}^{2} (-1)^i$. Trace $[(g^j)^* | H^{i,0}] = (p-1)\ \chi(X, \mathcal{O}_X)$ by making use of the Holomorphic Lefschetz Formula (3.3). This contribution I call the *holomorphic Lefschetz number* of θ_α. Likewise I compute the contribution of θ_α to $c_2(X)$, by making use of the ordinary Lefschetz Formula (3.1) – I call this the topological Lefschetz number of θ_α. I want to make use of the description of g around θ_α as given in Remark (4.2). In order to apply it notice the following observations which also will be of later use:

(4.4) LEMMA: *If C is a nodal curve on X, $g(C) = C$.*

PROOF: Since $g \in Aut_*(X)$, $(g(C), C) = (C, C) = -2$, but $g(C)$ is irreducible, hence $g(C) = C$. □

So I find the holomorphic Lefschetz number of θ_α from (4.2) and (3.3). First of all $b_\alpha = k_\alpha \cdot p$ and we find

$$k_\alpha \cdot \left\{ \sum_{a+b\neq 0 \bmod p} \frac{1}{(1-\rho^a)(1-\rho^b)} + \sum_{k=1}^{p-1} \frac{1}{(1-\rho^k)} + \sum_{k=1}^{p-1} \frac{2\rho^k}{(1-\rho^k)^2} \right\},$$

where $\rho = \exp\left(\frac{2\pi_i}{p}\right)$ and where the last two terms come from a nodal curve that is fixed under g. After some manipulation one gets:

(4.4.1) The holomorphic Lefschetz number of a fibre of type ${}_mI_{k\cdot p}$ equals

$$-\frac{k}{2}(p-1)(p-2).$$

The contribution to c_2 is easier to compute. I find:

(4.4.2) The topological Lefschetz number of a fibre of type ${}_mI_{k\cdot p}$ equals k.p.

So if both fibres θ_1 and θ_2 are of type ${}_mI_b (b\neq 0)$ the sum of the holomorphic Lefschetz numbers on the one hand is non-positive, by (4.4.1) and on the other hand equals

$$(p-1)\ \chi(X, \mathcal{O}_X) = (p-1)\ \frac{c_1^2(X)+c_2(X)}{12} = (p-1)\ \frac{c_2(X)}{12},$$

since X is elliptic – so is non-negative. This leads to $c_2(X)=0$, contrary to the assumptions. If both fibres θ_1 and θ_2 are of type ${}_mI_0$ by (4.2) the isolated fixed points are regular if they occur and any fixed curve F is smooth elliptic and has $F^2=0$ – if it occurs. The contradiction comes now from an application of (3.5).

If one, say θ_1 is of type ${}_{m_1}I_0$ and θ_2 is of type ${}_{m_2}I_b (b\neq 0)$ two possibilities occur: if $p\neq 2$, θ_1 is fixed and has zero holomorphic Lefschetz number, whereas θ_2 has non-positive Lefschetz number and one gets a contradiction as before. If $p=2$ and θ_1 is not a fixed curve there are 2 transversal fixed points on them, giving the Lefschetz number 2. $[\frac{1}{4}]=\frac{1}{2}$, whereas θ_2 has zero Lefschetz number, by (4.4.1). So the total Lefschetz number would be $\frac{1}{2}$ – which is not an integer. Contradiction!

The only possibility that is left is that one fibre, say θ_1 is multiple and the other fibre, θ_2, is non-multiple. In this situation on θ_2 there only may lie regular fixed points and no fixed curves may lie on it. This enormously restricts the possible types of θ_2. Namely, by (4.4) g preserves each nodal curve, so g must be the identity on it as soon as this nodal curve meets three other nodal curves. Going down Kodaira's list [7], Theorem 6.2 we find the possibilities:

a) type I_b $(b\geq 1)$, excluded by (4.2)
b) type II: a rational curve with one cusp – and g has at most 2 regular fixed points
c) type III: two rational curves touching at one point – and g has at most 3 regular fixed points.

So I let $\mu_1(\leq 3)$ be the number of regular fixed points on θ_2 and I assume θ_1 is multiple of type ${}_mI_b$. In case $b=0$ either θ_1 is fixed *or* g has 2 regular fixed points on θ_1. In both cases (3.5) gives a contradiction. In case $b\neq 0$, $b=k.p$ and the Lefschetz numbers of θ_1 are given in (4.4.1) and (4.4.2). Since each regular fixed point has holomorphic Lefschetz number $\sum_{k=0}^{p-1} \frac{1}{(1-\rho^k)(1-\rho^{-k})} = \frac{1}{12}[p^2-1]$ and topological Lefschetz number 1, I find: $(p-1)\,\chi(\mathcal{O}_X) = \frac{1}{12}(p-1)\,c_2(X) = \frac{1}{12}\mu_1(p^2-1) - \frac{1}{2}k(p-1)(p-2)$ and $\mu_1 + k.p = c_2(X)$. Comparing both equalities one finds a contradiction:

If $p=2$: $\mu_1 = k$ and $c_2 = 3k \leq 9$, but $\frac{c_2}{12} \in \mathbf{Z}$, hence $c_2 = 0$;

If $p=3$: $\mu_1 = 3k$ and $c_2 = 6k = 2\mu_1 \leq 6$ and similarly $c_2 = 0$;

If $p \geq 5$, one has $k \equiv 0 \bmod p$, say $k = n.p$ and

$$\mu_1 = 7np - 12n \geq 27n \geq 27, \text{ but } \mu_1 \leq 3.$$

CASE 2: $p_g(X) = 0$ and $g \in Aut_*(X, \mathbf{Z})$.

In this case $p_g(X) = q(X) = 0$. Observe that there must be *at least two singular fibres.* Indeed, recalling the canonical bundle formula for an elliptic fibration $f\colon X \to C$ with multiple fibres $m_j\theta_j (j = 1, \ldots, k)$ (cf. (4.3.1)) if one defines for every natural number n the non-negative integers n_j and r_j such that

$$m(m_j - 1) = n_j m_j + r_j \text{ with } 0 \leq r_j < m_j (j = 1, \ldots, k)$$

the m-th tensor power of the canonical bundle reads:

$$K_X^{\otimes m} = f^*(\mathcal{L}^{\otimes m} \otimes \mathcal{M}) \otimes \left(\sum_{j=1}^{k} r_j\theta_j\right) \tag{4.4.3}$$

with

$$\deg \mathcal{M} = \sum_{j=1}^{k} n_j \leq m\left[\sum_{j=1}^{k} \{1 - 1/m_j\}\right]$$

and

$$\deg \mathcal{L} = 2\pi(C) - 2 + \chi(\mathcal{O}_X)$$

as before. So, if $p_g(X) = q(X) = 0$, and $C = \mathbf{P}^1$, $k \leq 1$, we have

$$\deg(\mathcal{L}^{\otimes m} \otimes \mathcal{M}) \leq m\left(-1 + \left[1 - \frac{1}{m_1}\right]\right) < 0,$$

and $P_m(X) = \dim h^0(K_X^{\otimes m}) = \dim H^0(\mathcal{L}^{\otimes m} \otimes \mathcal{M}) = 0$ for all $m > 0$. So X

would be rational or ruled by [8], Theorem 54.

On the other hand g *must fix any multiple fibre*: Indeed, g at most permutes those fibres that have equal multiplicity. But, if $\theta_1 \neq \theta_2$ both have multiplicity a and $g(\theta_1) = \theta_2$ on the one hand $[\theta_1] = [\theta_2]$, i.e. the Chern class of the line bundle defined by $\theta_1 - \theta_2$ is zero. On the other hand $\theta_1 - \theta_2$ is a *non-zero* torsion element in Pic(X). Since $q(X) = 0$, taking Chern classes is injective, hence $[\theta_1 - \theta_2] \neq 0$. This contradiction shows that g has to fix any multiple fibre.

These two facts together imply that X has precisely two multiple fibres fixed under g.

One now obtains a contradiction as in Case 1. The reasoning however is much simpler and is therefore omitted. □

(4.5) THEOREM: *Let X be a minimal honestly elliptic Kähler surface with $c_2(X) > 0$. Then $Aut_*(X) = \{1\}$, unless $p_g(X) = q(X) = 0$. In the last case $Aut_*(X, \mathbf{Z}) = \{1\}$.*

PROOF: I refer to the notations and assumptions made just before (4.3). Proposition (4.3) implies that g is a rigid fibering automorphism. Since p is prime only the direct, indirect and trigonal cases need to be considered. In fact, the *direct case* can be dealt with as follows: (4.1) shows that all isolated fixed points are regular and all fixed curves F are smooth elliptic curves with $F^2 = 0$. So $c_2(X) = 0$, by (3.5), contrary to the assumptions. The *trigonal case* can be dealt with very easily: only multiple fibres of type ${}_3I_0$ may occur, by (4.1), hence $c_2(X) = 0$. Finally I have to exclude the *indirect case*. If $p_g(X) > 0$, $|K_X|$ has base locus only in the multiple fibres, hence (3.4) implies that fixed curves occur among curves contained in the multiple fibres, so this is ruled out by (4.1). In case $p_g(X) = 0$, also $q(X) = 0$ and $c_2(X) = 12$ one is in the same situation as in [13], §4 and the same proof as given there applies and gives a contradiction in this case as well. (Notice that the Enriques surface indeed is a special case of what I call "honestly elliptic surface" – and it is to be expected that at least part of Ueno's complicated proof should apply to surfaces having the same numerical invariants as Enriques surfaces.) □

§5 Honestly elliptic surfaces with zero Euler number

For an elliptic surface X the Euler number $e(X)$ is the sum of the Euler numbers of the singular fibres of an elliptic fibration f: $X \to C$. For any singular fibre θ of f one has the inequality $e(\theta) \geq 0$ with equality if and only if θ is a multiple of a smooth elliptic curve, i.e. of type ${}_mI_0$ in Kodaira's list [7], Theorem 6.2 (compare the table in [7] on p. 14). So $e(X) = 0$ if and only if the singular fibres of f are of type ${}_mI_0$. I will call such fibrations *elliptic quasi-bundles* and if such an elliptic quasi-bundle has no singular fibres at all

I call it *elliptic bundle*. All smooth fibres of an elliptic quasi-bundle are isomorphic and if $f: X \to C$ is an elliptic quasi-bundle with fibre F there exists:

(i) An elliptic bundle $f': X' \to C'$

(ii) A finite abelian group G of fibre preserving automorphisms of X' inducing on C' a group H such that $X = X'/G$, $C = C'/H$ and f' induces f.

In case f' gives X' the structure of an elliptic principal bundle with fibre F' and if all $g \in G$ commute with the F'-action, recall from (2.3) that the resulting fibration f is called an (F', G)-Seifert fibration, or more shortly: *Torus Seifert fibration.* They are characterized among the elliptic quasi-bundles as those having non-zero holomorphic vector fields parallel to the fibres or those with $q(X) = \pi(C) + 1$. I need some information about those with $p_g(X) = 0$.

(5.1) LEMMA: *Any minimal Kähler surface with $p_g(X) = 0$ and $q(X) = 1$ is algebraic and of the form $X = (B \times F)/G$, where G is a group of automorphisms of B as well as F, $E = B/G$ is elliptic, F/G is rational, and one of the following is true*:

(a) *If X has no vector fields $f: X \to E$ is an elliptic quasi-bundle with fibre F, having at least one singular fibre*;

(b) *Otherwise $f: X \to F/G = \mathbf{P}^1$ is a Torus-Seifert fibration with fibre B.*

PROOF: Since $p_g(X) = 0$ and $b_1(X)$ is even, X is algebraic by [8], Theorem 10 and one may apply [2], Th. VI.13. Observe that $f: X \to E$ is an elliptic bundle if and only if X has a vector field in which case both F and B are elliptic, so that this is subsumed under (b). □

Next I need some information about the possible multiple fibres in case (b) of the preceding Lemma:

(5.2) LEMMA: *If $f: X \to \mathbf{P}^1$ is a Torus-Seifert fibration and if X is honestly elliptic there are at least 3 multiple fibres and at least 4 if two double fibres occur.*

PROOF: Recall the formulas for the plurigenera for an elliptic fibration (4.4.3). With the notation employed there deg $\mathscr{L} = -2$ if $C = \mathbf{P}^1$, hence $\deg(\mathscr{L}^{\otimes m} \otimes \mathscr{M}) \leq m \left[-2 + \sum_{i=1}^{k} \{1 - 1/m_i\}\right]$, which is negative if $k \leq 2$ or if $k = 3$ and $2 = m_1 = m_2 \leq m_3$. Hence $P_m(X) = 0$ in these cases and X would be ruled, by ([8], Theorem 54). □

From now on I assume that $f: X \to C$ is an elliptic quasi-bundle, with the description of (5.1) in case $p_g(X) = 0$; I let $g \in Aut_*(X)$ if f is a Torus-Seifert fibration over a curve of genus ≥ 2 and $g \in Aut_*(X, \mathbf{Z})$ otherwise. I want to prove that $g \in Aut_0(X)$. For a Torus-Seifert fibration this means that g induces a constant translation in each fibre; in the other cases it means $g = \mathrm{id}$.

(5.3) PROPOSITION: *g preserves the fibre structure and induces the identity on the base curve.*

PROOF: The first clause follows from the fact that $(g(\theta), \theta) = (\theta, \theta) = 0$ for any smooth fibre together with the fact that $g(\theta)$ is an irreducible curve. The second clause one proves as follows: Let h be the automorphism of C induced by g. Since $f\colon X \to C$ is a quasi-bundle $H^1(X, \mathbf{C})$ has $H^1(C, \mathbf{C})$ as a direct factor, hence $h^* \in Aut_*(C) = Aut_0(C)$ by Hurwitz result [6]. So for $\pi(C) \geq 2$, $h = \mathrm{id}$ and I am done. For $\pi(C) = 1$, h is a translation and one fixed point is needed; for $\pi(C) = 0$ three fixed points are needed.

First I state a lemma:

LEMMA: *Let $f\colon X \to C$ be an elliptic fibration and $g \in Aut_*(X, \mathbf{Z})$. Suppose that $m\theta$ $(m \geq 2)$ is a multiple fibre with $g(\theta) \neq \theta$. Then necessarily, $m = 2$ and $C = \mathbf{P}^1$. If moreover θ' is a double fibre distinct from θ then $g(\theta') = \theta$.*

PROOF: Since $g \in Aut_*(X, \mathbf{Z})$ one has that $g(\theta) - \theta \in \mathrm{Pic}^0(X)$, hence $g(g(\theta) - \theta) \equiv g(\theta) - \theta$, where "$\equiv$" means linear equivalence. Hence $2g(\theta) \equiv g^2(\theta) + \theta$. This implies that $2g(\theta)$ moves linearly, so firstly $2g(\theta)$ must be a fibre and secondly-since this fibre moves linearly- C must be $\mathbf{P}^1$.

Now $\theta' - \theta$ is 2-torsion in $\mathrm{Pic}(X)$, hence preserved by g, i.e. $\theta + g(\theta') \equiv \theta' + g(\theta)$. Restricting to the curve θ I find that $(\theta + g(\theta')) \mid \theta$ is zero in $\mathrm{Pic}(\theta)$ hence $g(\theta') \mid \theta$ must be non-zero in $\mathrm{Pic}(\theta)$, i.e. $g(\theta') = \theta$.

Using this lemma I treat the possibilities separately:

CASE 1: C is elliptic.
In this case $p_g(X) = 0$ or 1. In the first case there must be at least one multiple fibre because of (5.1.a) and in the second case there also must be at least one multiple fibre, since otherwise $f\colon X \to C$ would be a principal fibre bundle over an elliptic curve, hence a torus. On the other hand, the Lemma implies that g fixes all multiple fibres, so I'm done.

CASE 2: C is $\mathbf{P}^1$.

(a) There are only multiple fibres of multiplicity ≥ 3.
These must be fixed, by the lemma, and since there are at least 3 multiple fibres by (5.2), I'm done.

(b) There is exactly one double fibre.
It must be fixed and moreover the remaining multiple fibres must be fixed by the Lemma. Again an application of (5.9) concludes the argument in this case.

(c) There are at least three double fibres. They must be all fixed, because of the Lemma.

(d) There are exactly two double fibres.

In this case there must be at least two other multiple fibres, by (5.2). I may assume that g is an involution permuting the double fibres θ_1 and θ_2 and fixing the multiple fibres θ_3 and θ_4(I may assume that no other multiple fibres are present). After some calculation (which I delete) it turns out that $H_1(X, \mathbf{Z})$ contains a 2-torsion element represented by $\tau_1 - \tau_2$ where τ_j are small positively oriented loops in C around the points $f(\theta_j)$ $(j = 1, 2)$ with $h(\tau_1) = \tau_2$. Since $g_*(\tau_1) = (\tau_2 - \tau_1) + \tau_1 \neq \tau_1$ in $H_1(X, \mathbf{Z})$ this gives a contradiction. □

Now I want to finish the proof in the elliptic quasi-bundle case. So I let $f: X \to C$ be an elliptic quasi-bundle and as before $f': X' \to C'$ is a covering elliptic bundle. Since g induces the identity on C it can be lifted to a fibre preserving automorphism g' of X' inducing the identity on C'.

Step 1: g' induces a translation in each fibre.

Indeed, this already holds for g: in case $\pi(C) \leq 1$, using (5.1) either f can be replaced by a Torus-Seifert fibration or f is a quasi-bundle over an elliptic curve having at least one singular fibre. In the first case $H^1(F)$ embeds in $H^1(X)$, so g induces the identity on $H^1(F)$ and g induces a translation in each fibre. In the second case or if $\pi(C) \geq 2$ and g would not induce a translation there would be a fixed curve meeting each fibre. Each component F_j of this fixed curve has $\pi(F_j) \geq 2$, so $e(F) < 0$, contradicting the Lefschetz Fixed Point Formula (3.1), which in this case reads:

$$\sum_j e(F_j) = c_2(X) \; (=0).$$

Step 2: It is sufficient to prove that g' induces a constant translation in each fibre in the Torus–Seifert case, and $g' = \mathrm{id}$ in the other case. Moreover, it is sufficient to prove this for fibrations having a holomorphic section.

The first clause of this assertion is clear. For the second clause one needs to observe firstly that all Kählerian fibre bundles over a curve C and with fibre F form a complex analytic family $\mathcal{V}(C, F)$ parametrized by $H^1(C, \mathcal{O}(F))$ (this is a consequence of [7], Theorems 11.9 and 11.4) and secondly that g' induces an automorphism in each member of $\mathcal{V}(C', F')$ which is a constant translation, resp. the identity if and only g' is a constant translation, resp. the identity.

We now drop the accents and treat separately the two cases where $X \to C$ is a holomorphic principal fibre bundle having a section (i.e. $X = C \times F$) and where $X \to C$ is a non-principal holomorphic fibre bundle with a section.

Step 3: The case where $X = C \times F$.

Let $s = C \times \{0\}$ the zero-section. Since s maps onto a non-zero class, when forming quotients of X by a group of fibre preserving automorphisms, one may assume that s and $g(s)$ define the same class. Let ϕ: $C \to F$ be defined by $g(c, f) = (c, f + \phi(c))$ and let Γ_ϕ be the graph of ϕ. Then $(s, g(s)) = (\Gamma_\phi, C \times (0)) = \deg(\phi)$, hence is zero, since $(s, g(s)) = (s, s) = 0$. This exactly means that g induces a constant translation in each fibre.

Step 4: The case $X = (D \times F)/H$, where $H \subset \text{Aut } F \cap \text{Aut } D$ acts freely on $D \times F$, but H is not a group of translations of F.

In this case f: $X \to D/H = C$ is the elliptic fibration and there are finitely many multi-sections $s_1, \ldots, s_m$: an n-section corresponds to a set of n branch points forming the fibre of the map μ: $F \to F/H = \mathbf{P}^1$. Clearly g lifts to $D \times F$ and the previous argument shows that g induces a constant translation on F. All I have to do is to find a fixed point. Now observe that g permutes the branch points of μ, say $P_1, \ldots, P_m \in F$, and let $e_i = e(P_i)$ be the branching order. The Riemann–Hurwitz formula:

$$2\pi(F) - 2 = 0 = |H|.\left[-2 + \sum_{i=1}^{m} (1 - 1/e_i)\right]$$

implies that the only possibilities are: $m = 4$ and $(e_1, e_2, e_3, e_4) = (2, 2, 2, 2)$; $m = 3$, $(e_1, e_2, e_3) = (3, 3, 3)$, $(2, 4, 4)$ or $(2, 3, 6)$.

Only in the first two cases g might permute the multi-sections. If for instance $m = 4$ and degree$(\mu) = 2$ there are 4 sections s_1, s_2, s_3, s_4 and since f^*: $\text{Pic}^0(C) \xrightarrow{\sim} \text{Pic}^0(X)$ these are easily seen to define a torsion subgroup of $H^2(X, \mathbf{Z})$ isomorphic to $\mathbf{Z}_2 \oplus \mathbf{Z}_2$. So g must fix them, since these type of torsion elements remain fixed: g is the lifting of a fibering automorphism g_1 of a quotient β: $X \to X'$ of X and sections s_j correspond to multisections σ_j with $\beta^* \sigma_j = s_j$ in cohomology. Hence, if $g(s_i) = s_j$, $g_1(\sigma_i) = \sigma_j$. If $i \neq j$ this would imply that $(g_1)^*$ would fix the non-trivial torsion element $(\sigma_i - \sigma_j)$. □

REFERENCES

[1] M. Atiyah and I. Singer: The index of elliptic operators III. Ann. of Math. *87* (1968) 546–604.

[2] A. Beauville: Surfaces algébriques complexes, Astérisque *54*, Soc. Math. France, 1978.

[3] A. Beauville: L'application canonique pour les surfaces de type général. Inv. Math. *55* (1979) 121–140.

[4] S. Bochner and D. Montgomery: Groups on analytic manifolds, Ann. of Math. *48* (1946) 659–669.

[5] H. Cartan: Quotient d'un espace analytique par un groupe d'automorphismes, in: Algebraic Geometry, a symposium in Honor of S. Lefschetz, p. 90–102. Princeton University Press, Princeton 1957.

[6] A. HURWITZ: Uber algebraische Gebilde mit eindeutigen Transformationen in sich, Math. Ann. *41*, (1893) 403–442.
[7] K. KODAIRA: On compact complex surfaces, II–III, Ann. Math. *77* (1963), 563–626, *78* (1963) 1–40.
[8] K. KODAIRA: On the structure of compact complex surfaces, I, IV, Amer. J. Math. *86* (1964), 751–798, *90* (1968) 1048–1066.
[9] D. LIEBERMAN: Compactness of the Chow scheme: applications to automorphisms and deformations of Kähler manifolds, Sém. Norguet 1976, Lecture Notes 670, Springer Verlag, 1978.
[10] Y. MIYAOKA: Extension Theorems for Kähler metrics, Proc. Jap. Acad. Vol. *50* (1974) 407–410.
[11] Y. MIYAOKA: On the Chern numbers of surfaces of general type, Inv. Math. *42* (1977) 225–237.
[12] C. PETERS: Holomorphic automorphisms of compact Kähler surfaces and induced actions in cohomology, Inv. Math. *52* (1979) 143–148.
[13] K. UENO: A remark on automorphisms of Enriques surfaces, Journal of Fac. Sc. Univ. of Tokyo, Sec. IA, Vol. *23* no. 1 (1976) 149–165.

Mathematisch Instituut
Rijksuniversiteit Leiden
Wassenaarseweg 80, Leiden (Netherlands)

Variétés de dimension 3

A CHARACTERIZATION OF AN ABELIAN VARIETY

Yujiro Kawamata

The ground field is assumed to be the complex number field $\mathbf{C}$. Our theory is birational geometry, i.e., we shall handle only birational invariants and obtain properties which can be stated only up to birational equivalences.

Let X be a non-singular projective variety. Define $P_m(X) = \dim H^0(X, mK_X)$ for $m = 1, 2, 3, \ldots, p_g(X) = P_1(X)$ and $q(X) = \dim H^0(X, \Omega_X^1)$. Let $\kappa(X)$ be the Kodaira dimension of X, which can take values $-\infty, 0, 1, \ldots, \dim X$.

MAIN THEOREM: *Let $\alpha : X \to A(X)$ be the Albanese map of X. If $\kappa(X) = 0$, then α is surjective and has connected fibers.*

COROLLARY: *X is birational to an abelian variety iff $\kappa(X) = 0$ and $q(X) =$* $\dim X$.

The main theorem follows from the following two theorems and one proposition.

THEOREM 1 (joint work with E. Viehweg): *Let $f : X \to A$ be a surjective morphism onto an abelian variety A such that* $\dim A = \dim X$. *If $\kappa(X) = 0$, then f is birationally equivalent to an étale morphism.*

THEOREM 2: *Let $f : X \to Y$ be a surjective morphism of non-singular projective varieties with connected fibers. If $\kappa(X) \geq 0$ and $\kappa(Y) =$* $\dim Y \geq 1$, *then $\kappa(X) \geq 1$.*

PROPOSITION: *Let X be a non-singular projective variety which is generically finite over an abelian variety A. Then $\kappa(X) \geq 0$ and after a change of birational model of X, if necessary, there exists a surjective morphism $\pi : X' \to \bar{X}$ from an étale cover X' of X to a non-singular projective variety $\bar{X}$ such that*

(1) *The general fiber of π is isogenous to an abelian subvariety B of A,*
(2) *$\bar{X}$ is generically finite over A/B,*
(3) $\kappa(X) = \kappa(\bar{X}) = \dim \bar{X}$.

REMARK: In the situation of theorem 2, we can show that $\kappa(X) = \kappa(Y) + \kappa(Z)$, where Z is a general fiber of f. This fact is an easy consequence of theorem 2.

The proof of theorem 2 is obtained from the following

THEOREM 3: *Let $f: X \to Y$ be a surjective morphism of non-singular projective varieties with connected fibers, D a divisor with normal crossing on Y, such that:*

(1) *When we write $Y_0 = Y - D$ and $X_0 = f^{-1}(Y_0)$, the restriction $f_0 = f|_{X_0}$ is smooth,*

(2) *The local monodromies around D acting on $R^n f_* \mathbf{C}_{X_0}$ are unipotent, where $n = \dim X - \dim Y$.*

Then, the sheaf $f_* K_{X/Y}$ is locally free and semi-positive.

DEFINITION: Let $\pi: V \to X$ be a vector bundle over X. V is said to be *semi-positive*, if for any morphism $\varphi: C \to X$ from any curve C and for any quotient line bundle Q of $\varphi^* V$, we have $\deg Q \geq 0$.

FACT 1 (Fujita): Let V be a semi-positive vector bundle on X and let L be an ample line bundle on X. Then $V \otimes L$ is an ample vector bundle.

FACT 2 (Kodaira): Let Y be a non-singular projective variety and let L be an ample line bundle on Y. If $\kappa(Y) = \dim Y$, then there exists a positive integer m such that $H^0(Y, mK - L) \neq 0$.

The proof of theorem 3 uses the main result of Schmid: Variation of Hodge structures, Inv. Math. 22.

Department of Mathematics
Faculty of Science
University of Tokyo
Hongo, Bunkyo, Tokyo (Japan)

CANONICAL 3-FOLDS

Miles Reid

It seems to me that one
cannot get a good view
of the sky carrying a
platter on one's head.
Ssu-Ma Ch'ien

§-∞. Introduction

This is a report on the theory of canonical models of 3-folds of f.g. general type, aiming to generalise both the theory of Du Val surface singularities and some theoretical aspects of the global theory of canonical models of surfaces. The heart of the approach is the definition of *canonical singularities*, which generalises the adjunction-theoretic characterisation of Du Val singularities.

The introductory §0 discusses some of the ways in which canonical 3-fold singularities differ from canonical surface singularities – these are the points which any eventual theory will have to cover. §1 discusses some of the formal consequences of the definition of canonical singularities. §2 outlines a theory which gives some hold on the classification of canonical singularities. §3 and §4 describe important classes of canonical singularities which can be tackled by means of the toric geometry of Mumford and Kempf; this is a key technique in algebraic geometry which extends the range of computability in a spectacular way, and I refer to Danilov[15] for an extremely attractive treatment.

§4 contains in passing an implicit list of singularities which can reasonably be called "simple elliptic" 3-fold singularities.

§5 contains a formula for the plurigenera of 3-folds of f.g. general type, independent of the preceding partial classification of canonical singularities. §6 contains remarks and further problems arising out of the preceding sections.

In acknowledgement, I must plead guilty to shameless exploitation of my research student Nick Shepherd–Barron; several of the key ideas in this paper originated with him, and in particular the beautiful connection between rational Gorenstein 3-fold singularities and elliptic Gorenstein surface singularities (Theorem 2.6) was prompted by his determined and original attempts to prove Theorem 2.2.

All varieties and maps are defined over the complex ground field $k = \mathbf{C}$. A linguistic novelty introduced here is a *free* linear system to mean a linear system free from fixed components and base points.

Contents

§0. Du Val singularities 274
§1. Definition of canonical singularities 276
Appendix to §1. Weil divisors, divisorial sheaves and $\omega_X^{[r]}$ 281
§2. Inductive treatment of 3-fold rational Gorenstein points 285
§3. Toric and quotient singularities 291
§4. Hypersurface singularities and quasi-homogeneity 296
§5. The plurigenus formula 302
§6. Open problems and concluding remarks 305
References 309

§0. Du Val singularities

DEFINITION (0.1): A (non-singular, projective) variety V of dimension n is of *f.g. general type* if the canonical ring $R(V) = \bigoplus_{m \geq 0} H^0(V, \omega_V^{\otimes m})$ is finitely generated over k, and of the maximum transcendence degree $n+1$. In this case $X = \operatorname{Proj} R(V)$ is birationally equivalent to V, and is called the *canonical model* of V.

In this section X will be called a canonical variety, or $P \in X$ a canonical singularity, if X is the canonical model of some V (compare Definition 1.1 and Proposition 1.2, (II)). The question studied here is this: what do canonical 3-folds look like? Global properties will be considered in §5, but most of the paper will be concerned with the local question, or the study of canonical 3-fold singularities.

For surfaces it is well known that any canonical singularity is analytically isomorphic to a Du Val singularity, that is to the hypersurface singularity in $\mathbf{A}^3$ defined by one of the following polynomial equations:

Table (0.2)

A_n: $x^2 + y^2 + z^{n+1}$, for $n \geq 1$;
D_n: $x^2 + y^2 z + z^{n-1}$, for $n \geq 4$;
E_6: $x^2 + y^3 + z^4$
E_7: $x^2 + y^3 + yz^3$
E_8: $x^2 + y^3 + z^5$.

THEOREM (0.3): *The following are 4 characterisations of Du Val singularities:*

(I) **Adjunction-theoretic.** *$P \in X$ is a normal Gorenstein point such that if $s \in \omega_X$ is a local generator, then for any resolution $f\colon Y \to X$, $s \in \omega_Y$, that is, s remains regular on $f^{-1}P$ when considered as a differential on Y (in other words, $P \in X$ is a rational Gorenstein point).*

(II) **Inductive.** *$P \in X$ is normal, and for any chain $X_n \to \cdots \to X_0 = X$ of length $n \geqq 0$, in which $s_i : X_i \to X_{i-1}$ is the blow-up of the maximal ideal m_{P_i} for some closed point $P_i \in X_{i-1}$, X_n has at most isolated double points*; (in particular, $P \in X$ is a hypersurface double point, and the tangent cone is a plane conic, which may be irreducible, a line pair, or a double line).

This is in fact a powerful algorithmic method for determining whether a given singularity is a Du Val point.

(III) **Quotient singularities.** *$P \in X$ is isomorphic to a quotient singularity $\mathbf{A}^2/G$, where $G \subset SL(2, k)$ is a finite group.*

(IV) **Quasi-homogeneous hypersurfaces.** *$P \in X$ is isomorphic to an isolated hypersurface singularity $0 \in X \subset \mathbf{A}^3$ defined by $f(x) = 0$, with f quasi-homogeneous with respect to some weighting α, with*

$$\alpha\left(\frac{x_1x_2x_3}{f}\right) = \sum \alpha(x_i) - \alpha(f) > 0.$$

(For example, $x^2 + y^3 + z^5$ is quasi-homogeneous of weight 1, where $\alpha(xyz) = \frac{1}{2} + \frac{1}{3} + \frac{1}{5} > 1$.)

The four sections 1–4 of this paper correspond to the four parts of this theorem, and are concerned with trying to generalise the statements to higher dimension; the fact that all four characterisations lead to the same surface singularities is however a miracle which we cannot hope to see continue in the higher-dimensional case. The four characterisations in the theorem go back in the main to Du Val [1], although the first invariably appears in Artin's dual cohomological form $R^1f_*\mathcal{O}_Y = 0$ (see [2] and [3]); see also Durfee [5].

The following is a list of the difficulties of classifying canonical singularities; these arise as much from known properties of canonical 3-folds as from technicalities of proof.

REMARK (0.4):

(i) It is quite inadequate to restrict attention to hypersurface singularities; indeed, the quotient singularities $X = \mathbf{A}^3/G$, with $G \subset SL(3, k)$ give examples of Gorenstein canonical points having embedding dimension $\dim T_{X,p} = \dim m_p/m_p^2$ arbitrarily large. In fact "typical" examples of 3-fold rational Gorenstein points are given by the affine cone over a del Pezzo surface.

(ii) The Weil divisor class K_X need not be a Cartier divisor; formally we know that rK_X must be an ample Cartier divisor for some $r \geqq 1$. I define the smallest such r to be the *index* of X; there are canonical quotient singularities, first discovered by Shepherd-Barron, having arbitrary index (§3).

There is a cyclic covering trick (Corollary 1.9) which reduces canonical singularities of local index $r > 1$ to the $r = 1$ case. It is therefore sufficient for some purposes to concentrate on the case that ω_X is locally free.

(iii) Canonical 3-folds have in general 1-dimensional singular loci; in fact it seems to me perverse to distinguish isolated singularities, since even in very simple cases non-isolated singularities will unavoidably appear in the course of resolving isolated singularities.

(iv) As a technical difficulty, there is no very simple reason why canonical singularities should be Cohen–Macaulay in higher dimensions[1]; the 3-fold case has recently been settled by Shepherd–Barron [36].

(v) Even for hypersurface singularities the simple and elegant inductive criterion (II) above cannot extend as such to the higher dimensional case. For example consider the two hypersurface singularities

$$k = 2:\ x^2 + y^4 + z^4 + t^4 = 0;$$

and $$k = 1:\ x^2 + y^3 + z^6 + t^6 = 0.$$

These are weighted cones over del Pezzo surfaces of degree k, and are easily seen to be canonical (Proposition 4.5). However, on blowing up the maximal ideal of the origin in the first we get a non-normal variety; and in the second we get a normal 3-fold having a curve of singularities whose surface sections are simple elliptic singularities. Thus the blow-up need not be canonical, so that an inductive criterion analogous to (II) must be more complicated.

§1. Definition of canonical singularities

The appendix to this section deals with Weil divisors and divisorial sheaves on a a normal variety X, and introduces the divisorial sheaves $\omega_X^{[r]} = \mathcal{O}_X(rK_X)$ of regular r-differentials on X.

I have written this paper consistently using the language of the sheaves $\omega_X^{[r]}$ rather than the equivalent language of Weil divisors K_X; the reader who wishes to translate some of the definitions or arguments back into the language of Weil divisors will benefit immensely from the exercise.

DEFINITION (1.1): A quasi-projective variety X is said to have *canonical singularities* if it is normal, and if the following 2 conditions hold:

(i) for some integer $r \geqq 1$ $\omega_X^{[r]}$ is locally free;

(ii) for some resolution $f\colon Y \to X$, and r as in (i), $f_*\omega_Y^{\otimes r} = \omega_X^{[r]}$.

Observe that these conditions are local on X; if they hold in a

[1] This has been proved in all dimensions by O. Gabber and R. Elkik.

neighbourhood of $P \in X$, I will say that X is canonical at P, or that $P \in X$ is a *canonical singularity*; the smallest r for which (i) holds at P is called the *index* of $P \in X$.

If furthermore X is projective and

(iii) $\omega_X^{[r]}$ is ample,

then X is said to be a *canonical variety*.

PROPOSITION (1.2):

(I) *$P \in X$ is a canonical singularity if and only if for some integer $r \geqq 1$ $\omega_X^{[r]}$ is generated by a section $s \in \omega_X^{[r]}$, such that $s \in \omega_Y^{\otimes r}$ for a resolution $f\colon Y \to X$; that is, the r-differential s, when considered as a rational r-differential on Y, remains regular on a neighbourhood of $f^{-1}P$.*

(II) *X is a canonical variety if and only if it is the canonical model of a variety of f.g. general type.*

REMARK (1.3): Assuming (i), (ii) is equivalent to

(ii′) for every proper birational morphism $f\colon Y \to X$, and every $s \geqq 1$, $f_*\omega_Y^{[s]} = \omega_X^{[s]}$.

In particular, canonical singularities satisfy "Kempf's condition" $f_*\omega_Y = \omega_X$ for a resolution $f\colon Y \to X$, so that according to Kempf's duality argument ([6], p. 44), canonical singularities are rational if and only if they are Cohen–Macaulay. However, not all rational singularities are canonical, since (ii′) is stronger than Kempf's condition.

Question (1.4) Does (ii′) imply (i)?[1]

To emphasize that the condition in (I) of Proposition 1.2 is readily calculable once we know a resolution, let me give two examples, which give a foretaste of results of §3 and §4.

Example (1.5) (i) Let $X \subset \mathbf{A}^{n+1}$ be a hypersurface with an ordinary k-fold point at the origin 0; *then $0 \in X$ is canonical if and only if $k \leqq n$.*

(ii) Let $X = \mathbf{A}^n/\boldsymbol{\mu}_k$, where $\boldsymbol{\mu}_k$ is the cyclic group of k-th roots of 1, acting by $\epsilon\colon (x_1, \dots, x_n) \mapsto (\epsilon x_1, \dots, \epsilon x_n)$; X is isomorphic to the affine cone over the k-fold Veronese embedding of $\mathbf{P}^{n-1}$; *then X is canonical if and only if $k \leqq n$, and its index is the denominator of n/k.*

Computation. (i) Let X be given by $f = f(x_0, \dots, x_n) = 0$, so that ω_X is generated by

$$s = \operatorname{Res}_X\left(\frac{dx_0 \wedge \cdots \wedge dx_n}{f}\right) = \frac{dx_1 \wedge \cdots \wedge dx_n}{(\partial f/\partial x_0)}.$$

[1] This has been answered negatively by Pinkham; the counter-example is one of a type of singularities studied by Laufer [35].

A typical piece of the blow-up of $\mathbf{A}^{n+1}$ has coordinates $y_0, \ldots, y_n$ with $x_0 = y_0$, $x_i = y_0 y_i$, and in this piece the non-singular proper transform X' is given by $f' = \frac{1}{y_0^k} f(y_0, y_0 y_i) = 0$. In terms of the generator $t = \operatorname{Res}\left(\frac{dy_0 \wedge \cdots \wedge dy_n}{f'}\right) \in \omega_{X'}$ we have $s = y_0^{n-k} t$, Q.E.D.

(ii) Among the coordinate functions on X I pick out the invariant monomials $u_1 = x_1^k$, $u_i = x_1^{k-1} x_i$; let $\frac{n}{k} = \frac{b}{a}$ with a and b coprime positive integers. Write

$$s = (dx_1 \wedge \cdots \wedge dx_n)^a = (\text{const.}) \cdot \frac{(du_1 \wedge \cdots \wedge du_n)^a}{u_1^{b(k-1)}};$$

this is a rational differential on X having no zeroes or poles, and is thus a generator of $\omega_X^{[a]}$.

This cone also becomes non-singular after a single blow-up; coordinates on a typical piece of X' are $v_1, \ldots, v_n$, with $u_1 = v_1$, $u_i = v_1 v_i$. Since $du_1 \wedge \cdots \wedge du_n = v_1^{n-1}(dv_1 \wedge \cdots \wedge dv_n)$ we have

$$s = (\text{const.}) \cdot v_1^\alpha \cdot (dv_1 \wedge \cdots \wedge dv_n)^a,$$

where $\alpha = a(n-1) - b(k-1) = b - a$. Thus s remains regular on X' if and only if $b \geqq a$, that is $k \leqq n$. Q.E.D.

The proofs of Remark 1.3, of (I) and of the "only if" part of (II) in Proposition 1.2 are purely formal, and are left to the reader as interesting exercises.

PROOF OF "IF" PART OF (II): Let V be a variety of f.g. general type. There exists an r such that the graded ring

$$R(V)^{(r)} = \bigoplus_{n \geqq 0} H^0(V, \omega_V^{\otimes nr})$$

is generated by its elements of the least degree r. Blowing up the base locus of $|rK_V|$ on V, I may assume that

$$|rK_V| = |D| + F,$$

with F fixed and $|D|$ free.

Let $\varphi_D = \varphi_{rK_V}\colon V \to X$ be the morphism defined by $|D|$; I will show that $\varphi_D(F)$ has no components of codimension 1, so that for some open set $X^0 \subset X$ having complement of codimension $\geqq 2$, $\varphi_D\colon \varphi_D^{-1}(X^0) \xrightarrow{\approx} X^0$, providing an isomorphism of $\mathcal{O}_X(1)$ with $\omega_X^{[r]}$ over X^0. Thus $\omega_X^{[r]} = \mathcal{O}_X(1) = \varphi_{D*}\omega_V^{\otimes r}$, so that X is canonical. There only remains to prove the following assertion.

LEMMA (1.6): *For every component Γ of F, $\varphi_D(\Gamma)$ has codimension $\geqq 2$.*

PROOF: Write $n = \dim V$. By hypothesis, for every $m > 0$,

$$H^0(\mathcal{O}_V(mD + \Gamma)) \longrightarrow H^0(\mathcal{O}_\Gamma(mD + \Gamma))$$

is the zero map. An easy argument using the Leray spectral sequence for φ_D shows that $h^1(\mathcal{O}_V(mD))$, and with it $h^0(\mathcal{O}_\Gamma(mD + \Gamma))$ is bounded by (const.)m^{n-2}; thus the Iitaka dimension $\kappa(\Gamma, D_\Gamma) \leqq n-2$, Q.E.D.

The remainder of this section is devoted to some easy but important formal consequences of the definition of canonical singularities.

PROPOSITION (1.7): *Suppose that $\varphi\colon Y \to X$ is a proper morphism, with X and Y normal varieties, and suppose that φ is etale in codimension* 1 *on Y. Then*

(I) *if X has canonical singularities, so does Y.*

Suppose furthermore that ω_X is locally free; then

(II) *if Y has canonical singularities, so does X.*

PROOF: Form a commutative diagram

$$\begin{array}{ccc} \tilde{Y} & \xrightarrow{\tilde{\varphi}} & \tilde{X} \\ {\scriptstyle g}\downarrow & & \downarrow{\scriptstyle f} \\ Y & \xrightarrow{\varphi} & X, \end{array} \qquad (*)$$

with f and g resolutions. Then if $s \in \omega_X^{[r]}$ is a generator, so is $\varphi^* s \in \omega_Y^{[r]}$. For (I) note that if f^*s is regular on $\tilde{X}$ then $g^*(\varphi^* s) = \tilde{\varphi}^* f^* s$ is regular on $\tilde{Y}$.

(II) follows by computing $v_\gamma(s)$ for v_γ a valuation of $k(X)$ in terms of a valuation v_δ of $k(Y)$ lying over v_γ, and having ramification index e:

$$v_\gamma(s) = \frac{1}{e}(v_\delta(\varphi^* s) - r(e-1)); \qquad (**)$$

thus if $r = 1$ and Y is canonical, then $v_\delta(\varphi^* s) \geqq 0$, so that the integer $v_\gamma(s) \geqq 0$.

REMARK (1.8): In fact X is canonical if and only if, for every r such that $\omega_X^{[r]}$ is locally free, generated by t, say, $v_\delta(t) \geqq r(e-1)$ for every valuation v_δ of $k(Y)$, where e is the ramification index of v_δ in the field extension $k(Y)/k(X)$. This criterion is of course useless in practice, since one has no hope of finding every relevant valuation v_δ without first resolving X.

A consequence of (II) is that all Gorenstein quotient singularities are canonical. From (I) we get a cyclic covering trick[1], which reduces the study of canonical singularities with $r > 1$ to the $r = 1$ case.

COROLLARY (1.9): *Let $P \in X$ be a canonical point of index r; then there exists a finite cover $Y \to X$ which is Galois with group* **Z**/r, *and which is etale in*

[1] Due independently to J. Wahl.

codimension 1, such that Y is canonical with ω_Y locally free; the construction is defined locally and uniquely determined up to local analytic isomorphism.

PROOF: By means of a local generator, identify $\omega_X^{[r]}$ with $\mathcal{O}_X$, and define in the obvious way an algebra structure on $\mathcal{A} = \mathcal{O}_X \oplus \omega_X^{[1]} \oplus \cdots \oplus \omega_X^{[r-1]}$; the finite $\mathbf{Z}/r$-cover $Y = \mathrm{Spec}_X \mathcal{A} \to X$ is unramified in codimension 1 because $\omega_X^{[a]} \otimes \omega_X^{[b]} \to \omega_X^{[a+b]}$ is an isomorphism in codimension 1; Y is therefore non-singular in codimension 1, and is normal because each of the sheaves $\omega_X^{[a]}$ is divisorial, so that $\mathcal{A}$ is saturated in the sense of (iv) of Proposition 2 of the Appendix. Q.E.D.

Example (1.10). The Galois tower

$$\begin{array}{ccccc} & \nearrow & \mathbf{A}^3/\boldsymbol{\mu}_2 & \searrow & \\ \mathbf{A}^3 & & & & X = \mathbf{A}^3/\boldsymbol{\mu}_6 \\ & \searrow & \mathbf{A}^3/\boldsymbol{\mu}_3 & \nearrow & \end{array}$$

(where the actions of the $\boldsymbol{\mu}_k$ are as in Example 1.5, (ii)) shows that a group action which splits as a direct sum, and such that each factor has canonical quotient, need not have canonical quotient. Furthermore, since both X and $\mathbf{A}^3/\boldsymbol{\mu}_2$ have the same index 2, the hypothesis made before (II) in Proposition 1.6 cannot be replaced by "index X = index Y".

Problem (1.11). §2 will give some ideas towards the classification of canonical 3-fold singularities of index 1; together with Corollary 1.9 this puts an upper bound on the problem of classification of singularities of any index r, which can be obtained as quotients of index 1 singularities by a cyclic group $\boldsymbol{\mu}_r$ whose representation on ω_Y is faithful. However, I have no very precise idea as to when such a quotient $Y/\boldsymbol{\mu}_r$ will be canonical, apart from Remark 1.8 and the useful examples in §3.

Canonical varieties satisfy a trivial but important compatibility with taking hyperplane sections. The reader will excuse the following rather obscene digression.

LEMMA (1.12) (Seidenberg): *Let $X \subset \mathbf{P}^N$ be a quasi-projective scheme over any field. If X satisfies Serre's condition S_r then so does its general hyperplane section; if X is a normal variety then so is its general hyperplane section.*

PROOF: Let

$$\begin{array}{ccc} & Z \subset X \times \check{\mathbf{P}}^N & \\ {}^{p_1}\swarrow & & \searrow^{p_2} \\ X & & \check{\mathbf{P}}^N \end{array} \qquad (***)$$

be the incidence relation $Z = \{(x, h) \mid x \in h\}$. $p_1 : Z \to X$ is a $\mathbf{P}^{N-1}$-fibre bundle, so that X satisfies S_r implies Z satisfies S_r. If $\eta \in \check{\mathbf{P}}^N$ is the generic point then the generic fibre Z_η of p_2 is also S_r, because the local rings of Z_η are particular local rings of Z; I conclude by E.G.A. IV$_3$, Proposition (9.9.2), (viii). For the final part one uses Serre's criterion for normality and the trivial Bertini theorem.

THEOREM (1.13): *If X has canonical singularities then so has its general hyperplane section.*

PROOF: Return to the incidence diagram (***) above. Since p_1 is a $\mathbf{P}^{N-1}$-fibre bundle, Z is canonical; the general hyperplane section Y of X is the general fibre of p_2: $Z \to \check{\mathbf{P}}^N$. Now let f: $\tilde{Z} \to Z$ be any resolution; the general fibre $\tilde{Y}$ of $g = p_2 \circ f$ is a resolution of Y (by Bertini's theorem), and Y is canonical by standard adjunction considerations.

COROLLARY (1.14): *Let X be a 3-fold with canonical singularities. Then with the exception of at most a finite number of "dissident" points $P \in X$, every point has an analytic neighbourhood which is (non-singular or) isomorphic to a Du Val surface singularity $\times\, \mathbf{A}^1$.*

The point is just that the Du Val singularities have no moduli.

The results of §5 on global properties of canonical 3-folds are consequences of this result, and do not depend on the attempts to classify canonical 3-fold singularities in §§2–4.

Appendix to §1; Weil divisors, divisorial sheaves and $\omega_X^{[r]}$

This section is intended to be complementary to Hartshorne's book Algebraic Geometry, II.6 and III.7.

Let X be a quasi-projective variety defined over a field k, and let $k(X)$ be its function field. Until further notice X is assumed normal. A *prime divisor* of X is an irreducible subvariety of codimension 1.

THEOREM (1): (i) *For every prime divisor Γ the local ring $\mathcal{O}_{X,\Gamma}$ is a discrete valuation ring, with valuation v_Γ: $k(X) \to \mathbf{Z} \cup \{-\infty\}$;*

(ii) $\mathcal{O}_X = \cap\, \mathcal{O}_{X,\Gamma}$, *in the following 2 senses:*

(a) *for all $P \in X$,* $\mathcal{O}_{X,P} = \bigcap_{P \in \Gamma} \mathcal{O}_{X,\Gamma}$;

(b) *for all open $U \subset X$,* $\Gamma(U, \mathcal{O}_X) = \bigcap_{\Gamma \in U} \mathcal{O}_{X,\Gamma}$.

PROOF: [8], p. 124.

Sections of $\mathcal{O}_X$ are thus rational functions $f \in k(X)$ which are regular along each prime divisor; this is an algebraic form of Hartog's lemma.

Let $\mathcal{L}$ be a coherent sheaf of $\mathcal{O}_X$-modules which is torsion-free and of rank 1. The generic stalk $\mathcal{L} \otimes k(X)$ is a 1-dimensional vector space over

$k(X)$, so that choosing a basis identifies $\mathscr{L}$ with a subsheaf of the constant sheaf $k(X)$.

PROPOSITION (2): *Equivalent conditions*:

(i) $\mathscr{L} = \mathscr{L}^{**}$, *where for a sheaf of* $\mathcal{O}_X$*-modules* $\mathscr{F}$, $\mathscr{F}^*$ *denotes the dual* $\mathscr{F}^* = Hom_{\mathcal{O}_X}(\mathscr{F}, \mathcal{O}_X)$;

(ii) $\mathscr{L} = \cap \mathscr{L}_\Gamma$ *in the sense of (ii) above*;

(iii) $\mathrm{Ass}(k(X)/\mathscr{L}) = \{\text{Prime divisors of } X\}$;

(iv) *for every inclusion* $\mathscr{L} \subset \mathscr{M}$ *where* $\mathscr{M}$ *is a torsion-free* $\mathcal{O}_X$*-module and* $\mathrm{Supp}(\mathscr{M}/\mathscr{L})$ *has codimension* $\geqq 2$, $\mathscr{L} = \mathscr{M}$;

(v) *if* $X^0 \subset X$ *is a non-singular open subvariety such that* $X \backslash X^0$ *has codimension* $\geqq 2$, *then* $\mathscr{L}_{|X^0}$ *is invertible, and* $\mathscr{L} = j_*(\mathscr{L}_{|X^0})$, *where* j *denotes the inclusion.*

PROOF: [7], §4, no. 2, Theorem 2.

A sheaf satisfying these conditions is called *divisorial*; a *Weil divisor* is defined as a formal sum $D = \Sigma\, n_\Gamma \Gamma$, with Γ prime divisors, $n_\Gamma \in \mathbf{Z}$, and almost all $n_\Gamma = 0$. If D is a Weil divisor, the subsheaf $\mathcal{O}_X(D) \subset k(X)$ is defined by

$$\Gamma(U, \mathcal{O}_X(D)) = \{f \in k(X) \mid v_\Gamma(f) \geqq -n_\Gamma \text{ for all } \Gamma \in U\}.$$

Exactly as for Cartier divisors one has:

THEOREM (3): *The correspondence* $D \mapsto \mathcal{O}_X(D)$ *defines a bijection*

$$\left\{\begin{matrix}\text{Weil divisors}\\ \text{on } X\end{matrix}\right\} \overset{\text{bij}}{\longleftrightarrow} \left\{\begin{matrix}\text{divisorial sub-}\\ \text{sheaves } \mathscr{L} \subset k(X)\end{matrix}\right\} / \Gamma(X, \mathcal{O}_X)^*.$$

PROOF: [7], §1, no. 3, Theorem 2.

LEMMA (4): *Let* $P \in X$, *and let* D *be a Weil divisor on* X. *Then equivalent conditions*:

(i) $\mathcal{O}_X(D)$ *is invertible at* P;

(ii) *there exists an* $f \in k(X)$ *such that* $v_\Gamma(f) = -n_\Gamma$ *for every prime divisor* Γ *with* $P \in \Gamma$;

(iii) *there exists a neighbourhood* U *of* P, *and a section* $s \in \Gamma(U, \mathcal{O}_X(D))$ *such that* s *generates* $\mathcal{O}_X(D)$ *over an open subset* $V \subset U$ *with codimension* $(U \backslash V) \geqq 2$.

PROOF: Trivial.

Such a divisor D is called *principal* at P, or a *Cartier divisor* at P. Locally one can always choose a Cartier divisor $E \geqq D$, so that $\mathcal{O}_X(D)$ has an expression $\mathcal{O}_X(D) = \mathscr{I}_{E-D} \cdot \mathcal{O}_X(E)$ as a product of a divisorial ideal sheaf $\mathscr{I}_{E-D}$ with an invertible sheaf $\mathcal{O}_X(E)$; this holds globally if, as here, X is assumed to be quasi-projective.

REMARK (5): It may well happen that $\mathcal{O}_X(D_1)\otimes\mathcal{O}_X(D_2)\neq\mathcal{O}_X(D_1+D_2)$; the left-hand side may have torsion, and it may not map onto the right-hand side either: let X be the quadric cone $X=(x^2-yz=0)\subset\mathbf{A}^3$, and let D be the line $(x=y=0)$; then $\mathcal{O}_X(-D)=\mathcal{I}_D$, and $\mathcal{O}_X(-D)\otimes\mathcal{O}_X(-D)\twoheadrightarrow\mathcal{I}_D^2$, which is clearly not the same as the principal ideal $y\cdot\mathcal{O}_X=\mathcal{I}_{2D}$.

It will however always be true that

$$(\mathcal{O}_X(D_1)\otimes O_X(D_2))^{**}=\mathcal{O}_X(D_1+D_2),$$

and this process of taking the product, and then the double dual, is similar to the procedure of taking the "symbolic power" of a prime ideal in the theory of primary decomposition.

In the remainder of this appendix I will show how to define a divisorial sheaf $\omega_X=\mathcal{O}_X(K_X)$, and set $\omega_X^{[r]}=\mathcal{O}_X(rK_X)=(\omega_X^{\otimes r})^{**}$.

Suppose now that $X\subset\mathbf{P}^N$ is an irreducible n-dimensional variety, *not necessarily normal.* Set

$$\omega_X=\operatorname{Ext}^{N-n}_{\mathcal{O}_{\mathbf{P}^N}}(\mathcal{O}_X,\omega_{\mathbf{P}^N}),$$

where $\omega_{\mathbf{P}^N}=\Omega^N_{\mathbf{P}^N}\cong\mathcal{O}_{\mathbf{P}^N}(-N-1)$; compare [9], p. 1.

Now let $X^0\subset X$ denote the non-singular locus.

PROPOSITION (6): $\omega_X|X^0=\Omega^n_X|X^0$.

PROOF: [9], p. 14 (the two sides can be calculated by means of an identical adjunction procedure).

THEOREM (7): *ω_X is a torsion-free sheaf of rank* 1, *satisfying the saturation condition* (iv) *of Proposition* 2; *in particular, if X is normal, ω_X is a divisorial sheaf.*

PROOF (compare [9], p. 8): (a) ω_X has rank 1 at the generic point, according to Proposition 6.

(b) ω_X is torsion-free; for if $\mathcal{F}\subset\omega_X$ is a torsion part, $\dim(\operatorname{Supp}\mathcal{F})\leqq n-1$, so that $H^n(\mathcal{F})=0$, and hence dually $\operatorname{Hom}(\mathcal{F},\omega_X)=0$, so that $\mathcal{F}=0$.

(c) Let $\omega_X\subset\mathcal{F}$, with $\dim(\operatorname{Supp}\mathcal{F}/\omega_X)\leqq n-2$; then $H^{n-1}(\mathcal{F}/\omega_X)=0$, and hence $H^n(\omega_X)=H^n(\mathcal{F})\cong k$. By duality I obtain a non-zero element of $\operatorname{Hom}(\mathcal{F},\omega_X)$, which provides a splitting of the exact sequence

$$0\longrightarrow\omega_X\longrightarrow\mathcal{F}\longrightarrow\mathcal{F}/\omega_X\longrightarrow 0,$$

and since $\mathcal{F}$ was supposed torsion free, $\mathcal{F}/\omega_X=0$. Q.E.D.

COROLLARY (8): *Suppose that X is normal; then*

$$\omega_X=(\Omega^n_X)^{**}=j_*(\omega_{X^0})=\Omega_X=\mathcal{O}_X(K_X),$$

where (i) *$j\colon X^0\hookrightarrow X$ is the inclusion of the smooth locus, or more generally of any smooth part of X having complement of codimension $\geqq 2$;*

(ii) Ω_X *is the sheaf of Zariski differentials regular in codimension* 1 (see [10], Proposition 8.7):

$$\Gamma(U, \Omega_X) = \{s \in \Omega^n_{k(X)} \mid s \in \Omega^n_{X,\Gamma} \text{ for all prime divisors } \Gamma \subset U\};$$

and (iii) K_X *is the Weil divisor class corresponding to the sheaf* ω_X *under Theorem* 3 *above, or the class constructed below.*

For each prime divisor Γ of X the stalk $\Omega^n_{X,\Gamma}$ is the $\mathcal{O}_{X,\Gamma}$-module generated by $s_\Gamma = dx_1 \wedge \cdots \wedge dx_n$, where x_1 is any local parameter of $\mathcal{O}_{X,\Gamma}$, and $x_2, \ldots, x_n \in \mathcal{O}_{X,\Gamma}$ are elements whose residues in $k(\Gamma)$ form a separating transcendence basis. Thus for any rational differential $s \in \Omega^n_{k(X)}$ there is a unique integer $v_\Gamma(s)$ such that

$$s = (\text{unit}) \cdot x_1^{v_\Gamma(s)} \cdot s_\Gamma,$$

and the *divisor* of s is the finite sum

$$(s) = \sum v_\Gamma(s)\Gamma;$$

then $K_X \sim (s)$.

As a coda to this appendix I include two remarks on the non-normal case.

Firstly, the structure sheaf $\mathcal{O}_X$ has a natural saturation in the sense of condition (iv), Proposition 2, consisting of rational functions $f \in k(X)$ which belong to $\mathcal{O}_X$ in codimension 1; it is natural to call this the S_2-isation, $S_2(\mathcal{O}_X)$,

$$\mathcal{O}_X \subset S_2(\mathcal{O}_X) \subset \tilde{\mathcal{O}}_X,$$

since $S_2(X) = \operatorname{Spec}(S_2(\mathcal{O}_X)) \xrightarrow{\pi} X$ is the unique finite morphism which is an isomorphism in codimension 1, and such that $S_2(X)$ satisfies Serre's condition S_2. A consequence of Theorem 7 is that ω_X is an $S_2(\mathcal{O}_X)$-module and coincides with $\pi_*\omega_{S_2(X)}$. Thus in discussing ω_X there is little loss of generality in assuming that X satisfies S_2.

Secondly, there is a sense in which the computation of ω_C for a curve C in terms of Rosenlicht differentials (see [11], p. 76) determines ω_X on any quasi-projective variety. To be precise, X has a linear section which is a reduced curve C (Lemma 1.12), and such that at each point $P \in C$ the equations $x_1, \ldots, x_{n-1}$ of the linear section $C \subset X$ form a regular sequence. In a neighbourhood of $P \in X$ one can then construct an isomorphism

$$\omega_X \otimes \mathcal{O}_C \xrightarrow{\approx} \omega_C, \tag{*}$$

denoted $s \longmapsto \operatorname{Res}_C\left(\dfrac{s}{\prod x_i}\right)$,

and constructed as follows: let $X = X_n \supset X_{n-1} \supset \cdots \supset X_1 = C$ be the chain of local divisors defined in turn by $x_1, \ldots, x_{n-1}$; each X_{i-1} is a Cartier divisor in X_i, locally defined by $x_{n-i} = 0$. Then the construction of [9], p. 7 provides a "residue" isomorphism

$$\omega_{X_i}(X_{i-1}) \otimes \mathcal{O}_{X_{i-1}} \xrightarrow{\approx} \omega_{X_{i-1}},$$

which can be fitted together into an isomorphism

$$\mathrm{Res}_C\colon \omega_X(D) \otimes \mathcal{O}_C \xrightarrow{\approx} \omega_C,$$

where D is the Cartier divisor defined by $x_1 \cdot x_2 \ldots \cdot x_{n-1}$. (*) is then obtained by composing this with $\omega_X \to \omega_X(D)$, given by $s \mapsto \dfrac{s}{\prod x_i}$. It is then easily seen that ω_X is uniquely determined as the sheaf of rational differentials on X whose residue (in the sense of (*)) belongs to ω_C for every sufficiently general section C.

§2. Inductive treatment of 3-fold rational Gorenstein points

DEFINITION (2.1): A point $P \in X$ of a 3-fold is called a *compound Du Val point* if for some section H through P, $P \in H$ is a Du Val singularity. Equivalently, $P \in X$ is *cDV* if it is locally analytically isomorphic to the hypersurface singularity given by

$$f + tg = 0,$$

where $f \in k[x, y, z]$ is one of the polynomials listed in Table 0.2, and $g \in k[x, y, z, t]$ is arbitrary.

A cDV point point may be isolated or otherwise. It will be shown below that it must be canonical.

As pointed out in Remark 0.4 (v), the blow-up of a canonical 3-fold point need not be normal, and if it is normal need not be canonical. However, if $f\colon Y \to X$ is any proper birational morphism with Y normal and X canonical of index 1, and if $\omega_Y = f^*\omega_X$ then Y is also canonical (this is an easy consequence of Proposition 1.2 (I)). For general $f\colon Y \to X$ $f^*\omega_X = \omega_Y(-\Delta)$ (or $K_Y = f^*K_X + \Delta$), with $\Delta = \Delta(f) \geqq 0$ a Weil divisor, the *discrepancy* of f; a prime divisor of $f^{-1}p$ which occurs in Δ with strictly positive coefficient is called *discrepant*, and one not occuring in Δ is called *crepant*.

THEOREM (2.2): *If $P \in X$ is a canonical point of index* 1 *which is not cDV then there exists a proper birational morphism $f\colon Y \to X$ with*

(i) *f is crepant, that is $f^*\omega_X = \omega_Y$, and*

(ii) *$f^{-1}P$ contains at least one prime divisor of Y.*

This theorem will be proved here using the fact [36] that $P \in X$ is

Cohen–Macaulay; on the other hand, Shepherd–Barron (who gave a proof under extra conditions) points out that the result implies that $P \in X$ is Cohen–Macaulay, using the Grauert and Riemenschneider vanishing theorem [27].

The fact that the inductive process must terminate follows from this easy result:

LEMMA (2.3): *Let $P \in X$ be a canonical point of index 1; as $f: Y \to X$ runs through all proper birational morphisms to X, the number of crepant prime divisors of Y is bounded.*

PROOF: Let $\pi: \tilde{X} \to X$ be some resolution; then by Hironaka's resolution theorem every $Y \to X$ can be housed

$$\begin{array}{ccc} & \tilde{\tilde{X}} & \\ \swarrow & & \searrow \\ \tilde{X} & & Y \\ \searrow & & \swarrow \\ & X & \end{array}$$

under a blow-up $\tilde{\tilde{X}}$ of $\tilde{X}$. Every crepant prime divisor of Y must then lie under a crepant prime divisor of $\tilde{\tilde{X}}$. But the exceptional divisors of the blow-ups in $\tilde{\tilde{X}} \to \tilde{X}$ are certainly discrepant, so that the crepant prime divisors of Y can be mapped injectively to those of the fixed $\tilde{X}$. Q.E.D.

Recall that a variety X is called *Gorenstein* if it is Cohen–Macaulay and the sheaf ω_X is locally free. I will assume from now on (see [36]) that my index 1 point $P \in X$ is Cohen–Macaulay; this assumption will always be used in the form that a hyperplane section H through P having an isolated singularity is normal. To say that $P \in X$ is Cohen–Macaulay and canonical of index 1 is equivalent to saying that $P \in X$ is rational Gorenstein.

DEFINITION (2.4): A Gorenstein point $P \in X$ of an n-dimensional variety X is *rational* (respectively *elliptic*) if for a resolution $f: Y \to X$ we have $f_*\omega_Y = \omega_X$ (respectively $f_*\omega_Y = m_P \cdot \omega_X$, where m_P is the ideal of P).

(This is equivalent via duality to the cohomological assertion $R^{n-1}f_*\mathcal{O}_Y = 0$ (respectively, is a 1-dimensional k-vector space at P)).

It is convenient to make intrinsic (and generalise slightly) the notion of "a general hyperplane section through P":

DEFINITION (2.5): Let $(\mathcal{O}_{X,P}, m_P)$ be the local ring of a point $P \in X$ of a k-scheme, and let $V \subset m_P$ be a finite-dimensional k-vector space which maps onto m_P/m_P^2 (equivalently, by Nakayama's lemma, V generates the $\mathcal{O}_{X,P}$-ideal m_P); by a ***general hyperplane section through P*** is meant the subscheme $H \subset X_0$ defined in a suitable open neighbourhood X_0 of P by the ideal $\mathcal{O}_X \cdot v$, where $v \in V$ is a sufficiently general element (that is, v is a k-point of a certain dense Zariksi open $U \subset V$).

THEOREM (2.6): (I) *If* $P \in X$ *is a rational Gorenstein point (with* $n = \dim X \geqq 2$*) then for a general hyperplane section* H *through* P, $P \in H$ *is elliptic or rational Gorenstein*;

(II) *if there exists a hyperplane section* H *through* P *such that* $P \in H$ *is rational Gorenstein then* $P \in X$ *is rational Gorenstein*; *in particular cDV points are canonical.*

PROOF: The fact that the Cohen–Macaulay condition passes to and from a hyperplane section is obvious; the fact that ω_X is locally free if and only if ω_H is locally free follows from the residue isomorphism

$$\omega_X(H) \otimes \mathcal{O}_H \xrightarrow{\approx} \omega_H,$$

so that if ω_X is generated by s at P, ω_H is generated by $\operatorname{Res}_H\left(\frac{s}{h}\right)$, where $h \in \mathcal{O}_{X,P}$ is the local equation of H.

For (I), let $\sigma: X_1 \to X$ be the blow-up of $P \in X$, and let $g: Y \to X_1$ be any resolution; by construction of the blow-up $m_P \cdot \mathcal{O}_{X_1}$ is an invertible sheaf of ideals and the same continues to hold for Y, so that $m_P \cdot \mathcal{O}_Y = \mathcal{O}_Y(-E)$. Under these conditions the Cartier divisor E on Y is called the *strong geometric fundamental cycle* of the resolution $f = g \circ \sigma: Y \to X$.

As the hyperplane section H through P runs through any linear system whose local equations generate m_P, $f^*H = L + E$, where L runs through a linear system on Y which is free near $f^{-1}P$. Thus by Bertini's theorem a general L is a resolution of the corresponding $P \in H$:

$$\begin{array}{ccc} f^*H = L + E & \subset & Y \\ \downarrow & & \downarrow{\scriptstyle f} \\ H & \subset & X. \end{array}$$

Since X is canonical, the generator $s \in \omega_X$ remains regular on Y; $\frac{s}{h}$ generates the sheaf $\omega_X(H)$. At any point of Y, h factorises as $h = \ell \cdot e$, where ℓ is a local equation for L, and e one for E. Thus if $a \in m_{X,P}$, $\frac{as}{h} = \frac{as}{\ell e}$, and since a vanishes along E, $\frac{as}{h}$ is a regular section of $\omega_Y(L)$. It follows that for any element $\bar{a} \in m_{H,P}$, the product $\bar{a} \cdot \operatorname{Res}_H\left(\frac{s}{h}\right)$ of $\bar{a}$ with a generator of ω_H remains regular on L, $\bar{a} \cdot \operatorname{Res}_H\left(\frac{s}{h}\right) = \operatorname{Res}_L\left(\frac{as}{h}\right) \in \omega_L$. Thus $m_P \cdot \omega_H \subset f_* \omega_L \subset \omega_H$. Q.E.D.

(II) follows from one of the main results of Elkik [14], Theorem 4, p. 146, once I observe that X is a flat deformation of the variety $H \times \mathbb{A}^1$, which has rational singularities.

LEMMA (2.7): *Let X be an affine variety, and H a hyperplane section of X; then there exists a flat family $\mathfrak{X} \to \mathbf{A}^1$ having fibres $X_t \simeq X$ if $t \neq 0$, and $X_0 \simeq H \times \mathbf{A}^1$.*

PROOF: If $X \subset \mathbf{A}^N$ is given by the ideal $I = I(X) \subset k[T_1, \ldots, T_N]$, with H the hyperplane $T_N = 0$, let φ: $k[T_1, \ldots, T_N] \to k[S_1, \ldots, S_{N+1}]$ be given by $T_i \mapsto S_i$ for $i \leqq N-1$, $T_N \mapsto S_N S_{N+1}$; it is then easy to check that the ideal $J \subset k[S_1, \ldots, S_{N+1}]$ generated by $\varphi(I)$ defines a variety $\mathfrak{X} \subset \mathbf{A}^{N+1}$, with a morphism to $\mathbf{A}^1$ given by S_{N+1}, having the required property. Q.E.D.

It seems worthwhile to illustrate Theorem 2.6 with a summary of the low-dimensional cases.

Table (2.8):

dim.	rational Gor.	elliptic Gor.
1	non-sing. point	node or cusp
2	Du Val point	Laufer–Reid
3	this paper	??
4	???	

Theorem 2.6 is extremely strong, due to the fact that elliptic Gorenstein surface singularities form an extremely well-defined and tightly controlled class of singularities; see [13], where they are called "minimally elliptic", or my unpublished manuscript [12]. The following is a summary of some results of [12] and [13]; (see especially [13], Theorem 3.13, p. 1270 and Theorem 3.15, p. 1275).

PROPOSITION (2.9): *One can attach a natural number $k = -Z^2$, $k \geqq 1$ to each elliptic Gorenstein surface point $P \in S$, in such a way that*

(i) *if $k \geqq 2$ then $k = \operatorname{mult}_P S$;*

(ii) *if $k \geqq 3$ then $k =$ minimal embedding dimension $= \dim m_P/m_P^2$; if $k \geqq 3$ then the blow-up $S_1 \to S$ of (the reduced point) P in S is a normal surface having only Du Val singularities.*

If $k = 2$ then $P \in S$ is isomorphic to a hypersurface given by $x^2 + f(y, z) = 0$, with f a sum of monomials $y^a z^b$ of degree $a + b \geqq 4$; if α is the weighting $\alpha(x) = 2$, $\alpha(y) = \alpha(z) = 1$ then the α-blow-up (see §4) $S_1 \to S$ is a normal surface having only Du Val points.

If $k = 1$ then $P \in S$ is given by $x^2 + y^3 + f(y, z)$, where f is a sum of monomials yz^a with $a \geqq 4$ and z^a with $a \geqq 6$; if α is the weighting $\alpha(x) = 3$, $\alpha(y) = 2$, $\alpha(z) = 1$ then the α-blow-up $S_1 \to S$ is a normal surface having at most 1 Du Val point.

The given blow-up σ: $S_1 \to S$ has the following effect on the canonical sheaf: $\omega_{S_1} = \sigma^* \omega_S(-Z_1)$, where Z_1 is the geometric fundamental cycle for σ; that is, Z_1 is a Cartier divisor, and for $k \geqq 2$ $m_P \cdot \mathcal{O}_{S_1} = \mathcal{O}_{S_1}(-Z_1)$, so that Z_1 is the strong geometric fundamental cycle. If $k = 1$ then there is a point $Q \in Z_1$, non-singular on Z_1 and on S_1 such that $m_P \cdot \mathcal{O}_{S_1} = m_Q^2 \cap \mathcal{I}_{Z_1} = m_Q \cdot \mathcal{O}_{S_1}(-Z_1)$.

The assertions about the weighted blow-up of the $k=2$ or $k=1$ points are not in [12] or [13]; morally they should be proved by relating the weighting α to the higher adjunction ideals $\mathcal{I}_n \subset \mathcal{O}_S$ (that is, the ideals $\mathcal{I}_n$ such that $f_*\omega_Y^{\otimes n} = \mathcal{I}_n \cdot \omega_S^{\otimes n}$, where $f\colon Y \to S$ is a resolution), and proving general results about the "relative canonical model" $\mathrm{Proj}_S(\bigoplus \mathcal{I}_n)$. However, as a practical alternative they can be proved case-by-case by performing the α-blow-up on each of the $k=2$ or $k=1$ points, listed in [12] or [13], p. 1290; this amounts to making a projective transformation, one affine piece of which is given by setting

$$k=2:\quad x = z^2x_1,\ y = zy_1,\ z = z,$$
$$k=1:\quad x = z^3x_1,\ y = z^2y_1,\ z = z,$$

and deleting the unwanted factor z^4 or z^6 from the resulting equation. For the $k=2$ points this can also be described as the ordinary blow-up followed by normalisation.

The assertions about the canonical sheaf and the fundamental cycles follow easily from similar results for the minimal resolution (see [12], p.3.4, and compare [13], Lemma 3.12, p. 1268).

COROLLARY (2.10): *To a rational Gorenstein 3-fold point $P \in X$ one can attach a natural number $k \geqq 0$ such that*

$$k = 0 \Leftrightarrow P \in X \text{ is a cDV point} \Leftrightarrow \begin{cases} \text{the general section } H \text{ through} \\ P \text{ has a Du Val point } P \in H; \end{cases}$$

$k \geqq 1$: the general section H through P has an elliptic Gorenstein point $P \in H$ with invariant k. In particular,

(i) *if $k \geqq 2$ then $k = \mathrm{mult}_P X$;*

(ii) *if $k \geqq 3$ then $k+1 =$ minimal embedding dimension $= \dim m_P/m_P^2$.*

If $k=2$ then $P \in X$ is isomorphic to a hypersurface given by $x^2 + f(y,z,t) = 0$, with f a sum of monomials of degree $\geqq 4$; if $k=1$ then $P \in X$ is given by $x^2 + y^3 + f(y,z,t) = 0$, where $f = yf_1(z,t) + f_2(z,t)$ and f_1 (respectively f_2) is a sum of monomials z^at^b of degree $a+b \geqq 4$ (respectively $\geqq 6$).

The next result is a precise form of Theorem 2.2.

THEOREM (2.11): *Let $P \in X$ be a rational Gorenstein point with invariant $k \geqq 1$, and let $\sigma\colon X_1 \to X$ be defined as follows: if $k \geqq 3$, $\sigma\colon X_1 \to X$ is the blow-up of (the reduced point) P. If $k = 2$ or 1, choose coordinates so that $P \in X$ is the hypersurface point in $\mathbf{A}^4$ given by an equation as in the last sentence of Corollary 2.10; let α be the weighting*

$$k=2:\quad \alpha(x) = 2,\ \alpha(y) = \alpha(z) = \alpha(t) = 1$$
$$\text{or}\quad k=1:\quad \alpha(x) = 3,\ \alpha(y) = 2,\ \alpha(z) = \alpha(t) = 1,$$

and let $\sigma\colon X_1 \to X$ be the α-blow-up (see §4).

Then X_1 is normal and Cohen–Macaulay, and $\sigma^*\omega_X = \omega_{X_1}$, so that X_1 is again rational Gorenstein.

PROOF: The blow-up $X_1 \to X$ has a geometric fundamental cycle E_1 which is a Cartier divisor; in case $k \geqq 2$, E_1 is a strong fundamental cycle, because σ dominates the blow-up of m_P. If $k = 1$ the reader can check by writing down the equations of the α-blow-up that E_1 is still a Cartier divisor, although now only a weak geometric fundamental cycle (that is, $\mathcal{O}_{X_1}(-E_1) = (m_P \cdot \mathcal{O}_{X_1})^{**}$).

If $H \subset X$ is a sufficiently general section through P then $\sigma^* H = H_1 + E_1$, where H_1 is a Cartier divisor, and the restriction $\sigma\colon H_1 \to H$ is the standard blow-up of the elliptic Gorenstein point $P \in H$ referred to in Proposition 2.9. Because H_1 is relatively very ample and is itself normal (by Proposition 2.9) it follows that X_1 is normal except possibly at a finite number of points. I now want to prove that X_1 is Cohen–Macaulay at each point of E_1; this is obvious for the $k = 2$ or 1 points, since X_1 remains a hypersurface. For $k \geqq 3$, $P \in X$ is a Gorenstein point having embedding dimension $k + 1$ and multiplicity k; it follows from the main result of Sally [20] that the tangent cone E_1 is Cohen–Macaulay, and hence so is X_1.

The assertion $\sigma^* \omega_X = \omega_{X_1}$ is now a simple consequence of the last paragraph of Proposition 2.9 and the technique of proof used in (I) of Theorem 2.6. The equation $h \in m_P$ of the general section H through P splits locally on X_1 as $h = h_1 \cdot e$, where h_1 defines H_1, and e defines E_1. Now $H_1 \to H$ is the standard blow-up, and the restriction to H_1 of the cycle E_1 is the fundamental cycle referred to in Proposition 2.9. Now let $s \in \omega_X$ be a local generator near P; the generator $\operatorname{Res}_H \left(\frac{s}{h}\right) \in \omega_H$, whcn considered as a rational differential on H_1, generates $\omega_{H_1}(Z_1)$, according to Proposition 2.9. Thus by the adjunction formula $\frac{s}{h}$ must generate $\omega_{X_1}(Z_1 + H_1)$ in a neighbourhood of H_1; but $Z_1 + H_1 = \sigma^* H$ is the divisor defined on X_1 by h, so that s must generate ω_{X_1} in a neighbourhood of H_1. Since H_1 meets every component of $\sigma^{-1}P$, s can have no zeroes on X_1, proving the theorem.

COROLLARY (2.12): *If X is a 3-fold with rational Gorenstein singularities then there exists a partial resolution $f\colon Y \to X$ which is proper and birational, such that*

(i) *f is crepant, $f^* \omega_X = \omega_Y$;*

(ii) *Y has only cDV singularities.*

I do not wish at present to go into the various interesting questions concerned with resolving *cDV* points; for many purposes it seems natural to leave them alone! However, merely the existence of a crepant $Y \to X$ with Y having only hypersurface singularities implies that the local invariant $(-c_2 \cdot \Delta)$ defined in §5 is zero for $P \in X$ rational Gorenstein (see Corollary 5.6).

PROPOSITION (2.13): *Let $P \in X$ be a rational Gorenstein point with invariant $k \geqq 1$; let $T = T_{X,P}$ be the projectivised tangent cone if $k \geqq 3$, or the α-tangent cone if $k = 2$ or 1. Then T is a (generalised)* del Pezzo surface, *in the sense that it satisfies the following host of conditions:*

(i) *T is a 2-dimensional Gorenstein scheme;*

(ii) *the dual invertible sheaf to ω_T is ample, $\omega_T^* = \mathcal{O}_T(1)$;*

(iii) *$h^1(\mathcal{O}_T(m)) = 0$ for all m, and*

$$h^0(\mathcal{O}_T(m)) = \begin{cases} 0 \text{ if } m < 0 \\ 1 + k\binom{m}{2} \text{ for } m \geqq 0; \end{cases}$$

(iv) *form the graded ring $R = R(T, \omega_T^*) = \bigoplus_{m \geqq 0} H^0(\mathcal{O}_T(m))$;*

then if $k \geqq 3$, R is generated by its elements of degree 1. If $k = 2$ (respectively $k = 1$) then

$$R = k[x, y, z, t]/f$$

where x, y, z, t and f have the weights 2, 1, 1, 1 *and* 4 (*respectively* 3, 2, 1, 1 *and* 6).

(v) *the reduced irreducible components of T are projectively normal surfaces of degree $a - 1$ or a in $\mathbf{P}^a$, and in particular are either rational or elliptic ruled surfaces.*

Sketch proof. The affine tangent cone remains Gorenstein according to Sally [20]; then T is Gorenstein with $\omega_T = \mathcal{O}_T(m)$ for some m, as follows from the main theorem of Goto and Watanabe [22]. The fact that $m = -1$ then follows from the adjunction formula: $T \subset X_1$, with $\mathcal{O}_X(T) \otimes \mathcal{O}_T = \mathcal{O}_T(-1)$, and $\omega_{X_1} = \sigma^* \omega_X$, so that $\omega_{X_1} \otimes \mathcal{O}_T = \mathcal{O}_T$.

The remaining assertions depend on similar assertions for the tangent cone to an elliptic Gorenstein surface singularity, which follow by considering the minimal resolution, as in [12] or [13]; in particular it is easily seen that every component of the projectivised tangent cone to an elliptic Gorenstein point is a normal rational or elliptic curve, or a nodal or cuspidal rational curve embedded normally. Q.E.D.

COROLLARY (2.14): *Let $P \in X$ be a rational Gorenstein point; then there exists a resolution $f: Y \to X$ such that $f^{-1}P$ is a union of rational and ruled surfaces.*

For the partial resolution $f: Y \to X$ of Corollary 2.12, $f^{-1}P$ is a union of rational and elliptic ruled surfaces by (v) of Proposition 2.13; but Sing Y may contain curves of positive genus above P.

§3. Toric and quotient singularities

In this section I review some notions of toric geometry, and give criteria for toric varieties to be canonical; for more details of the definitions and properties of differentials on toric varieties see [15]. Toric methods have appeared implicitly in the last section in the form of weighted blow-ups, and they will play a crucial part in §4; a more immediate aim is the proof of the following result, which was suggested by some examples of Shepherd–Barron, who also proved the theorem in a particular case.

THEOREM (3.1): *Let $G \subset GL(n, k)$ be a finite group acting linearly on $\mathbf{A}^n$. Suppose that G has no quasi-reflections, so that the map $\mathbf{A}^n \to \mathbf{A}^n/G = X$ is etale in codimension 1. Then X is canonical if and only if for every element $g \in G$ of order r, and ϵ any primitive rth root of* 1, *the diagonal form of the action of g is*

$$g\colon x_i \longrightarrow \epsilon^{a_i} x_i, \text{ such that } 0 \leqq a_i < r,$$

with $\Sigma\, a_i \geqq r$.

REMARK (3.2): X is Gorenstein if and only if $\Sigma\, a_i \equiv 0 \bmod r$, in which case it is already canonical by Proposition 1.7, (II). (This is a theorem of Watanabe and Khinich, [31] and [32]).

By Remark 1.7, the condition for X to be canonical can be expressed in terms of the ramification of valuations v_δ in the field extension $k(\mathbf{A}^n)/k(X)$; standard ramification theory (see for example [7], p. 284, together with the fact that the ramification group must be cyclic in characteristic 0) then reduces the condition to the cyclic subgroup $R_{v_\delta} \subset G$. Thus for the proof of Theorem 3.1, which I defer to the end of this section (Theorem 3.9), I can assume that G is an Abelian group acting diagonally.

Let $\bar{M} = \mathbf{Z}^n$, and for $m = (m_1, \ldots, m_n) \in \bar{M}$ write x^m for the monomial $x^m = \Pi x_i^{m_i} \in k[\mathbf{A}^n]$. The action of a diagonal group G on $\mathbf{A}^n$ is given by a homomorphism

$$\alpha\colon G \longrightarrow \mathrm{Hom}_{Ab}(\bar{M}, \mathbf{G}_m),$$

so that $g \in G$ acts as $x^m \mapsto \alpha_g(m) \cdot x^m$. The invariant monomials are x^m, with $m \in M$, where $M \subset \bar{M}$ is the sublattice of finite index

$$M = \bigcap_g \mathrm{Ker}\, \alpha_g \subset \bar{M}.$$

Let $\sigma \subset M_{\mathbf{R}}$ be the first quadrant $\sigma = \{m \mid m_i \geqq 0 \text{ for each } i\}$. Then $\mathbf{A}^n = \mathrm{Spec}\, k[\sigma \cap \bar{M}]$, and $X = \mathrm{Spec}\, k[\sigma \cap M]$, with $\mathbf{A}^n \to X$ corresponding to the inclusion of the exponent semigroups $\sigma \cap M \subset \sigma \cap \bar{M}$.

Quotients $\mathbf{A}^n/G$ by an Abelian group acting without quasi-reflections correspond precisely to simplicial toric varieties: if $\sigma \subset M_{\mathbf{R}}$ is a rational simplicial cone, then there exists a unique overlattice $\bar{M} \supset M$ such that $\sigma = \langle e_1, \ldots, e_n \rangle$, with $\{e_i\}$ a basis of $\bar{M}$, and such that the following condition holds:

$$\text{for every } i,\ \bar{M} = M + \sum_{j \neq i} \mathbf{Z} \cdot e_j \qquad (*)$$

this condition corresponds to the fact that G has no quasi-reflections.

Now let M be any lattice of rank n, and let $\sigma \subset M_{\mathbf{R}}$ be a cone spanning $M_{\mathbf{R}}$; set $X = \operatorname{Spec} k[\sigma \cap M]$. X contains a big torus, $\mathbf{T} \subset X$, with $\mathbf{T} = \operatorname{Spec} k[M] \cong \mathbf{G}_m^n$. For every wall $\tau \subset \sigma$, $\sigma - \tau$ is the half-space of $M_{\mathbf{R}}$ containing σ and bounded by τ, and the corresponding variety $X^\tau = \operatorname{Spec} k[(\sigma - \tau) \cap M]$ is isomorphic to $\mathbf{A}^1 \times \mathbf{G}_m^{n-1}$, with $\mathbf{T} \subset X^\tau \subset X$. The complement $X \backslash \mathbf{T}$ is a union of prime divisors Γ_τ, and the generic point of Γ_τ corresponds to the last remaining coordinate hyperplane in $X^\tau \cong \mathbf{A}^1 \times \mathbf{G}_m^{n-1}$. Thus to check the regularity of a differential on X it is sufficient to know it on $\mathbf{T}$, and to check at each prime divisor $\Gamma_\tau \subset X^\tau$.

For $\{m_1, \ldots, m_n\} \subset M$ a linearly independent set, the rational differential

$$s = \frac{\pm 1}{[M : \bigoplus \mathbf{Z} \cdot m_i]} \cdot \frac{\mathrm{d}x^{m_1}}{x^{m_1}} \wedge \cdots \wedge \frac{\mathrm{d}x^{m_n}}{x^{m_n}}$$

does not depend on the choice of $\{m_1, \ldots, m_n\}$, and is a generator of the $\mathcal{O}_{\mathbf{T}}$-module $\omega_{\mathbf{T}}$.

LEMMA (3.3): *$x^m s^r \in \Gamma(X, \omega_X^{[r]})$ if and only if for every $\tau \in \operatorname{Walls}(\sigma)$ we have*

$$m \in r \cdot \operatorname{Int}((\sigma - \tau) \cap M).$$

PROOF: If I set $M_2 = \operatorname{Span}(\tau) \cap M$, and let $m_1 \in (\sigma - \tau) \cap M$ be a complementary element, then the semigroup $(\sigma - \tau) \cap M$ decomposes as $\langle m_1 \rangle \times M_2$. The discrete valuation ring $\mathcal{O}_{X, \Gamma_\tau}$ then has x^{m_1} as a local parameter, and x^m is a unit for $m \in M_2$. Thus taking a basis $m_2, \ldots, m_n$ of M_2, I can write

$$s = \frac{\mathrm{d}x^{m_1}}{x^{m_1}} \wedge \cdots \wedge \frac{\mathrm{d}x^{m_n}}{x^{m_n}};$$

thus $x^m s^r$ is regular along Γ_τ if and only if $m - r m_1 \in (\sigma - \tau)$. Q.E.D.

Write

$$r\text{-}\operatorname{Int}(\sigma \cap M) = \bigcap_{\tau \in \operatorname{Walls}(\sigma)} r \cdot \operatorname{Int}((\sigma - \tau) \cap M);$$

note that for $r = 1$, $1\text{-}\operatorname{Int}(\sigma \cap M) = \operatorname{Int}(\sigma) \cap M$. Let $\mathfrak{A}^{[r]} \subset k[\sigma \cap M]$ be the ideal generated by x^m, $m \in r\text{-}\operatorname{Int}(\sigma \cap M)$. Then the map

$$\mathfrak{A}^{[r]} \longrightarrow \Gamma(X, \omega_X^{[r]})$$

given by $x^m \longrightarrow x^m s^r$

is an isomorphism. Compare Danilov [15], §4.

COROLLARY (3.4): *$\omega_X^{[r]}$ is locally free if and only if the semigroup ideal $r\text{-}\operatorname{Int}(\sigma \cap M) \subset \sigma \cap M$ is principal.*

Let $\sigma' = \langle f_1, \ldots, f_n \rangle$ be any basic cone with $\sigma' \supset \sigma$; then $X' = \operatorname{Spec} k[\sigma' \cap M] \cong \mathbf{A}^n$, and has a birational morphism $X' \to X$ defined by the inclusion $\sigma \cap M \subset \sigma' \cap M$. A resolution $f: Y \to X$ can be made by glueing together such affine constructions as σ' runs through a fan of cones (see [15], II, §8). Since for a basic cone $\sigma' = \langle f_1, \ldots, f_n \rangle$, $r\text{-Int}(\sigma' \cap M)$ is the principal ideal generated by $r(f_1 + \cdots + f_n)$, I get the following result.

COROLLARY (3.5): *X satisfies the condition* (ii′) *of Remark* 1.3 *if and only if, for every* $r \geqq 1$, *and for every basic cone* $\sigma' = \langle f_1, \ldots, f_n \rangle$ *with* $\sigma' \supset \sigma$ *we have*

$$r\text{-Int}(\sigma \cap M) \subset r\text{-Int}(\sigma' \cap M) = r(f_1 + \cdots + f_n) + \sigma' \cap M.$$

Since for $r = 1$ this amounts to $\operatorname{Int}(\sigma \cap M) \subset \operatorname{Int}(\sigma' \cap M)$, which is trivially satisfied, the next result follows.

COROLLARY (3.6): *If X is toric then for every proper birational morphism* $f: Y \to X$, $f_* \omega_Y = \omega_X$. *In particular if X is Gorenstein then it is canonical.*

This also follows from the fact that toric varieties are Cohen–Macaulay ([15], §3) and rational ([15], §8) by using Kempf's duality argument ([6], p. 50).

Now assume that σ is simplicial[1], and let $M \subset \bar{M} = \mathbf{Z}^n$ be the overlattice in which σ becomes basic, $\sigma = \langle e_1, \ldots, e_n \rangle$, with $\{e_i\}$ a basis of $\bar{M}$ and condition (*) satisfied.

LEMMA (3.7): $\omega_X^{[r]}$ *is locally free if and only if* $r(e_1 + \cdots + e_n) \in M$.

PROOF: σ has walls τ_i given by $m_i = 0$. Furthermore, according to (*) each $\operatorname{Int}((\sigma - \tau_i) \cap M)$ contains an element $e_i + \sum_{j \neq i} a_j e_j$; by adding an element of $\tau_i \cap M$ I can even assume that each $a_j \geqq N$ for any chosen $N \in \mathbf{Z}$, so that for each i, $r\text{-Int}(\sigma \cap M)$ contains a vector $re_i + \sum_{j \neq i} b_j e_j$. If $m = \Sigma m_i e_i$ is a generator for the ideal $r\text{-Int}(\sigma \cap M)$ it follows that each $m_i \leqq r$; the inequality the other way is trivial, so that the only possible single generator of $r\text{-Int}(\sigma \cap M)$ is $r(e_1 + \cdots + e_n)$. Q.E.D.

---

[1] The general case is covered by the following result, kindly communicated by Danilov.

THEOREM: Let M be a lattice, $\sigma \subset M_{\mathbf{R}}$ a cone, and set $X = \operatorname{Spec} k[\sigma \cap M]$. Let N be the dual lattice, and $\check{\sigma} \subset N_{\mathbf{R}}$ the dual cone; write $1\text{-}Sk\,\check{\sigma}$ for the set of primitive $\alpha \in \check{\sigma} \cap N$ such that $\tau_\alpha = (\alpha = 0)$ is a wall of σ. Then

(I) $\omega_X^{[r]}$ is locally free for some r if and only if there exists $f \in M_{\mathbf{Q}}$ such that, considering f as a linear map $f: N \to \mathbf{Q}$,

$$f_{|1\text{-}Sk\check{\sigma}} = 1;$$

(II) X is canonical if and only if furthermore

$$f_{|\check{\sigma} \cap N \setminus \{0\}} \geqq 1.$$

The proof is similar to that of 3.7–3.9.

COROLLARY (3.8): *Suppose that $\omega_X^{[r]}$ is locally free; then X is canonical if and only if, for every basic cone $\sigma' = \langle f_1, \ldots, f_n \rangle \supset \sigma$, with $\{f_i\}$ a basis of M, we have*

$$(e_1 + \cdots + e_n) - (f_1 + \cdots + f_n) \in \sigma'.$$

PROOF: The condition in Corollary 3.5 can be rewritten

$$r(e_1 + \cdots + e_n) + \sigma \cap M \subset r(f_1 + \cdots + f_n) + \sigma' \cap M. \qquad \text{Q.E.D.}$$

Now let $N = M^*$ be the lattice dual to M; N consists of linear forms $\alpha(m) = \Sigma\, q_i m_i$, with $q_i \in \mathbf{Q}$ and $\alpha(m) \in \mathbf{Z}$ for every $m \in M$. Dual to σ we have the positive quadrant $\check{\sigma} = \{\alpha \mid q_i \geqq 0 \text{ for each } i\}$.

The following criterion is equivalent to Theorem 3.1 for Abelian G.

THEOREM (3.9): *X is canonical if and only if for every non-zero $\alpha \in \check{\sigma} \cap N$ we have $\alpha(e_1 + \cdots + e_n) = \Sigma\, q_i \geqq 1$.*

Of course this condition need only be tested on primitive α in the unit cube.

PROOF: Given any primitive vector $\alpha \in \check{\sigma} \cap N$, I can extend it to a basis $\alpha = f_1^*, \ldots, f_n^*$ of N lying in $\check{\sigma}$. The dual basis $f_1, \ldots, f_n$ spans a basic cone $\sigma' \supset \sigma$, and every such basic cone σ' arises in this way. But $\Sigma\, e_i - \Sigma\, f_i \in \sigma'$ is the assertion that for each i we have $f_i^*(\Sigma\, e_j - \Sigma\, f_j) \geqq 0$; that is, $f_i^*(e_1 + \cdots + e_n) \geqq 1$. Q.E.D.

EXAMPLE (3.10): The "Shepherd–Barron node"[1] $X_r = \mathbf{A}^3/\boldsymbol{\mu}_r$, where $\rho \in \boldsymbol{\mu}_r$ acts by

$$(x, y, z) \longmapsto (\rho x, \rho y, \rho^{r-1} z),$$

is a canonical singularity of index r; for $\rho = \epsilon^k$ acts with eigenvalues ϵ^k, ϵ^k, ϵ^{r-k}, and $k + k + (r - k) \geqq r$.

These singularities actually occur as the only singularities of a general weighted hypersurface $X_{dr(r-1)} \subset \mathbf{P}(1, 1, r-1, r, r)$; (see [17] for the techniques needed to justify this assertion). This is a 3-fold with canonical singularities, and $\omega_X = \mathcal{O}_X(k)$, with $k = dr^2 - dr - 3r - 1$; if $k \geqq 1$, X is a canonical 3-fold; we have

$$K_X^3 = dr(r-1)k^3/r^2(r-1) = dk^3/r,$$

and since $k \equiv -1 \bmod r$, the invariant K_X^3 defined in §5 is a rational number which can have arbitrary denominator.

PROBLEM (3.11): Give necessary and sufficient combinatorial conditions on a sequence of integers $(b_1, \ldots, b_k; a_1, \ldots, a_{k+n+1})$ for the general weighted

---

[1]Considered independently by J. Wahl.

complete intersection

$$X_{b_1,\ldots,b_k} \subset \mathbf{P}(a_1, \ldots, a_{k+n+1})$$

to have canonical singularities. This condition might resemble (***) in Theorem 4.5.

§4. Hypersurfaces and quasi-homogeneity

Let $X \subset \mathbf{A}^n$ be a hypersurface, $P \in X$, and let $x_1, \ldots, x_n$ be analytic coordinates on $\mathbf{A}^n$ around P; near P, $X \subset \mathbf{A}^n$ is given by an equation $g = g(x)$.

I will use the following notations: $M = \mathbf{Z}^n$, with $\{e_i\}$ the natural basis; $m \in M$ corresponds to, and is sometimes identified with, the monomial x^m, with $x^{e_i} = x_i$. The first quadrant is $\sigma \subset M_{\mathbf{R}}$, N is the dual lattice to M, $\check{\sigma} \subset N_{\mathbf{R}}$ is the dual first quadrant. For $m = \Sigma\, m_i e_i \in M$ and $\alpha = (q_1, \ldots, q_n) \in N_{\mathbf{Q}}$, $\alpha(m) = \Sigma\, m_i q_i$. I will abuse the notation by writing $\alpha(x^m) = \alpha(m)$ for a monomial x^m, and extend the definition of α to the whole of $k[x_1, \ldots, x_n]$ by setting, for $g = \Sigma\, a_m x^m$,

$$\alpha(g) = \inf\{\alpha(m) \mid a_m \neq 0\}.$$

For example, if $\alpha = e_i^*$ then $\alpha(g) = 1$ if and only if g vanishes along the coordinate hyperplane $x_i = 0$ to multiplicity 1.

THEOREM (4.1): *The following is a necessary condition for X to have canonical singularities:*

$$(*)\qquad \begin{cases} \textit{for all } P \in X, \\ \textit{for all analytic coordinates } x_1, \ldots, x_n \textit{ around } P, \\ \textit{for all } \alpha \in \check{\sigma} \cap N_{\mathbf{Q}}, \textit{ with } \alpha \neq q e_i^*, \\ \alpha\left(\dfrac{x_1 \ldots x_n}{g}\right) = \sum q_i - \alpha(g) > 0. \end{cases}$$

It is often useful to make an obvious normalisation, and to assume that $\Sigma\, q_i = 1$; I will occasionally assume without warning that $q_1 \geqq \cdots \geqq q_n$.

CONJECTURE (4.2): *The condition (*) in Theorem 4.1 is also sufficient, that is*

$$\left.\begin{matrix} P \in X \textit{ is} \\ \textit{non-rational} \end{matrix}\right\} \Leftrightarrow \begin{cases} \textit{there exist analytic coordinates,} \\ \textit{there exists } \alpha \in \check{\sigma} \cap N_{\mathbf{Q}},\ \alpha \neq e_i^*, \textit{ with } \alpha(e_1 + \cdots + e_n) = 1, \\ \textit{such that } \alpha(g) \geqq 1. \end{cases}$$

The condition certainly implies that X is "naïvely canonical" in the sense that X has multiplicity $\operatorname{mult}_Y X < r$ along every subvariety $Y \subset X$ of dimension $\dim Y = n - r$ (for any $r \geqq 2$); for at a general point of Y, Y can be given by $x_1 = \cdots = x_r = 0$, and setting $\alpha = \frac{1}{r}(e_1^* + \cdots + e_r^*)$, the condition $\alpha(g) < 1$ holds if and only if $\mathrm{mult}_Y X < r$.

The next result is a feeble approximation[1] to Conjecture 4.2.

PROPOSITION (4.3): *The hypersurface $X \subset \mathbf{A}^n$ defined by $g = \sum_{i=1}^{n} x_i^{a_i} = 0$ has a canonical singularity at 0 if and only if $\sum \frac{1}{a_i} > 1$.*

The proofs of Theorem 4.1 and Proposition 4.3 are both based on the notion of weighted blow-up, which is a particular case of the toric morphism defined by a subdivision of a fan (see [15], §5). Let $\alpha \in \check{\sigma} \cap N_{\mathbf{Q}}$, and let d be the least denominator of α, so that $d\alpha \in N$. For each $i = 1, \ldots, n$ (later for clarity I will take $i = 1$) $\sigma_i \subset M_{\mathbf{R}}$ denotes the cone

$$\sigma_i = \{m \mid \alpha(m) \geqq 0, \text{ and } m_j \geqq 0 \text{ for each } j \neq i\}.$$

If $\alpha = e_i^*$ then $\sigma_i = \sigma$, so that the construction will be trivial. Set $Z_i = \operatorname{Spec} k[\sigma_i \cap M]$, and let $\varphi_i\colon Z_i \to \mathbf{A}^n$ be the birational map corresponding to $\sigma \cap M \subset \sigma_i \cap M$; for $i \leqq 1 \leqq n$, the φ_i glue together into a projective morphism $\varphi\colon Z \to \mathbf{A}^n$, the α-blow-up of $\mathbf{A}^n$.

It is easy to give a toric description of the weighted projective space $\mathbf{P}(\alpha)$, and to check that Z is none other than the normalised graph of the rational map $\mathbf{A}^n \dashrightarrow \mathbf{P}(\alpha)$ which makes $\mathbf{A}^n \backslash \{0\}$ into a $\mathbf{G}_m$-bundle.

Write $E \subset Z$ for the exceptional locus $E = \varphi^{-1}0$ ($\varphi\colon E \to \mathbf{P}(\alpha)$ is a finite cover); in Z_1, $E_1 = E \cap Z_1$ is the stratum of Z_1 corresponding to the wall $\tau_1^{(0)} \subset \sigma_1$ given by $\alpha = 0$ (see [15], §2.5). If $m_0 \in M$ is such that $d\alpha(m_0) = 1$ then x^{m_0} is a local parameter of the discrete valuation ring $\mathcal{O}_{Z,E}$, and $v_E(x_i) = d\alpha(e_i)$. Note that the reduced E is not necessarily a Cartier divisor, although in Z_1, x_1 is a local equation for $d\alpha(x_1) \cdot E_1$.

Z contains 2 kinds of coordinate hyperplanes: the proper transforms $(x_i = 0) \subset \mathbf{A}^n$, which is $Z \backslash Z_i$, and E itself; the intersection $\bigcap_i Z_i = Z_{(0)}$ is a neighbourhood of the generic point of E, and any monomial x^m with $\alpha(m) = 0$ becomes a unit when restricted to $Z_{(0)}$.

Now let $d\alpha(g) = c$, and suppose that X is irreducible and not contained in any coordinate hyperplane of $\mathbf{A}^n$; if I write $g = (x^{m_0})^c g'$ then g' is a unit in $\mathcal{O}_{Z,E}$, and defines the proper transform X' of X in $Z_{(0)}$:

$$\varphi^* X = X' + cE.$$

X' will in general not be a Cartier divisor on the whole of Z.

PROOF OF THEOREM 4.1: I will assume that $\alpha(e_1 + \cdots + e_n) = 1$ but

[1] The method of proof given here also proves the conjecture if g is non-degenerate with respect to its Newton polygon, in the sense of Koushnirenko [38]; compare [37], Theorem 2.3.1.

$\alpha(g) \geqq 1$, that is $c \geqq d$, and deduce that X is not canonical. Let s be the usual basis of $\omega_{\mathbf{T}}$ (as in §3), so that $\omega_{\mathbf{A}^n}$ is based by $x_1 \ldots x_n \cdot s$ and $t = g^{-1} \cdot x_1 \ldots x_n \cdot s$ generates $\omega_{\mathbf{A}^n}(X)$. If I show that $\varphi^* t$ has E as a pole on Z when considered as a rational section of $\omega_Z(X')$, then X is not canonical; indeed, t will be of the form $t = u \cdot v^{-1}$, where $u \in \omega_Z(X')$ is a basis and $v \in \mathcal{O}_Z$ has a zero along E. The Poincaré residue of t is then a product of $\mathrm{Res}_{X'} u$, which bases $\omega_{X'}$ by the adjunction formula ([9], p. 7), and the restriction of v^{-1}, where $v \in \mathcal{O}_{X'}$ is a non-unit; note that this argument does not assume that X' is normal.

ω_Z is generated near E by $x^{m_0} \cdot s$, so that $x_1 \ldots x_n \cdot s$, considered as a differential on Z, has divisor of zeroes eE, where $e = d\alpha(x_1 \ldots x_n) - 1 = d - 1$; hence it generates $\omega_Z(-eE)$, and $\varphi^* t$ generates $\omega_Z(\varphi^* X - eE) = \omega_Z(X' + (c-e)E)$. Under the hypothesis $c \geqq d$,

$$c - e = c - (d-1) > 0,$$

showing that $\varphi^* t$ has a pole. Q.E.D.

PROOF OF PROPOSITION 4.3: Let $\alpha = \sum \frac{1}{a_i} e_i^*$, and carry out the α-blow-up as above. It will follow from Lemma 4.4 below that X' has the following two virtues:

(i) X' is normal;

(ii) there exists a resolution $f\colon Y \to X'$ with $f_* \omega_Y = \omega_{X'}$.

In view of (i) and the computation in the proof of Theorem 4.1, the generator of ω_X, which is the Poincaré residue of t, lifts to a regular differential on X'. Combining this with (ii), X is canonical. It remains to prove the following result.

LEMMA (4.4): *X' is toroidal.*

PROOF: For each $j \neq 1$ (for clarity I will later take $j = 2$), write

$$a_1 = b_j \gamma_j \text{ and } a_j = c_j \gamma_j,$$

with b_j and c_j coprime integers.

Write $z_j = x_1^{-b_j} x_j^{c_j}$, which is a coordinate function on Z_1. On Z_1, X' is given by

$$x_1^{-a_1} \cdot g = g' = 1 + \sum_{j \neq 1} z_j^{\gamma_j}.$$

It follows that at each point of X' one of the z_i is non-zero, say $z_2 \neq 0$.

Let $Z_{1;2} = \{P \in Z_1 \mid z_2 \neq 0\}$, and $X_2' = X' \cap Z_{1;2}$ be a typical piece of X'; then there is a decomposition $M = M_1 \times M_2$, with M_1 the 1-dimensional lattice based by $(-b_2, c_2, 0, \ldots, 0)$, and M_2 a complementary lattice, which induces a decomposition

$$Z_{1;2} \cong \mathbf{G}_m \times Y,$$

with Y the toric variety corresponding to $\sigma_1 \cap M_2$; z_2 is the coordinate in $\mathbf{G}_m$. Now $X_2' \subset Z_{1;2}$ is given by the equation

$$z_2^{\gamma_2} = 1 - \sum_{j \geqq 3} z_j^{\gamma_j},$$

and since on this piece z_2 never vanishes it follows that the restriction of the second projection $Z_{1;2} \to Y$ is an etale morphism $X_2 \to Y$. Q.E.D.

There is no doubt that Conjecture 4.2 is true, at least in the case $n = 4$. Here are two (related) ideas for its proof. Firstly one can make a list (in hierarchical order) of all g which satisfy (*), and show that this list satisfies the inductive property analogous to Theorem 2.6. Indeed, it is easily seen that for any g satisfying (*) and not defining a cDV point, one of the α-blow-ups used in the proof of Theorem 2.6 is appropriate ($\alpha = (1, 1, 1, 1)$ or $(2, 1, 1, 1)$ or $(3, 2, 1, 1)$), and leads to a variety X' which again has only hypersurface singularities; to prove the conjecture one has to show that for any g in our list the singularities of X' are either cDV points or points occurring earlier in the list. Although this is a perfectly feasible program, I have only scratched the surface; apart from the fact that the effort involved in making the list seems to be about 10 times that required for the analogous lists of elliptic surface singularities (see the tables in [12] and [13]), a more serious difficulty is that there do not seem to be any checks to eliminate errors—in the surface case the equations of the singularities and the shape of the resolution (a configuration of curves on a non-singular surface) both fit into nicely controllable hierarchical patterns.

The second possible proof of Conjecture 4.2 is to try to prove directly that in making the appropriate α-blow-up $X' \to X$, condition (*) for X implies (*) for X'; for some fixed set of coordinates on X, and the coordinates on X' resulting from the toric description of the α-blow-up, this is trivial. What is therefore required is some theoretical understanding of which analytic changes of coordinates on X' are relevant to (*), and which of these come from changes of coordinates on X.

I conclude this section with a discussion of the combinatoric condition in (*). This condition is formally similar to the numerical condition for stability of a hypersurface $X \subset \mathbf{P}^n$ of given degree under the action of $PGL(n)$ (see [23], pp. 48 and 80, and also [24]). As in that theory, it should be possible, for combinatorial reasons, to write down a *finite* set $\{\alpha_i\}_{i \in I}$ of $\alpha_i \in \check{\sigma} \cap N_{\mathbf{Q}}$ with $\alpha_i(e_1 + \cdots + e_n) = 1$ which have the same effect as *all* α, that is:

$$(**) \quad \begin{cases} \text{for all } \alpha \in \check{\sigma} \cap N_{\mathbf{Q}} \text{ with } \alpha(e_1 + \cdots + e_n) = 1 \\ \text{there exists an } i \in I \text{ such that} \\ \alpha(m) \geqq 1 \Rightarrow \alpha_i(m) \geqq 1 \text{ for all } m \in \sigma \cap M. \end{cases}$$

For $n = 2$ and 3 this blessèd purpose is accomplished by the sets

$$A_2 = \{\tfrac{1}{2}(1, 1)\}.$$

and $$A_3 = \{\tfrac{1}{3}(1, 1, 1), \tfrac{1}{4}(2, 1, 1), \tfrac{1}{6}(3, 2, 1)\} \cup \{\tfrac{1}{2}(1, 1, 0)\};$$

this explains the significance of these weightings for surface rational and elliptic singularities; the first component of A_3 is easily characterised as the set of solutions of $\sum_{i=1}^{3} \frac{1}{a_i} = 1$.

THEOREM (4.5): *For $n = 4$, the following set A_4 satisfies (**):*

$$A_4 = A_4' \cup \{(q_1, q_2, q_3, 0) \mid (q_1, q_2, q_3) \in A_3\},$$

where $\quad A_4' = \{\alpha \in \check{\sigma} \cap N_{\mathbf{Q}} \mid \alpha(e_1 + \cdots + e_4) = 1,$ *and (***) holds*$\},$

where (***) is the condition

(***) $\begin{cases} \textit{for each } i \textit{ there is a monomial } x_i^{a_i} \cdot r_i \textit{ with } \alpha(x_i^{a_i} \cdot r_i) = 1, \\ \textit{with } a_i \in \mathbf{Z},\ a_i \geqq 2, \textit{ and with } r_i \textit{ a monomial of degree} \leqq 1; \\ \textit{and for each } j \textit{ not all of these are divisible by } x_j. \end{cases}$

In terms of the geometry of the lattice, (***) means that the tetrahedron $(\alpha = 1) \cap \sigma \subset M_{\mathbf{R}}$ has "near-vertices" $x_i^{a_i}$ or $x_i^{a_i} x_j$.

The idea of the proof is that if $\gamma \in \check{\sigma} \cap N_{\mathbf{Q}}$ also has $\gamma(e_1 + \cdots + e_n) = 1$, and if there are *no* monomials with $\alpha(x_i^{a_i} \cdot r) = 1$ and with $\gamma(r) < 1$, then I can replace α by $\alpha' = (1 - \lambda)\alpha + \lambda\gamma$ (with small λ), and have

$$\alpha(m) \geqq 1 \Rightarrow \alpha'(m) \geqq 1; \qquad (0)$$

as λ increases we eventually acquire a new monomial $x_i^{a_i} \cdot r$ with $\alpha'(x_i^{a_i} r) = 1$, $\gamma(r) < 1$. Repeating this with a different γ satisfying $\gamma(x_i^{a_i} r) = 1$, we eventually get an α' still satisfying (0), but now forced to have an assortment of solutions of $\alpha'(x_i^{a_i} \cdot r) = 1$ for certain very restricted r, which forces α' to have a solution with $\deg(r) \leqq 1$.

A_4' is a finite set, since the a_i satisfy $a_i q_i \leqq 1$, and hence

$$1 = \sum q_i \leqq \sum \frac{1}{a_i},$$

and for each solution of this inequality (essentially a finite set) there are at most 4^4 $(= 256)$ hyperplanes in $M_{\mathbf{R}}$ spanned by some choice of the monomials $x_i^{a_i}$ or $x_i^{a_i} x_j$, one for each i; some of these hyperplanes will of course fail to pass through $(1, 1, 1, 1)$. A_4' seems to be quite large, with (apparently) 95 elements.

For $\alpha \in A_4'$ I can write $\alpha = \frac{1}{d}(b_1, b_2, b_3, b_4)$, with $b_i \in \mathbf{Z}$ and d the least common denominator of the $q_i = b_i/d$; the condition on α guarantees that a sufficiently general hypersurface

$$X_d \subset \mathbf{P}(b_1, b_2, b_3, b_4),\ \alpha \in A_4'$$

has singularities "not much worse" than those of $\mathbf{P}$ itself (see [17]), and ensures that X_d is a $K3$ surface with at worst Du Val points. A_4' should thus also provide a complete list of all weighted hypersurfaces which are $K3$ surfaces; these can be constructed from a non-singular $K3$ surface S, together with a rational divisor class $h \in \operatorname{Pic} S \otimes \mathbf{Q}$, by a construction due to Demazure [25]. These projective surfaces have corresponding affine cones, the general weighted hypersurface of weight α, which correspond to "simple elliptic" 3-fold singularities.

Thus Conjecture 4.2 implies that one of the beautiful features of the hierarchy of surface singularities carries over in some form to higher dimensions: lurking on the fringe of the rational singularities there are simple elliptic ones.

Examples (4.6): $\frac{1}{42}(21, 14, 6, 1)$; $\frac{1}{5}(2, 1, 1, 1)$; $\frac{1}{42}(21, 14, 4, 3)$; $\frac{1}{14}(7, 4, 2, 1)$.

$X_{42} \subset \mathbf{P}(21, 14, 6, 1)$ has Du Val points A_6, A_2 and A_1 at the transverse intersection of X_{42} with the 1-dimensional singular strata of $\mathbf{P}$. On the $K3$ resolution, t^{42} defines the following divisor (all the components are rational non-singular with self-intersection -2):

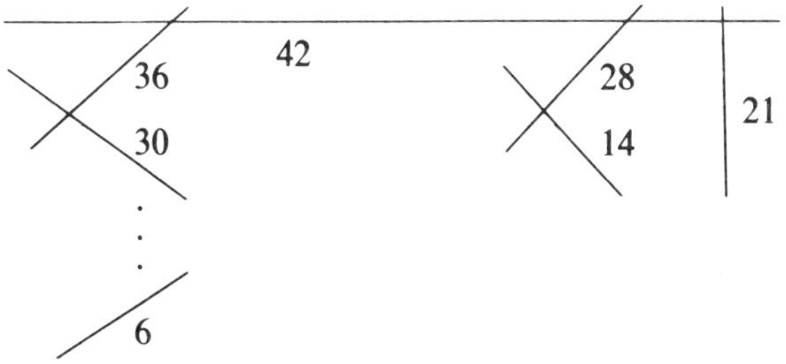

$X_5 \subset \mathbf{P}(2, 1, 1, 1)$ is a divisor in the cone on the Veronese $\mathbf{P}$, having a simple node at the vertex; this example occurs in Saint–Donat [26].

I have included the last two examples to show that things can get quite complicated: the hypersurface $X_{42} \subset \mathbf{P}(21, 14, 4, 3)$ given by $x^2 + y^3 + yz^7 + z^9t^2 + t^{14}$ has the following singularities:

$$\left.\begin{array}{ll} z = t = 0 & 1 \times A_6 \\ y = z = 0 & 2 \times A_2 \\ x = t = 0 & 1 \times A_1 \end{array}\right\} \text{transverse intersections of singularities of } \mathbf{P}$$

and an A_3 point at $(0, 0, 1, 0)$. The corresponding desingularisation and Demazure divisor is:

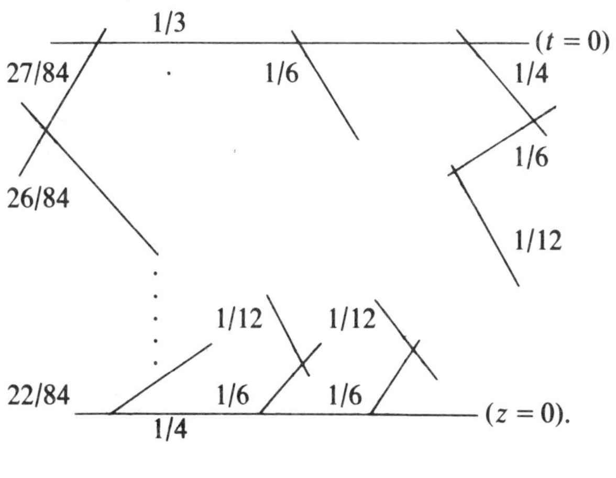

§5. The plurigenus formula

Let X be a canonical 3-fold of index r. The fact that $\omega_X^{[r]}$ is ample, together with the easy Corollary 1.14, allows us to construct a resolution $f\colon Y \to X$ satisfying the following two mild conditions.

DEFINITION (5.1): A resolution $f\colon Y \to X$ is *0-minimal* if $f^*\omega_X^{[r]} = \omega_Y^{\otimes r}(-\Delta_r)$, with $\Delta_r \geqq 0$ a divisor on Y such that $f(\Delta_r) \subset X$ is a finite set.

DEFINITION (5.2): A resolution $f\colon Y \to X$ is *elegant* if for $s = 1, \ldots, r-1$ (hence for all s) the subsheaf $f'\omega_X^{[s]} = \operatorname{Im} f^*\omega_X^{[s]} \subset \omega_Y^{\otimes s}$ is invertible.

REMARK (5.3): For a torsion-free sheaf $\mathcal{F}$ on X, $f'\mathcal{F} = f^*\mathcal{F}/\text{Torsion}$ is the sheaf denoted $\mathcal{F}\circ f$ by Grauert and Riemenschneider ([27], p. 267), who pointed out that in general torsion turns up in taking the sheaf-of-$\mathcal{O}$-module theoretical f^*, defined by setting the stalk $(f^*\mathcal{F})_P = \mathcal{F}_{fP} \otimes_{\mathcal{O}_{X,fP}} \mathcal{O}_{Y,P}$ for $P \in Y$. There is of course no problem in taking f^* if $\mathcal{F}$ is locally free.

For an elegant resolution, it follows that $f'\omega_X^{[n]} = \omega_Y^{\otimes n}(-\Delta_n)$ for each $n \geqq 0$, with $\Delta_n \geqq 0$ of the form

$$\Delta_n = m\Delta_r + \Delta_i,$$

where $n = mr + i$, $0 \leq i \leq r-1$.

Since $\omega_X^{[r]}$ is invertible and ample, $\omega_X^{[mr+i]} = \omega^{[i]} \otimes (\omega^{[r]})^{\otimes m}$ is generated by its global sections for all sufficiently large m; the same is therefore true of $f'\omega_X^{[mr+i]}$, so that elegance is equivalent to demanding that for all sufficiently large n

$$|nK_Y| = |D_n| + \Delta_n,$$

with Δ_n fixed and $|D_n|$ free.

PROPOSITION (5.4): *X has an elegant 0-minimal resolution $f\colon Y \to X$.*

PROOF: Both of these conditions are very easy to satisfy. Firstly, in order that Y be elegant it is necessary and sufficient that Y dominates each of the blow-ups of the divisorial sheaves $\omega_X^{[i]}$ for $i = 1, \ldots, r-1$. By the blow-up of a divisorial sheaf $\mathcal{L}$ is intended the following: express $\mathcal{L} \cong \mathcal{O}_X(D)$ in the form $\mathcal{O}_X(D) = \mathcal{I}_{E-D} \cdot \mathcal{O}_X(E)$, where $\mathcal{O}_X(E)$ is invertible, and $\mathcal{I}_{E-D}$ is a divisorial ideal sheaf (as in the Appendix to § 1); the blow-up of $\mathcal{L}$ is the blow-up of the sheaf of ideals $\mathcal{I}_{E-D}$, which is obviously independent of the choice of D and E. Since each of the $\omega_X^{[i]}$ is invertible outside finitely many dissident points of X, the condition that Y dominates the blow-up of each of them does not affect zero-minimality.

A 0-minimal resolution can be obtained by any sequence of steps $Y = X_n \to \cdots \to X_0 = X$ which leads to a non-singular Y, such that each step $s_i: X_{i+1} \to X_i$ satisfies one of the following two conditions:

(i) s_i is an isomorphism above all but a finite number of points of X;

(ii) s_i is the blow-up with centre $C_i \subset \operatorname{Sing} X_i$ a reduced curve C_i which lies over a 1-dimensional component of the singular locus of X. In case (ii) s_i is necessarily a blow-up of a curve of singularities which is generically (Du Val point) $\times \mathbf{A}^1$; s_i will be crepant (Definition 2.1) outside a finite number of points of X_i. Q.E.D.

THEOREM (5.5): *The following formula for the plurigenera of Y holds for all $n \equiv 1 \bmod r$, $n \geqq 2$, and for all sufficiently large n:*

$$P_n(Y) = P_n(X) = \frac{K_X^3}{12} \cdot (2n-1)n(n-1) - (2n-1)\chi(\mathcal{O}_X) + \ell(n). \qquad (*)$$

Here $\ell(n) \in \mathbf{Q}$ is linear with periodic adjustments, and $K_X^3 \in \mathbf{Q}$ are invariants of X defined by their appearance in (*); K_X^3 is also determined by $r^3 \cdot K_X^3 = (rK_X)^3$, where the right-hand side makes sense because rK_X is a Cartier divisor.

It has already been pointed out in Example 3.10 that K_X^3 can have arbitrary denominator. Further information on the invariants appearing in (*), and a discussion of its significance, will be given after the proof.

PROOF: If $r \mid n$ then $f'\omega_X^{[n]} = f^*\omega_X^{[n]}$ is the inverse image of an ample sheaf under a birational morphism, and so is quasi-positive ([27], p. 265); if $n \gg 0$ then as already observed $f'\omega_X^{[n]}$ is generated by its global sections, and taking n bigger still it defines a birational map. Thus by the vanishing theorem of Grauert and Riemenschneider ([27], p. 273),

$$H^p(Y, f'\omega_X^{[n]} \otimes \omega_Y) = 0$$

for all $p > 0$, and for all n with $r \mid n$ or $n \gg 0$.

Thus $P_{n+1}(Y) = \chi(Y, f'\omega_X^{[n]} \otimes \omega_Y)$, which can be computed by the Hirzebruch–Riemann–Roch formula. Let $n = mr + i$, with $0 \leqq i \leqq r-1$. Thus

$$\begin{aligned} P_{n+1} = \chi &= \kappa_3[\operatorname{ch}(D) \cdot \operatorname{Td}(X)] \\ &= \kappa_3\left[\left(1 + D + \frac{1}{2}D^2 + \frac{1}{6}D^3\right) \cdot \left(1 + \frac{1}{2}c_1 + \frac{1}{12}(c_1^2 + c_2) + \frac{1}{24}c_1c_2\right)\right], \end{aligned}$$

where the Chern classes are those of Y, and

$$D = mf^*(rK_X) + (i+1)K_Y - \Delta_i. \qquad (1)$$

This will simplify to (*), using

$$c_1 = -K_Y, \qquad (2)$$

$$\frac{1}{24}c_1c_2 = \chi(\mathcal{O}_Y) = \chi(\mathcal{O}_X), \qquad (3)$$

$$(f^*(rK_X)) \cdot \Delta_i = 0, \tag{4}$$

and

$$rK_Y = f^*(rK_X) + \Delta_r; \tag{5}$$

here (3) holds because X is Cohen–Macaulay [36], and so $R^if_*\mathcal{O}_Y = 0$ for $i > 0$, and (4) because $f(\Delta_i) \subset X$ is a finite set for each i.

Thus

$$P_{n+1} = \frac{1}{6}D^3 - \frac{1}{4}K_YD^2 + \frac{1}{12}c_1^2D + \frac{1}{12}c_2D + \frac{1}{24}c_1c_2. \tag{6}$$

Using (5),

$$D = (n+1)K_Y - m\Delta_r - \Delta_i, \tag{7}$$

so that the last two terms in (6) are

$$\frac{1}{12}c_2D + \frac{1}{24}c_1c_2 = -\frac{1}{24}(2n+1)\chi(\mathcal{O}_X) - \frac{1}{12}[mc_2\Delta_r + c_2\Delta_i]. \tag{8}$$

The first three terms can be rewritten using (5) and (7) in terms of $f^*(rK_X)$ and the Δ:

$$\begin{aligned} &\frac{1}{12}D(2D-K_Y)(D-K_Y)\\ &\quad= \frac{1}{12r^3}\{(n+1)f^*rK_X - r\Delta_i\}\{(2n+1)f^*rK_X - 2r\Delta_i - \Delta_r\}\\ &\quad\times\{nf^*rK_X - r\Delta_i - \Delta_r\}\\ &\quad= \frac{1}{12}(2n+1)(n+1)nK_X^3 - \frac{1}{12r^2}\Delta_i\{2r\Delta_i + \Delta_r\}\{r\Delta_i + \Delta_r\}; \end{aligned} \tag{9}$$

the final equality has involved (4).

(6), (8) and (9) imply (*), together with the following formula for $\ell(n)$:

$$\ell(n+1) = \frac{1}{12}m(-c_2\Delta_r) - \frac{1}{12}c_2\Delta_i - \frac{1}{12}\left(2\Delta_i^3 + \frac{3}{r}\Delta_i^2\Delta_r + \frac{1}{r^2}\Delta_i\Delta_r^2\right),$$

where $n = mr + i$, $0 \leqq i \leqq r-1$. In particular, if $r \mid n$,

$$\ell(n+1) = \frac{1}{12}\frac{n}{r}(-c_2\Delta_r),$$

so that $\frac{n}{r}(-c_2\Delta_r)$ is the linear part of ℓ.

The fact that P_n is an integer implies varies congruences modulo $12r$ on the invariants appearing in (*). In particular, the denominator of K_X^3 divides r.

The divisors Δ_i (for $i = 1, \ldots, r$) occur naturally as unions of connected components $\Delta_i(P)$ with $f(\Delta_i(P)) = P$, lying over finitely many points $P \in X$. A consequence of their appearance in the formula (*) for the birationally invariant plurigenera of Y is the following result.

COROLLARY (5.6): *The quantities* $-c_2\Delta_r(P)$ *and*

$$-\left(c_2\Delta_i + 2\Delta_i^3 + \frac{3}{r}\Delta_i^2\,\Delta_r + \frac{1}{r^2}\Delta_i\Delta_r^2\right)(P) \quad (\text{for } i = 1, \ldots, r-1)$$

are invariants of the canonical singularity $P \in X$, independent of the resolution $f\colon Y \to X$.

A similar argument based on calculating $H^0(\mathcal{O}_X(n_1H) \otimes \omega_X^{[n_2]})$, where H is an ample divisor and $n_1 \gg n_2 \gg 0$, and using the birational invariance of logarithmic differentials, proves Corollary 5.6 for a local canonical singularity $P \in X$, without assuming that $P \in X$ is isomorphic to a point of a global canonical variety.

It is obvious that for a hypersurface rational point the single invariant $-c_2\Delta = 0$; and Corollary 2.12 implies that this continues to hold for all rational Gorenstein 3-fold points.

PROBLEM (5.7): (i) For a canonical point $P \in X$ of index r, relate the invariants of Corollary 5.6 to the following numerical functions of the $\mathcal{O}_{X,P}$-modules $\omega_X^{[i]}$:

(a) the Hilbert functions $H(n, i) = \dim_k \omega_X^{[i]}/m_P^n\omega_X^{[i]}$;

(b) the lengths $r(i, j)$ and $s(i, j)$ of the kernels and cokernels of $\omega_X^{[i]} \otimes \omega_X^{[j]} \to \omega_X^{[i+j]}$.

(ii) Calculate these invariants for the quotient singularities $\mathbf{A}^3/\boldsymbol{\mu}_r$; (it's quite likely that these are representative of all index r points).

(iii) Topological interpretation?

(iv) Is it true that $\ell(n) \geqq 0$?

§6. Open problems and concluding remarks

6.1. *Is the canonical ring finitely generated?*

Wilson has shown that on a non-singular 3-fold V with $\kappa(V) \neq -\infty$, K_V is ample if and only if $K_VC > 0$ for every curve $C \subset V$. On the other hand we have the adjunction formula

$$K_VC + \deg N_{V|C} = 2p_a(C) - 2, \tag{*}$$

where $N_{V|C}$ is the normal sheaf; by the Riemann–Roch theorem

$$h^0(N_{V|C}) - h^1(N_{V|C}) = -K_VC.$$

By deformation theory, if $K_VC < 0$, C should then move in a positive-dimensional family; C will thus lie in a surface F, which it would be highly desirable[1] to contract by a birational modification of V. The techniques for

[1] The possibility of carrying out this contraction has been proved by S. Mori in a precise form; unfortunately, this is as yet only the first step (and not the inductive step) in the direction of finite generation.

such modifications have been pioneered by Kulikov [28], and simplified by Persson and Pinkham [29].

6.2. Now suppose that $K_V C = 0$; the following remark is partly suggested by a conversation with Bombieri: if K_V is ample on $V \backslash C$, but $\mathcal{O}_C(K_V) \in \operatorname{Pic} C$ is not a torsion class, then $R(V)$ is *not* finitely generated. Compare Zariski [30], p. 562.

CONJECTURE: If K_V is ample on $V \backslash C$ and $K_V C = 0$ then $p_a C = 0$.

For S a surface of general type, $p_a C \geqq 1$ implies that $K_S C > 0$ (without using minimal models). Using the index theorem and minimal models, $K_S C \geqq \sqrt{C^2}$, so that the first term in (*) cannot be too small.

6.3. *The adjunction sequence.*

The following is a local version of the problem of finite generation. If $f\colon Y \to X$ is a resolution of a variety X (suppose *either* that X is normal, *or* that ω_X is invertible), define the *adjunction sequence* to be the sequence of subsheaves $f_*\omega_Y^{\otimes n} \subset \omega_X^{[n]}$; if ω_X is invertible, $f_*\omega_Y^{\otimes n} = \mathcal{J}_n \cdot \omega_X^{\otimes n}$, where $\mathcal{J}_n$ is the *n-th higher adjunction ideal.*

Problem. Is the $\mathcal{O}_X$-algebra $\bigoplus_{n \geqq 0} f_*\omega_Y^{\otimes n}$ finitely generated?

This is equivalent to knowing that the ring $R(Y, f^*\mathcal{O}_X(k) \otimes \omega_Y)$ is f.g. for $k \gg 0$. If this is true then

$$\operatorname{Proj} R(Y, f^*\mathcal{O}_X(k) \otimes \omega_Y) = \operatorname{Proj}_X \Big(\bigoplus_{n \geqq 0} f_*\omega_Y^{\otimes n}\Big) \to X$$

is called the *relative canonical model* of Y, or the *canonical blow-up* of X.

6.4. For simple types of hypersurface singularities one expects the sequence of ideals $\{\mathcal{J}_n\}$ to be defined by weighting conditions as in §4. The following conjecture would extend to 3-fold hypersurface singularities the most fundamental properties of elliptic surface singularities:

CONJECTURE: *Let* $0 \in X \subset \mathbf{A}^n$ *be an elliptic singularity* (*Definition* 2.4); *then there exist coordinates* x_i *on* $\mathbf{A}^4$, *and an* $\alpha \in A'_4$ (*Theorem* 4.5) *which is uniquely determined by any of the following statements*:

(i) $\alpha(g) = 1$; *where g is the defining equation of* X;

(ii) $\mathcal{J}_n = \left\{ f \in \mathcal{O}_X \;\middle|\; \alpha(f) \geqq \frac{n}{d} \right\}$, *where d is the least denominator of* α;

(iii) *the* α*-blow-up* $X_1 \to X$ *is a variety with canonical singularities along* $f^{-1}0$.

6.5. The varieties of f.g. general type for which the canonical model is Cohen–Macaulay have the following property: after making a cyclic cover of degree r ramified in a general element of $|mrK_V|$, $m \gg 0$, one can make a birational modification W such that W has only cDV points and $|nK_W|$ is free for all $n \gg 0$; in particular, $K_W C \geqq 0$ for every curve $C \subset W$.

6.6. One does not expect to get a unique minimal non-singular model of a 3-fold; instead, one could ask for a class of "nice resolutions" of the canonical model X. One might hope to index nice resolutions by some kind of combinatorial data, and characterising canonical points as subvarieties (for example hypersurfaces) in toric varieties might be a first step in this direction. However, one should not merely restrict to complete intersections in toric varieties, since this would exclude many interesting varieties which are Weil divisors but not Cartier divisors—the weighted blow-up of a hypersurface is a case in point. The following is a rather vague hope.

CONJECTURE: *Every canonical singularity is isomorphic to a toric section $P \in X \subset A$, defined by an ideal I_X with $\alpha(A) > \alpha(I_X)$ for a class of weightings α.*

Here a *toric section* (quasi-complete intersection in a toric space) is an irreducible subvariety $X \subset A$ of codimension r, such that r equations $f_1, \dots, f_r$ define X outside the coordinate hyperplanes: $X \cap \mathbf{T} = (f_1 = \cdots = f_r = 0) \subset \mathbf{T}$. The notion of weighting awaits clarification.

6.7. The simplest kind of normal 3-fold singularity which is not Cohen–Macaulay would be a *fake cDV point* $P \in X$, that is a point $P \in X$ for which a general section H through P has a non-normal isolated singularity $P \in H$, whose normalisation $P \in S$ is a Du Val point; the existence of such points is related to the deformation theory of $P \in H$.

CONJECTURE: *Let $P \in S$ be a Du Val point, and let $P \in H$ be an isolated singularity with $\mathcal{O}_{H,P} \subset \mathcal{O}_{S,P}$ of finite codimension. Then any deformation of H arises from deforming the normalisation S or from moving $\mathcal{O}_{H,P}$ inside $\mathcal{O}_{S,P}$ as a subvector space of fixed codimension. In particular H is not a section of a normal 3-fold.*

For example, the simplest such $P \in H$, $H = \operatorname{Spec} k[x^2, x^3, y, xy]$ (obtained by pinching out the vector $(x^2 = y = 0) \subset \mathbf{A}^2$) is rigid.

There is now some evidence for this conjecture: by [36] a fake cDV singularity cannot be canonical; as kindly pointed out by Jonathan Wahl, Mumford's Theorem in IV of [39] shows that a fake cDV point cannot be isolated. However, the deformation theory is much harder to deal with.

6.8. One can continue Theorem 2.6 (I) to the assertion that if a Gorenstein point $P \in X$ satisfies $m_P^r \cdot \omega_X \subset f_* \omega_Y$, then the general section $P \in H$ will satisfy $m_P^{r+1} \cdot \omega_H \subset f_* \omega_L$. By an induction this can be chased down to the curve section: if $P \in X$ is a rational Gorenstein point of an n-fold then the general curve section through P, $P \in C$ say, satisfies

$$m_P^{n-1} \subset \mathscr{C},$$

where $\mathscr{C} = [\mathcal{O}_{\tilde{C}} : \mathcal{O}_C]$ is the conductor. In this context Theorem 2.11 is partly explained by Shepherd–Barron's remark that if $P \in C$ is a Gorenstein curve point such that $m_P^2 \subset \mathscr{C}$, then either $m_P^2 = \mathscr{C}$, or $P \in C$ is a very special curve such as the plane curve given by $(x^2 + y^n = 0,\ n \leqq 5)$.

6.9. The reader will have observed that despite my ideological commitment to replacing the cohomological arguments involving $R^i f_*$'s by adjunction-theoretic argument, I have basely betrayed my principles in the proof of (II) of Theorem 2.6. I would like to know if a proof of this result could be given on the lines of the proof of (I).

6.10. The combinatorics involved in resolving a rational Gorenstein 3-fold point as in §2 deserves further study. It is easily seen that on blowing up a point with invariant $k \geqq 3$ the invariant of any resulting point is at most k. Thus the resolution of these singularities consist of trees of del Pezzo surfaces. It is not clear as yet if there are any restrictions of the branching of these trees, say arising from some kind of index theorem.

6.11. The existence of canonical 3-folds of arbitrary index means that there can be no bound n such that the canonical ring of every 3-fold of f.g. type is generated by elements of degree $\leqq n$; it is not clear whether K_X^3 can be arbitrarily small, although it becomes fairly small for some complete intersections (see Problem 3.11). Most probably there should exist some bound depending only on r such that for every canonical 3-fold of index r $|mrK_X|$ is very ample. See [18], [19] and [41].

6.12. This problem is suggested by a remark of Beauville's: if ω_X is ample on a non-singular 3-fold X then it is a consequence of Yau's inequalities [33] that

$$\chi(\mathcal{O}_X) = \tfrac{1}{24} c_1 c_2 \leqq \tfrac{1}{64} c_1^3 < 0.$$

CONJECTURE: *Let X be a canonical 3-fold with Gorenstein singularities, and let $f\colon Y \to X$ be a 0-minimal resolution. Then $c_2(Y)$ is quasi-positive in the sense that for every prime divisor $\Gamma \subset Y$*

$$c_2(Y) \cdot \Gamma \geqq 0,$$

with equality if and only if $\dim f(\Gamma) \leqq 1$.

This might be a consequence of some kind of index theorem for 3-folds.

REFERENCES

[1] P. DU VAL: On the singularities which do not affect the condition of adjunction. Proc. Camb. Phil. Soc. *30* (1934) 453–491.

[2] M. ARTIN: Some numerical criteria for contractibility of curves on an algebraic surface. Amer. J. Math. *84* (1962) 485–496.

[3] M. ARTIN: On isolated rational singularities of surfaces. Amer. J. Math. *88* (1966) 129–136

[4] M. ARTIN: Coverings of the rational double points in characteristic *p*, in Complex analysis and algebraic geometry. W.L. Baily, Jr. and T. Shioda Eds., Iwanami Shoten and C.U.P., 1977.

[5] A. DURFEE: Fifteen characterisations of rational double points and simple critical points. Ens. Math.(2) *25*: 1–2, (1979) 131–163.

[6] KEMPF et al.: Toroidal embeddings I. Lecture Notes in Math. *339* (1973).

[7] N. BOURBAKI: Algèbre commutative, Chap. VII, Hermann, Paris, 1965.

[8] H. MATSUMURA: Commutative algebra. Benjamin, N.Y., 1970.

[9] A. ALTMAN and S. KLEIMAN: An introduction to Grothendieck duality theory. Lecture Notes in Math. *146* (1970).

[10] O. ZARISKI: An introduction to the theory of algebraic surfaces. Lecture Notes in Math. *83* (1969).

[11] J.-P. SERRE: Groupes algébriques et corps de classes. Hermann, Paris, 1959.

[12] M. REID: Elliptic Gorenstein singularities. Manuscript (1975).

[13] H. LAUFER: Minimally elliptic singularities. Amer. J. Math. *99* (1977) 1257–1295.

[14] R. ELKIK: Singularitiés rationelles et déformations. Invent. Math. *47* (1978) 139–147.

[15] V. DANILOV: The geometry of toric varieties. Uspekhi Mat. Nauk *33*:2 (1978) 85–134 = Russian Math. Surveys *33*:2 (1978) 97–154.

[16] ZARISKI and SAMUEL: Commutative algebra, Vol. 1, Springer Graduate Texts.

[17] I. DOLGACHEV: Weighted projective varieties. Manuscript (1976).

[18] P. WILSON: The pluricanonical map on varieties of general type. Bull. London Math. Soc. *12* (1980) 103–107.

[19] X. BENVENISTE: Variétés de dimension 3, de type général, tel que le système linéaire defini par un multiple du diviseur canonique soit sans point-base. C. R. Acad. Sc. Paris, *289*, Sér. A (1979) 691–694.

[20] J. D. SALLY: Tangent cones at Gorenstein singularities. Compositio Math. *40*:7 (1980).

[21] J. D. SALLY: Good embedding dimensions for Gorenstein singularities. Preprint.

[22] S. GOTO and K. WATANABE: On graded rings, I. J. Math. Soc. Japan *30* (1978) 179–213.

[23] D. MUMFORD: Geometric invariant theory. Springer Ergebnisse.

[24] D. MUMFORD: Stability of projective varieties. Ens. Math.(2), *23*: 1–2, (1977) 39–110.

[25] M. DEMAZURE: Anneaux gradués normaux, in Séminaire Demazure-Giraud-Teissier, Singularités des surfaces, Ecole Polytechnique, Palaiseau (1979).

[26] B. SAINT-DONAT: Projective models of K3 surfaces. Amer. J. Math. *96* (1974) 602–639.

[27] H. GRAUERT and O. RIEMENSCHNEIDER: Verschwindungssätze für analytische Kohomologiegruppen auf komplexen Räumen. Invent. Math. *11* (1970) 263–292.

[28] V. S. KULIKOV: Degenerations of K3 surfaces and Enriques surfaces. Izv. Akad. Nauk SSSR Ser. Mat. *41* (1977) = Math. USSR Izv. *11* (1977).

[29] U. PERSSON and H. PINKHAM: Degeneration of surfaces with trivial canonical bundle. Ann. of Math. (1980).

[30] O. ZARISKI: The theorem of Riemann-Roch for high multiples of an effective divisor on an algebraic surface. Ann. Math. *76* (1962) 560–615.

[31] K. WATANABE: Certain invariant subrings are Gorenstein, I and II Osaka J. Math. *11* (1974) 1–8 and 379–388.

[32] V. A. KHINICH: When is a ring of invariants of a Gorenstein ring also Gorenstein? Izv. Akad. Nauk SSSR Ser. Mat. *40* (1976) 50–56 = Math. USSR Izv. *10* (1976) 47–54.

[33] S.-T. YAU: Calabi's conjecture and some new results in algebraic geometry. Proc. Natl. Acad. Sci. USA *74* (1977) 1798–1799.

[34] S.S-T. YAU: 3 papers on surface singularities. Amer. J. Math. *101* (1979) 761–884.

[35] H. LAUFER: On $\mathbf{CP}^1$ as an exceptional set, SUNY preprint.

[36] N. SHEPHERD-BARRON: Canonical 3-fold singularities are Cohen–Macaulay, Warwick preprint.

[37] M. MERLE and B. TEISSIER: Conditions d'adjonction, d'après Du Val. in Séminaire sur les Singularitiés des surfaces, Springer-Verlag Lecture Notes 777 (1980).

[38] A. G. KOUSHNIRENKO: Polyèdres de Newton et nombre de Milnor, Invent. Math. *32* (1976) 1–32.

[39] D. MUMFORD: Footnotes to the work of C. P. Ramanujam, in C. P. Ramanujam - A Tribute, Tata Inst. Studies in Math., *8* Bombay (1978) 247–262.

[40] M. REID: The Du Val surface singularities A_n, D_n, E_6, E_7, E_8. To appear.

[41] P. WILSON: Algebraic varieties of general type, Kyoto R.I.M.S. preprint.

Math. Inst.,
Univ. of Warwick,
Coventry CV4 7AL,
England.

BIRATIONAL GEOMETRY OF ALGEBRAIC THREEFOLDS

Kenji Ueno

In the present notes, by an algebraic variety V we mean that V is an irreducible complete algebraic variety defined over the complex number field $\mathbf{C}$. A non-singular algebraic variety is called an algebraic manifold. A compact complex manifold M is called a *Moishezon manifold*, if we have tr. deg $\mathbf{C}(M) = \dim M$, where $\mathbf{C}(M)$ is the meromorphic function field of M. A Moishezon manifold has the structure of a proper smooth algebraic space over $\mathbf{C}$ and it is bimeromorphically equivalent to a projective manifold. Hence from the view point of the birational geometry, we need not distinguish Moishezon manifolds from algebraic manifolds.

Let D be a Cartier divisor (a line bundle) on a normal algebraic variety V. If the complete linear system $|mD|$ is not empty, we can define a rational mapping $\Phi_{mD}: V \to P^{\dim|mD|}$ associated with the complete linear system $|mD|$. The D-dimension $\kappa(D, V)$ is defined by

$$\kappa(D, V) = \begin{cases} -\infty, \text{ if } |mD| = \emptyset \text{ for all positive integers,} \\ \max\limits_{m\geq 1} \dim \Phi_{mD}(V), \text{ otherwise.} \end{cases}$$

Let V be an algebraic manifold and K_V a canonical divisor (the canonical line bundle) of V. Then $\kappa(K_V, V)$ is called the *Kodaira dimension* of V and is written as $\kappa(V)$. The Kodaira dimension is a birational invariant. Hence for a singular algebraic variety V, the Kodaira dimension $\kappa(V)$ of V is defined as $\kappa(V) = \kappa(V^*)$ where V^* is a non-singular model of V.

By definition $\kappa(V) = -\infty$ if and only if $P_m(V) = h^0(V, mK_V) = 0$ for all $m \geq 1$ and $\kappa(V) = 0$ if and only if $P_m(V) \leq 1$ for all $m \geq 1$ and there exists a positive integer m_0 such that $P_{m_0}(V) = 1$. Moreover $\kappa(V) > 0$ if and only if there exists a positive integer m such that $P_m(V) \geq 2$.

Using the notion of Kodaira dimension, we can classify all n-dimensional algebraic varieties into $n+2$ classes. When $n = 2$, the detailed study of these 4 classes was done by Italian algebraic geometers more than sixty years ago. Recently we obtained several important structure theorems on algebraic threefolds. In the present notes we shall briefly review these results and discuss some important unsolved problems.

In §1 we shall discuss differences between algebraic surfaces and algebraic manifolds of dimension $n \geq 3$. The main results on the classification theory of algebraic threefolds will be also given. In §2 we shall discuss algebraic threefolds with $\kappa = 0$. Several examples of such threefolds will be given. In §3, the symmetric powers of holomorphic forms will be discussed.

§1. Difference between surfaces and threefolds

In this section, we first discuss relatively minimal models of algebraic manifolds.

DEFINITION 1.1: Let V be an algebraic manifold (or more generally Moishezon manifold). Then V is called a relatively minimal model of V, if any birational morphism $f: V \to V'$ is an isomorphism.

The following theorem plays an essential role in the classification theory of surfaces.

THEOREM 1.2: *If a birational equivalence clase $\mathscr{C}$ of a surface S contains at least two distinct relatively minimal models, then $\mathscr{C}$ contains countably many relatively minimal models, and $\mathscr{C}$ is rational or ruled.*

On the other hand Moishezon [7] found a Moishezon threefold of general type which has at least two different relatively minimal models. Later, Francia [1] found examples of projective threefolds of general type which have finitely many distinct relatively minimal projective models. Recently Fujiki [2] found the following suprising fact.

THEOREM 1.3: *Let V be a relatively minimal Moishezon manifold of dimension ≥ 3. Suppose V contains a rational curve C (which may be singular). Then there exists a relatively minimal model which is not isomorphic to V.*

PROOF: For simplicity we shall consider the case $\dim V = 3$. A proof of the general case can be found in Fujiki [2]. If C is singular, we let $\pi_0: V_0 \to V$ be obtained by a finite succession of monoidal transformations along non-singular centers such that the proper transform C_0 of C is non-singular. The normal bundle of C_0 in V_0 has the form $\mathcal{O}_{\mathbf{P}^1}(a) \oplus \mathcal{O}_{\mathbf{P}^1}(b)$. If we blow-up V_0 at a point $p \in C_0$, the normal bundle of the proper transform of C_0 has the form $\mathcal{O}_{\mathbf{P}^1}(a-1) \oplus \mathcal{O}_{\mathbf{P}^1}(b-1)$. Hence, if necessary we blow-up V_0 at a finite number of distinct points on C_0, and we can assume that $a < 0$, $b < 0$, a,b are relatively prime if $a \neq b$, and if $a = b$, then $a = b = -1$. Assume $a < b$. We let $\pi: V_1 \to V_0$ be the blow-up of V_0 along C_0. The exceptional surface F_1 is a $\mathbf{P}^1$-bundle over $\mathbf{P}^1$ and has two sections. One of the section C_1 has the self-intersection number $a - b < 0$ in F_1. Then C_1 is a non-singular rational curve and its normal bundle in V_1 has the form $\mathcal{O}_{\mathbf{P}^1}(a-b) \oplus \mathcal{O}_{\mathbf{P}^1}(b)$. By a finite succession of this procedure, we obtain π_1: $V_1 \to V_0$ such that V_1 contains a non-singular rational curve C_1 with $\pi_1(C_1) = C_0$ whose normal bundle in V_1 is $\mathcal{O}_{\mathbf{P}^1}(-1) \oplus \mathcal{O}_{\mathbf{P}^1}(-1)$. Now we blow up V_1 along C_1 and obtain the morphism π_2: $V_2 \to V_1$. Let F_2 be the exceptional surface of the blowing up.

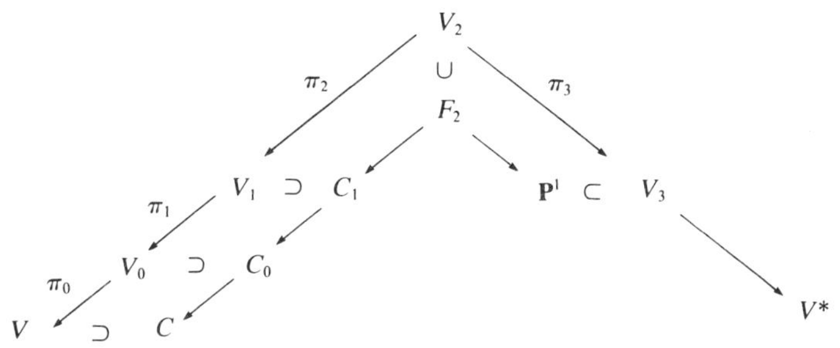

The restriction $\pi_2|_{F_2}: F_2 \to C_1$ gives the trivial $\mathbf{P}^1$-bundle over $C_1 = \mathbf{P}^1$. We let $g: F_2 \to \mathbf{P}^1$ be the projection in the other direction. Since the normal bundle of C_1 in V_1 is $\mathcal{O}_{\mathbf{P}^1}(-1) \oplus \mathcal{O}_{\mathbf{P}^1}(-1)$, the restriction of the line bundle $[F_2]$ on V_2 to each fibre of g is $\mathcal{O}_{\mathbf{P}^1}(-1)$. Hence we can blow down each fibre of g to a point and obtain π_3: $V_2 \to V_3$ such that $\pi_3|_{F_2} = g$. From the construction it is easy to show that there is *no* birational morphism $f: V_3 \to V$. If necessary we contract subvarieties of V_3 and obtain a relatively minimal model V^*. Then, V^* is not isomorphic to V. Q.E.D.

COROLLARY 1.4: *If a relatively minimal Moishezon manifold V contains a ruled surface (which may be singular), the birationally equivalent class of V contains continuously many distinct relatively minimal models.*

The important fact is that the statements of Theorem 1.3 and Corollary 1.4 have nothing to do with pluricanonical systems, hence, Kodaira dimensions. Therefore, the analogue to Theorem 1.2 does not hold for higher dimensional Moishezon manifolds. It is an interesting problem to find a "good" model of an algebraic manifold (Moishezon manifold) and Theorem 1.3 says that the notion of a relatively minimal model is too weak for that purpose.

Let us consider the following

LEMMA 1.5: *Let V be a Moishezon manifold with trivial canonical bundle. Then V is relatively minimal.*

PROOF: Suppose that there exists a non-isomorphic birational morphism $f: V \to V'$. Since $p_g(V') = p_g(V) = 1$, there exists a non-zero holomorphic n-form ω on V', where $n = \dim V$. Since $f^*\omega$ has no zero on

V, f is of maximal rank at each point. This contradicts the fact that f^{-1} is not a morphism. Q.E.D.

In §2 we shall give an example of an algebraic threefold with trivial canonical bundle which contains a rational surface, hence, it has continuously many relatively minimal models. But these relatively minimal models have in general non-trivial canonical bundles. Note that there exists a Moishezon threefold which has at least two distinct birational models with trivial canonical bundle.

There is another difference between surfaces and higher dimensional algebraic manifolds.

THEOREM 1.6.

1) *Let S be an algebraic surface with $\kappa \geq 0$. Then, there exist the birationally equivalent model (minimal model) S^* and a positive integer m_0 such that $|mm_0K_{S^*}|$ is free from base points and fixed components for any positive integer m.*

2) *Let k be an integer with $0 \leq k \leq n$ and $n \geq 3$. Then, there exists an algebraic manifold V of dimension n such that $\kappa(V) = k$ and for any birationally equivalent model V^* of V, $|mK_{V^*}|$ has fixed component if it is not empty.*

The proof of the second part can be found in Ueno [12]. For a surface S, if $\kappa(S) \geq 0$, then $P_{12}(S) \geq 1$. It is not known whether such fact holds for higher dimensional algebraic manifolds.

On the other hand, of course, algebraic threefolds enjoy many properties similar to those of surfaces. The following is a classification table of algebraic threefolds. In Angers, Viehweg gave the talk concerning the proof of the table below. The details can be found in Ueno [13], [14] and Viehweg [15], [16].

κ	g_1	structure
3	≥ 0	Φ_{mK} is birational for $m \gg 0$
2	≥ 0	elliptic threefold.
1	≥ 0	fibre space over a curve whose fibres are surfaces with $\kappa = 0$
0	3	abelian variety
	2	étale fibre bundle over an abelian surface whose fibre is an elliptic curve
	1	étale fibre bundle over an elliptic curve whose fibre is a surface with $\kappa = 0$
	0	?
$-\infty$	≥ 1	uniruled
	$= 0$	?

A part of the classification of algebraic threefolds with $\kappa = 0$ is generalized by Kawamata. In the talk of Angers, he gave an outline of the proof that the Albanese mapping is surjective for any algebraic manifold with $\kappa = 0$. Moreover, in this case he proved that the fibres of the Albanese mapping are connected. For the details see Kawamata [4], Kawamata-Viehweg [5].

§2. Algebraic threefolds of parabolic type

An algebraic manifold V is called of parabolic type if $\kappa(V) = 0$. In this section we shall consider algebraic threefolds of parabolic type more carefully. Let us define certain numerical invariants of algebraic manifolds. Put

$$g_k(V) = h^0(V, \Omega^k_V).$$

Then, the number $g_k(V)$ is a birational invariant of V. For an algebraic threefold V of parabolic type, from the above classification table it follows

$$g_1(V) \leq 3,$$
$$g_3(V) = p_g(V) = 1.$$

It is interesting to know which kind of triplet $(g_1(V), g_2(V), g_3(V))$ appears for an algebraic threefold V of parabolic type. As we shall see below, we know the answer if $g_1(V) \geq 1$. But if $g_1(V) = 0$, we do not know the complete answer. If there is a birational model V^* such that $K^m_{V^*}$ is trivial for an integer m, the answer is easy by virtue of the Riemann-Roch theorem. But we saw already that this is not necessarily the case for algebraic threefold (Theorem 1.6, 2)).

The following is the table of algebraic threefolds of parabolic type with $g_1 \geq 1$.

g_1	g_2	$g_3 = p_g$	structure
3	3	1	abelian threefold
2	1	0	elliptic bundle over an abelian surface
1	1	1	étale fibre bundle over an elliptic curve whose fibre is a certain abelian surface or K3 surface.
1	0	0	étale fibre bundle over an elliptic curve whose fibre is a surface with $\kappa = 0$.

To study the case $g_1 = 0$, we propose the following conjecture.

CONJECTURE: *Let V be an n-dimensional algebraic manifold with* $\kappa = 0$. *Then we have*

$$g_k(V) \leq \binom{n}{k} = g_k(A^n).$$

where A^n *is an n-dimensional abelian variety.*

In case $\dim V = 3$, the conjecture says that $g_2(V) \leq 3$. Assuming the conjecture, let us consider what kind of combination of (g_2, g_3) is possible if $g_1 = 0$. In the following table, the number $k)$ means that examples with such numerical invariants can be found in Example $k)$ below and the question mark means that there are no known examples.

g_1	g_2	g_3	$\chi(V, \mathcal{O})$	example
	3	1	3	?
	3	0	4	1)
	2	1	2	?
0	2	0	3	2)
	1	1	1	?
	1	0	2	3)
	0	1	0	4)
	0	0	1	5)

Example 1) $g_2 = 3$, $g_3 = 0$. Let V be a non-singular model of the quotient space of an abelian threefold by the standard involution. The manifold V is called a Kummer manifold and it is simply connected. Any birationally equivalent model of V has a non-zero effective bicanonical divisor. The manifold V has the following numerical invariants.

$$\kappa(V) = 0, \quad g_1(V) = 0, \quad g_2(V) = 3, \quad g_3(V) = 0, \quad P_2(V) = 1,$$

$\kappa(\Omega_V^1, V) = -\infty$, $\kappa(\Omega_V^2, V) = 2$. (For the definition of $\kappa(\Omega_V^1, V)$ and $\kappa(\Omega_V^2, V)$, see the next section.)

Example 2) $g_2 = 2$, $g_3 = 0$. Let E_ρ be the elliptic curve with period matrix $(1, \rho)$ where $\rho = \exp(2\pi\sqrt{-1}/3)$. Let V be a non-singular model of the quotient variety $(E\times E\times E)/G$ where G is the cyclic group of order 3 generated by the automorphism

$$g:(z_1, z_2, z_3)\mapsto(\rho z_1, \rho z_2, \rho^2 z_3)$$

The manifold V has a structure of fibre space over $\mathbf{P}^1$ whose general fibre is $E\times E$. Moreover, V is simply connected. We have the following numerical invariants.

$$\kappa(V)=0,\quad g_1(V)=0,\quad g_2(V)=2,\quad g_3(V)=0,\quad P_2(V)=0,$$
$$P_3(V)=1,\quad \kappa(\Omega_V^1, V)=-\infty,\quad \kappa(\Omega_V^2, V)=1.$$

Any birationally equivalent model of V has a non-zero effective tricanonical divisor.

Example 3) $g_2=1$, $g_3=0$. Let S be a $K3$ surface with involution h such that $h^*\omega=\omega$ for any non-zero holomorphic 2-form ω on S. (Examples of such $K3$ surfaces can be obtained from a certain double covering of a Kummer surface associated with the product of two elliptic curves. See [10].) We let V be a non-singular model of the quotient variety $(E\times S)/G$ where E is an elliptic curve with the standard involution ι and G is the cyclic group of order 2 generated by the automorphism

$$g:E\times S\longrightarrow E\times S$$
$$(z, w)\longmapsto(\iota(z), h(w)).$$

The algebraic threefold V is simply connected. We have the following numerical invariants.

$$\kappa(V)=0,\quad g_1(V)=0,\quad g_2(V)=1,\quad g_3(V)=0,\quad P_2(V)=1,$$
$$\kappa(\Omega_V^1, V)=-\infty,\quad \kappa(\Omega_V^2, V)=-\infty.$$

Moreover, V has a structure of fibre space over $\mathbf{P}^1$ whose general fibre is the $K3$ surface S.

Example 4) $g_2=0$, $g_3=1$. There are lots of examples in this class. We give three examples. The first example is a non-singular hypersurface of degree 5 in $\mathbf{P}^4$. The second example is a non-singular model V of a two-sheeted ramified covering $\pi:V'\to\mathbf{P}^3$ ramified along an irreducible surface of degree 8 with ordinary double points as its singularities. The manifold V is considered as a degeneration of a two-sheeted ramified covering over $\mathbf{P}^3$ ramified along a non-singular surface of degree 8. It is easily seen that our threefold V has no birationally equivalent model with trivial canonical bundle. These examples have the following numerical invariants.

$$\kappa(V)=0,\quad g_1(V)=0,\quad g_2(V)=0,\quad g_3(V)=1.$$

Next let us consider the elliptic curve E_ρ in Example 2). We let G be the cyclic group of order 3 generated by the analytic automorphism

$$g: E\times E\times E \mapsto E\times E\times E$$

$$(z_1, z_2, z_3)\mapsto(\rho z_1, \rho z_2, \rho z_3).$$

Let V be the non-singular model of the quotient space $(E\times E\times E)/G$ obtained by the canonical resolution of its singularities. (See [11], (16.10) p. 199). Then V is simply connected and V contains projective planes appearing by the resolution of singularities. Moreover, it is easy to show that the canonical bundle of V is trivial. The threefold V has the following numerical invariants.

$$\kappa(V)=0,\quad g_1(V)=0,\quad g_2(V)=0,\quad g_3(V)=1,$$

$$\kappa(\Omega_V^1, V)=-\infty,\quad \kappa(\Omega_V^2, V)=-\infty.$$

We note that Igusa [3] found an example of an algebraic threefold V in this class such that the product of three copies of an elliptic curve is an unramified covering of V and V has the following numerical invariants.

$$\kappa(V)=0,\quad g_1(V)=0,\quad g_2(V)=0,\quad g_3(V)=1,$$

$$\kappa(\Omega_V^1, V)=2,\quad \kappa(\Omega_V^2, V)=2.$$

Example 5. $g_2=0$, $g_3=0$. Let E be an elliptic curve with period matrix $(1,\tau)$. We consider analytic automorphisms of $E\times E\times E$ defined by

$$g_1:(z_1, z_2, z_3)\longrightarrow(-z_1, z_2+\tfrac{1}{2}, z_3+\tfrac{1}{2})$$

$$g_2:(z_1, z_2, z_3)\longrightarrow(z_1+\tfrac{1}{2}, -z_2, z_3+\tfrac{1}{2})$$

$$g_3:(z_1, z_2, z_3)\longrightarrow(z_1+\tfrac{1}{2}, z_2+\tfrac{1}{2}, -z_3)$$

The three automorphisms g_i, $i=1,2,3$ commute each other and form an abelian group G. Note that $g_i\cdot g_j$, $i\neq j$ has fixed varieties of dimension one. Let V be a non-singular model of the quotient variety $(E\times E\times E)/G$. Then V is of parabolic type and we have

$$g_1(V)=g_2(V)=g_3(V)=0,\quad P_2(V)=1,$$

$$\kappa(\Omega_V^1, V)=-\infty,\quad \kappa(\Omega_V^2, V)=-\infty.$$

The threefold V has a structure of fibre space over $\mathbf{P}^1$ whose general fibre is a Kummer surface. Moreover, V is simply connected.

It is interesting to know whether there exist algebraic threefolds of parabolic type with $(g_2, g_3) = (3, 1)$ or $(2, 1)$ or $(1, 1)$. Moreover, we note that if the above conjecture that $g_2(V) \leq 3$ for any algebraic threefold of parabolic type is true, then the algebraic fundamental group of an algebraic threefold of parabolic type with $g_1(V) = 0$ is finite except if $(g_2, g_3) = (0, 1)$.

§3. Symmetric power of holomorphic forms

To study holomorphic 2-forms on an algebraic threefold, it would be better to consider more generally the m-th symmetric power $S^m\Omega_V^2$ of the sheaf of germs of holomorphic 2-forms. The m-th symmetric power $S^m\Omega_V^1$ of the sheaf of germs of holomorphic 1-forms has been studied by F. Sakai [9]. In this section we shall consider symmetric powers of holomorphic 1-forms and 2-forms.

Let M be an n-dimensional algebraic manifold. Let $\mathbf{P}(\Omega_M^\nu)$ be the projective fibre space associated with the sheaf Ω_M^ν of germs of holomorphic ν-forms and $\mathcal{O}_{\mathbf{P}}(1)$ the tautological line bundle. Define $\kappa(\Omega_M^\nu, M)$ by

$$\kappa(\Omega_M^\nu, M) = \kappa(\mathcal{O}_{\mathbf{P}}(1), \mathbf{P}(\Omega_M^\nu)), \ \nu = 1, 2, \ldots, \dim M = n.$$

Then, $\kappa(\Omega_M^n, M)$ is nothing but the Kodaira dimension and $\kappa(\Omega_M^\nu, M)$ takes one of the values $-\infty,\ 0, 1, \ldots, n + \binom{n}{\nu} - 1$. $\kappa(\Omega_M^\nu, M)$ is a birational invariant. Moreover we have

LEMMA 3.1: *Let $\varphi: V \to W$ be a surjective morphism of algebraic manifolds. Then we have $\kappa(\Omega_V^\nu, V) \geq \kappa(\Omega_W^\nu, W)$ for $1 \leq \nu \leq \dim W$.*

F. Sakai introduced the notion of the cotangent dimension $\lambda(M)$. The relation between $\lambda(M)$ and $\kappa(\Omega_M^1, M)$ is given by

$$\lambda(M) = \begin{cases} \kappa(\Omega_M^1, M) - (\dim M - 1), & \text{if } \kappa(\Omega_M^1, M) \geq 0 \\ -\dim M, & \text{if } \kappa(\Omega_M^1, M) = -\infty. \end{cases}$$

In this section we shall use the notation $\kappa(\Omega_M^1, M)$ because of Lemma 3.1, which does not hold for $\lambda(M)$.

F. Sakai [9] and Kobayashi [6] showed $\lambda(S) \leq 0$ for every algebraic surface S of parabolic type. Moreover, Sakai conjectured $\lambda(V) \leq 0$ for every algebraic manifold V of parabolic type. In our terminology, $\kappa(\Omega_V^1, V) \leq \dim V - 1$ for every algebraic manifold of parabolic type. It is natural to generalize the conjecture and ask the following problem.

Problem. For every algebraic manifold V of parabolic type are the following inequalities true?

$$\kappa(\Omega_V^\nu, V) \le \binom{n}{\nu} - 1 = \kappa(\Omega_{A^n}^\nu, A^n),\ \nu = 1, 2, \ldots, n-1,$$

where $n = \dim V$ and A^n is an n-dimensional abelian variety.

The problem is a sort of a generalization of the conjecture in §2. We note that the answer to the problem is affirmative for each algebraic threefold of parabolic type with $g_1 \ge 1$ and all known examples with $g_1 = 0$.

Proposition 3.2: *Let $\varphi: V \to W$ be an elliptic fibre space (i.e. general fibres of φ are elliptic curves) with* $\dim W \ge 2$. *Assume that general fibres of φ have variable moduli. Then we have isomorphisms*

$$H^0(V, S^m\Omega_V^\nu) \xrightarrow{\sim} H^0(W, S^m\Omega_W^\nu),\quad \nu = 1, 2, \ldots, n-1,$$
$$m = 1, 2, 3, \ldots,$$

where $n = \dim V$.

Proof: For simplicity we assume $n = 3$ and prove the proposition for $\nu = 2$. Let $\varphi_0: V_0 \to W_0$ be the Zariski open set of our fibre space such that at any point of $\varphi^{-1}(W_0)$, φ is of maximal rank and for every point $x \in W - W_0$, there is a point on $\varphi^{-1}(x)$ at which φ is not of maximal rank. By the assumption, each fibre of φ_0 is an elliptic curve. We let $\tilde{T}: \tilde{W}_0 \to H = \{z \in \mathbf{C} \mid \mathrm{Im}(z) > 0\}$ be the period mapping associated with φ_0 where $\tilde{W}_0$ is the universal covering of W_0. Then we have

$$\tilde{T}(\gamma w) = \frac{a\tilde{T}(w) + b}{c\tilde{T}(w) + d},\quad A = \begin{pmatrix} a & b \\ c & d \end{pmatrix} \in SL_2(\mathbf{Z}),$$

for $\gamma \in \pi_1(W_0, w_0)$. The mapping $\gamma \mapsto A$ is a group homomorphism of $\pi_1(W_0, w_0)$ into $SL_2(\mathbf{Z})$. The semi-direct product $\Gamma = \pi_1(W_0, w_0) \times \mathbf{Z}^2$ operates on $W_0 \times \mathbf{C}$ as follows.

$$(\gamma, m_1, m_2): W_0 \times \mathbf{C} \to W_0 \times \mathbf{C},$$
$$(w, \zeta) \mapsto (\gamma(w), (c\tilde{T}(w) + d)^{-1}(\zeta + m_1\tilde{T}(w) + m_2)).$$

Then Γ operates on $W_0 \times \mathbf{C}$ properly discontinuously and freely. Let A_0 be the quotient manifold $(W_0 \times \mathbf{C})/\Gamma$. We have a surjective morphism $\pi_0: A_0 \to W_0$ such that every fibre of π_0 is an elliptic curve. Moreover, for each point $w \in W_0$ there exists a neighbourhood U such that $\pi_0^{-1}(U)$ and $\varphi_0^{-1}(U)$ are isomorphic as fibre spaces. Let (z_1, z_2) be local coordinates of U. Then, for each element $\omega \in H^0(V, S^m\Omega_V^2)$ the restriction $\omega|_{\varphi_0^{-1}(U)}$ can be expressed in the form

$$\sum_{a+b+c=m} A_{abc}(z, \zeta)(dz_1 \wedge dz_2)^a \cdot (dz_1 \wedge d\zeta)^b \cdot (dz_2 \wedge d\zeta)^c. \tag{3.1}$$

First we shall show that $A_{abc}(z, \zeta)$ is independent of the variable ζ. Note that the above expression is invariant under the translations

$$\zeta \longmapsto \zeta + m_1 T(z) + m_2, \quad m_1, m_2 \in \mathbf{Z}, \tag{3.2}$$

where $T(z)$ is a branch of T on U. Hence we have

$$A_{0,m,0}(z, \zeta + m_1 T(z) + m_2) = A_{0,m,0}(z, \zeta).$$

This implies that $A_{0,m,0}$ is independent of ζ. Similarly $A_{0,0,m}$ is independent of ζ. Next we have

$$A_{1,m-1,0}(z, \zeta) = A_{1,m-1,0}(z, \zeta + m_1 T(z) + m_2) + m_1 A_{0,m,0}(z) \frac{\partial T}{\partial z_2}. \tag{3.3}$$

This shows that $\dfrac{\partial A_{1,m-1,0}}{\partial \zeta}$ is invariant under the translations (3.1), hence independent of ζ. Therefore $A_{1,m-1,0}(z, \zeta)$ has the form $A(z) + B(z)\zeta$. Then, by (3.3) it follows that $B(z) = 0$. Thus, $A_{1,m-1,0}$ is independent of ζ. In this way we can prove that A_{abc} is a function of z_1, z_2 and is independent of ζ.

So far we need not have assumed that the moduli of general fibres are not constant. Now we will show that in the expression (3.1), only the term $A_{m,0,0}(z)(\mathrm{d}z_1 \wedge \mathrm{d}z_2)^m$ appears, if the moduli of general fibres are not constant. Let $\pi: \hat{W} \to W$ be a generically finite morphism and $\hat{\varphi}: \hat{V} \to \hat{W}$ a non-singular model of $V \times_W \hat{W} \to \hat{W}$. Then the pull-back $\hat{\omega}$ of ω onto $\hat{V}$ can be defined and is an element of $H^0(\hat{V}, S^m \Omega^2_{\hat{V}})$. Then, if we can show that $\hat{\omega}$ has an expression $B(w_1, w_2)(\mathrm{d}w_1 \wedge \mathrm{d}w_2)^m$ in $\hat{\varphi}^{-1}(\hat{U})$ where $\hat{U}$ is an open set in $\hat{W}$ with local coordinates (w_1, w_2) and $\pi(\hat{U}) \subset U$, then, in $\varphi^{-1}(U)$, ω has an expression $A_{m,o,o}(z_1, z_2)(\mathrm{d}z_1 \wedge \mathrm{d}z_2)^m$. Hence we can assume the following.

(1) $D = W - W_0$ is a divisor with normal crossings.

(2) There exist a point $p \in D$ and a neighbourhood U of p with local coordinates (u_1, u_2) such that in U, D is defined by $u_1 = 0$ and that the monodromy around $D \cap U$ is *unipotent*. (Since the moduli of general fibres are not constant, taking a suitable ramified covering of W, we can always assume this.)

(3) Put $U_0 = U - D \cap U$. By means of the toroidal embedding we can extend our family $\varphi_0^{-1}(U_0) \to U_0$ to a family $\psi_1: V_1 \to U$. Then the family $\psi_1: V_1 \to U$ is bimeromorphically equivalent to the original family $\varphi^{-1}(U) \to U$. (See Namikawa [8]).

Note that in $\varphi^{-1}(U_0)$, our form ω has the similar expression as (3.1).

$$\sum_{a+b+c=m} B_{a,b,c}(u_1, u_2)(\mathrm{d}u_1 \wedge \mathrm{d}u_2)^a \cdot (\mathrm{d}u_1 \wedge \mathrm{d}\zeta)^b \cdot (\mathrm{d}u_2 \wedge \mathrm{d}\zeta)^c, \tag{3.4}$$

where $B_{a,b,c}(u_1, u_2)$ is a holomorphic function on U_0. To show that the symmetric form (3.1) has the expression $A_{m,0,0}(z_1, z_2)(dz_1 \wedge dz_2)^m$, it is enough to show that in (3.4) the similar expression holds. In the toroidal relative compactification $\psi_1: V_1 \to U$, there is a dense open subset with local coordinates (w_1, w_2, w_3) such that on $\psi_1^{-1}(U_0)$ we have

$$w_1 = u_1, \quad w_2 = u_2, \quad w_3 = \exp(2\pi\sqrt{-1}\zeta),$$

and the morphism ψ_1 is given by

$$u_1 = w_1, \quad u_2 = w_2.$$

Hence by these coordinates, (3.4) has the form

$$\tau = B_{abc}(w_1, w_2)(dw_1 \wedge dw_2)^a \left(\frac{1}{2\pi\sqrt{-1}} dw_1 \wedge \frac{dw_3}{w_3}\right)^b \left(\frac{1}{2\pi\sqrt{-1}} dw_2 \wedge \frac{dw_3}{w_3}\right)^c.$$

Since $B_{abc}(w_1, w_2)$ is independent of ζ, if b or c is positive, then τ is not holomorphic. Therefore, τ has the form $B(w_1, w_2)(dw_1 \wedge dw_2)^m$. This implies

$$H^0(V, S^m\Omega_V^2) \subset \varphi^* H^0(W, S^m\Omega_W^2).$$

The converse inclusion is obvious. Q.E.D.

COROLLARY: *Let $\varphi: V \to W$ be the same as above. Then we have*

$$\kappa(\Omega_V^\nu, V) = \kappa(\Omega_W^\nu, W), \quad \nu = 1, 2, \ldots, \dim W.$$

This gives an affirmative answer to the problem when $\varphi: V \to W$ satisfies the above condition and V is an algebraic threefold of parabolic type. We obtain a similar result for fibrations of abelian varieties. This will be discussed elsewhere.

REFERENCES

[1] P. FRANCIA: Some remarks on minimal models I. to appear.

[2] A. FUJIKI: On the minimal models of compact complex manifolds. To appear.

[3] J. IGUSA: On the structure of a certain class of Kähler varieties. Amer. J. Math. *76* (1954) 669–678.

[4] Y. KAWAMATA: On a characterization of an abelian variety in the classification theory of algebraic varieties II. To appear.

[5] Y. KAWAMATA and E. VIEHWEG: On a characterization of an abelian variety in the classification theory of algebraic varieties. To appear.

[6] S. KOBAYASHI: The first Chern class and holomorphic symmetric tensor fields. To appear.